ADVANCES IN
Chromatography

VOLUME 37

Edited by

Phyllis R. Brown
UNIVERSITY OF RHODE ISLAND
KINGSTON, RHODE ISLAND

Eli Grushka
THE HEBREW UNIVERSITY OF JERUSALEM
JERUSALEM, ISRAEL

Marcel Dekker, Inc.   New York • Basel • Hong Kong

Coventry University

Library of Congress Cataloging-in-Publication Data
Main entry under title:

Advances in chromatography. v. 1-
  1965–
  New York, M. Dekker
    v. illus. 24 cm.
  Editors: v.1-J.C. Giddings and R.A. Keller.
  1. Chromatographic analysis-Addresses, essays, lectures.
I. Giddings, John Calvin, [date] ed.   II. Keller, Roy A., [date] ed.
QD271.A23 544.92 65-27435
ISBN: 0-8247-9804-X

The publisher offers discounts on this book when ordered in bulk quantities. For more information, write to Special Sales/Professional Marketing at the address below.

This book is printed on acid-free paper.

Copyright © 1997 by Marcel Dekker, Inc. All Rights Reserved.

Neither this book nor any part may be reproduced or transmitted in any form or by any means, electronic or mechanical, including photocopying, microfilming, and recording, or by any information storage and retrieval system, without permission in writing from the publisher.

Marcel Dekker, Inc.
270 Madison Avenue, New York, New York 10016

Current printing (last digit):
10 9 8 7 6 5 4 3 2 1

PRINTED IN THE UNITED STATES OF AMERICA

Contributors

Keith D. Bartle, Ph.D. Professor, School of Chemistry, University of Leeds, Leeds, United Kingdom

Phyllis R. Brown, Ph.D. Department of Chemistry, University of Rhode Island, Kingston, Rhode Island

Mark D. Burford, Ph.D., C. Chem., M.R.S.C. Research Fellow, School of Chemistry, University of Leeds, Leeds, United Kingdom

Kathryn M. De Antonis, Ph.D. Department of Chemistry, University of Rhode Island, Kingston, Rhode Island

Steven B. Hawthorne, Ph.D. Energy and Environmental Research Center, University of North Dakota, Grand Forks, North Dakota

John H. Knox, F.R.S., F.R.S.E., Ph.D., D.Sc Emeritus Professor, Department of Chemistry, University of Edinburgh, Edinburgh, United Kingdom

Cheng-Ming Liu, Ph.D. Bio-Rad Laboratories, Hercules, California

Linda B. McGown, Ph.D. Professor, Department of Chemistry, P. M. Gross Chemical Laboratory, Duke University, Durham, North Carolina

Zelda E. Penton, Ph.D. Senior Chemist, Varian Chromatography Systems, Walnut Creek, California

Roberto Rodriguez-Diaz, Ph.D. Research and Development Chemist, Laboratory Products Research and Development, Bio-Rad Laboratories, Hercules, California

Paul Ross, Ph.D. Development Manager, Research and Development, Hypersil, Runcorn, Cheshire, United Kingdom

Shahab A. Shamsi, Ph.D. Research Associate, Department of Chemistry, Louisiana State University, Baton Rouge, Louisiana

Muhammad A. Sharaf, Ph.D. Staff Scientist, Applied Biosystems Division, Perkin-Elmer Corporation, Foster City, California

Maria Brak Smalley, Ph.D.[*] Department of Chemistry, P. M. Gross Chemical Laboratory, Duke University, Durham, North Carolina

Isiah M. Warner, Ph.D. Chairman and Philip W. West Professor of Analytical and Environmental Chemistry, Louisiana State University, Baton Rouge, Louisiana

Tim Wehr, Ph.D. Manager, Capillary Electrophoresis Applications, Life Science Group, Bio-Rad Laboratories, Hercules, California

Crystal C. Williams, M.S. Board of Regents Fellow, Analytical Chemistry, Louisiana State University, Baton Rouge, Louisiana

---

[*]*Current affiliation:* Senior Scientist, Product Development/Quality Assurance, Johnson & Johnson Clinical Diagnostics, Rochester, New York

Contents of Volume 37

Contributors to Volume 37   iii
Contents of Other Volumes   ix

1. **Assessment of Chromatographic Peak Purity**   1

   Muhammad A. Sharaf

   - I. Introduction
   - II. The Detector's View of the World
   - III. Single-Channel Techniques
   - IV. Multichannel Techniques
   - V. Operational Considerations of Peak Purity Detection Strategies
   - VI. Statistical Considerations of Peak Purity Detection Strategies
   - VII. Concluding Remarks
     References

2. **Fluorescence Detectors in HPLC**   29

   Maria Brak Smalley and Linda B. McGown

   - I. Introduction
   - II. An Overview of Fluorescence Detection in HPLC
   - III. Steady-State Fluorescence Intensity and Spectral Detection
   - IV. Innovations in Detector Technology and Methodology
   - V. Dynamic-State Fluorescence Detection
   - VI. Conclusions
     References

3A. **Carbon-Based Packing Materials for Liquid Chromatography: Structure, Performance, and Retention Mechanisms**   73

   John H. Knox and Paul Ross

   - I. Introduction
   - II. Development and Production of Carbon-Based Packing Materials for HPLC
   - III. Structure of PGC/Hypercarb®
   - IV. Performance of PGC/Hypercarb®
   - V. Basis of Retention by Graphite in LC as Seen in 1988
   - VI. Retention Studies Post-1988
   - VII. Existing Theories of Retention in Reversed-Phase Chromatography and Their Application to Graphite
   - VIII. Correlation of Retention on Graphite with Measured or Calculated Physical Properties
   - IX. Conclusions
     References

3B. **Carbon-Based Packing Materials for Liquid Chromatography: Applications**   121

   Paul Ross and John H. Knox

   - I. Introduction
   - II. Separations of Geometric Isomers and Closely Related Compounds
   - III. Enantiomer Separations

IV. Separations of Sugars, Carbohydrates, and Glucuronides
  V. Residue Analysis
  VI. Separations of Ionized and Other Highly Polar Compounds
  Appendix 1: Separations of Geometric Isomers and Closely Related Compounds
  Appendix 2: Enantiomer Separations
  Appendix 3: Analysis of Sugars, Carbohydrates, and Glucuronides
  Appendix 4: Residue Analysis
  Appendix 5: Separations of Ionized and Highly Polar Compounds
  References

4. **Directly Coupled (On-Line) SFE–GC: Instrumentation and Applications** 163

   Mark D. Burford, Steven B. Hawthorne, and Keith D. Bartle

   I. Introduction
   II. Techniques for Coupling SFE–GC
   III. External Trapping of Analytes
   IV. Internal Accumulation of Analytes
   V. Construction of SFE–GC Instrumentation
   VI. Optimization of SFE–GC Chromatography
   VII. Quantitative Considerations for SFE–GC
   VIII. Optimization of Extraction Conditions for SFE–GC
   IX. SFE–GC Applications
   X. Conclusions
   References

5. **Sample Preparation for Gas Chromatography with Solid-Phase Extraction and Solid-Phase Microextraction** 205

   Zelda E. Penton

   I. Introduction
   II. Conventional Solid-Phase Extraction
   III. Solid-Phase Microextraction
   IV. Conclusions
   References

6. **Capillary Electrophoresis of Proteins** 237

   Tim Wehr, Robert Rodriguez-Diaz, and Cheng-Ming Liu

   I. Introduction
   II. Strategies for Reducing Protein–Wall Interactions
   III. Capillary Zone Electrophoresis
   IV. Capillary Isoelectric Focusing
   V. Sieving Separations
   VI. Clinical Applications
   References

7. **Chiral Micelle Polymers for Chiral Separations in Capillary Electrophoresis** 363

   Crystal C. Williams, Shahab A. Shamsi, and Isiah M. Warner

   I. Introduction
   II. Fundamentals of Capillary Electrophoresis
   III. MEKC Concepts
   IV. Polymeric Micelles
   V. Conclusion
   References

8. **Analysis of Derivatized Peptides Using High-Performance Liquid Chromatography and Capillary Electrophoresis** 425

   Kathryn M. De Antonis and Phyllis R. Brown

   I. Introduction
   II. Derivatization Chemistries
   III. Analysis of Derivatized Peptides
   IV. Conclusion
   References

Index   453

Contents of Other Volumes

Volumes 1-10 out of print

Volume 11

Quantitative Analysis by Gas Chromatography   *Josef Novák*
Polyamide Layer Chromatography   *Kung-Tsung Wang, Yau-Tang Lin, and Iris S. Y. Wang*
Specifically Adsorbing Silica Gels   *H. Bartels and P. Prijs*
Nondestructive Detection Methods in Paper and Thin-Layer Chromatography   *G. C. Barrett*

Volume 12

The Use of High-Pressure Liquid Chromatography in Pharmacology and Toxicology   *Phyllis R. Brown*
Chromatographic Separation and Molecular-Weight Distributions of Cellulose and Its Derivatives   *Leon Segal*
Practical Methods of High-Speed Liquid Chromatography   *Gary J. Fallick*
Measurement of Diffusion Coefficients by Gas-Chromatography Broadening Techniques: A Review   *Virgil R. Maynard and Eli Grushka*
Gas-Chromatography Analysis of Polychlorinated Diphenyls and Other Nonpesticide Organic Pollutants   *Joseph Sherma*

x / Contents of Other Volumes

High-Performance Electrometer Systems for Gas Chromatography
    Douglas H. Smith
Steam Carrier Gas-Solid Chromatography   *Akira Nonaka*

Volume 13   Out of Print

Volume 14

Nutrition: An Inviting Field to High-Pressure Liquid Chromatography
    Andrew J. Clifford
Polyelectrolyte Effects in Gel Chromatography   *Bengt Stenlund*
Chemically Bonded Phases in Chromatography   *Imrich Sebestian and István Halász*
Physicochemical Measurements Using Chromatography   *David C. Locke*
Gas-Liquid Chromatography in Drug Analysis   *W. J. A. VandenHeuvel and
    A. G. Zacchei*
The Investigation of Complex Association by Gas Chromatography and Related
    Chromatographic and Electrophoretic Methods   *C. L. de Ligny*
Gas-Liquid-Solid Chromatography   *Antonio De Corcia and Arnaldo Liberti*
Retention Indices in Gas Chromatography   *J. K. Haken*

Volume 15

Detection of Bacterial Metabolites in Spent Culture Media and Body Fluids by
    Electron Capture Gas-Liquid Chromatography   *John B. Brooks*
Signal and Resolution Enhancement Techniques in Chromatography
    Raymond Annino
The Analysis of Organic Water Pollutants by Gas Chromatography and Gas Chroma-
    tography-Mass Spectrometry   *Ronald A. Hites*
Hydrodynamic Chromatography and Flow-Induced Separations   *Hamish Small*
The Determination of Anticonvulsants in Biological Samples by Use of High-
    Pressure Liquid Chromatography   *Reginald F. Adams*
The Use of Microparticulate Reversed-Phase Packing in High-Pressure Liquid
    Chromatography of Compounds of Biological Interest   *John A. Montgomery,
    Thomas P. Johnston, H. Jeanette Thomas, James R. Piper, and
    Carroll Temple Jr.*
Gas-Chromatographic Analysis of the Soil Atmosphere   *K. A. Smith*
Kinematics of Gel Permeation Chromatography   *A. C. Ouano*
Some Clinical and Pharmacological Applications of High-Speed Liquid Chroma-
    tography   *J. Arly Nelson*

Volume 16   Out of Print

Volume 17

Progress in Photometric Methods of Quantitative Evaluation in TLO   *V. Pollak*
Ion-Exchange Packings for HPLC Separations: Care and Use   *Fredric M. Rabel*

Micropacked Columns in Gas Chromatography: An Evaluation *C. A. Cramers and J. A. Rijks*

Reversed-Phase Gas Chromatography and Emulsifier Characterization *J. K. Haken*

Template Chromatography *Herbert Schott and Ernst Bayer*

Recent Usage of Liquid Crystal Stationary Phases in Gas Chromatography *George M. Janini*

Current State of the Art in the Analysis of Catecholamines *Anté M. Krstulovic*

Volume 18

The Characterization of Long-Chain Fatty Acids and Their Derivatives by Chromatography *Marcel S. F. Lie Ken Jie*

Ion-Pair Chromatography on Normal- and Reversed-Phase Systems *Milton T. W. Hearn*

Current State of the Art in HPLC Analyses of Free Nucleotides, Nucleosides, and Bases in Biological Fluids *Phyllis R. Brown, Anté M. Krstulovic, and Richard A. Hartwick*

Resolution of Racemates by Ligand-Exchange Chromatography *Vadim A. Danankov*

The Analysis of Marijuana Cannabinoids and Their Metabolites in Biological Media by GC and/or GC-MS Techniques *Benjamin J. Gudzinowicz, Michael J. Gudzinowicz, Joanne Hologgitas, and James L. Driscoll*

Volume 19

Roles of High-Performance Liquid Chromatography in Nuclear Medicine *Steven How-Yan Wong*

Calibration of Separation Systems in Gel Permeation Chromatography for Polymer Characterization *Josef Janča*

Isomer-Specific Assay of 2,4-D Herbicide Products by HPLC: Regulaboratory Methodology *Timothy S. Stevens*

Hydrophobic Interaction Chromatography *Stellan Hjertén*

Liquid Chromatography with Programmed Composition of the Mobile Phase *Pavel Jandera and Jaroslav Churáček*

Chromatographic Separation of Aldosterone and Its Metabolites *David J. Morris and Ritsuko Tsai*

Volume 20

High-Performance Liquid Chromatography and Its Application to Protein Chemistry *Milton T. W. Hearn*

Chromatography of Vitamin D_3 and Metabolites *K. Thomas Koshy*

High-Performance Liquid Chromatography: Applications in a Children's Hospital *Steven J. Soldin*

The Silica Gel Surface and Its Interactions with Solvent and Solute in Liquid Chromatography *R. P. W. Scott*

New Developments in Capillary Columns for Gas Chromatography
Walter Jennings

Analysis of Fundamental Obstacles to the Size Exclusion Chromatography of Polymers of Ultrahigh Molecular Weight *J. Calvin Giddings*

Volume 21

High-Performance Liquid Chromatography/Mass Spectrometry (HPLC/MS) *David E. Grimes*

High-Performance Liquid Affinity Chromatography *Per-Olof Larsson, Magnus Glad, Lennart Hansson, Mats-Olle Månsson, Sten Ohlson, and Klaus Mosbach*

Dynamic Anion-Exchange Chromatography *Roger H. A. Sorel and Abram Hulshoff*

Capillary Columns in Liquid Chromatography *Daido Ishii and Toyohide Takeuchi*

Droplet Counter-Current Chromatography *Kurt Hostettmann*

Chromatographic Determination of Copolymer Composition *Sadao Mori*

High-Performance Liquid Chromatography of K Vitamins and Their Antagonists *Martin J. Shearer*

Problems of Quantitation in Trace Analysis by Gas Chromatography *Josef Novák*

Volume 22

High-Performance Liquid Chromatography and Mass Spectrometry of Neuropeptides in Biologic Tissue *Dominic M. Desiderio*

High-Performance Liquid Chromatography of Amino Acids: Ion-Exchange and Reversed-Phase Strategies *Robert F. Pfeifer and Dennis W. Hill*

Resolution of Racemates by High-Performance Liquid Chromatography *Vadium A. Davankov, Alexander A. Kurganov, and Alexander S. Bochkov*

High-Performance Liquid Chromatography of Metal Complexes *Hans Veening and Bennett R. Willeford*

Chromatography of Carotenoids and Retinoids *Richard F. Taylor*

High Performance Liquid Chromatography *Zbyslaw J. Petryka*

Small-Bore Columns in Liquid Chromatography *Raymond P. W. Scott*

Volume 23

Laser Spectroscopic Methods for Detection in Liquid Chromatography *Edward S. Yeung*

Low-Temperature High-Performance Liquid Chromatography for Separation of Thermally Labile Species *David E. Henderson and Daniel J. O'Connor*

Kinetic Analysis of Enzymatic Reactions Using High-Performance Liquid Chromatography *Donald L. Sloan*

Heparin-Sepharose Affinity Chromatography *Akhlaq A. Farooqui and Lloyd A. Horrocks*

Chromatopyrography *John Chih-An Hu*
Inverse Gas Chromatography *Seymour G. Gilbert*

Volume 24

Some Basic Statistical Methods for Chromatographic Data *Karen Kafadar and Keith R. Eberhardt*
Multifactor Optimization of HPLC Conditions *Stanley N. Deming, Julie G. Bower, and Keith D. Bower*
Statistical and Graphical Methods of Isocratic Solvent Selection for Optimal Separation in Liquid Chromatography *Haleem J. Issaq*
Electrochemical Detectors for Liquid Chromatography *Ante M. Krstulović Henri Colin, and Georges A. Guiochon*
Reversed-Flow Gas Chromatography Applied to Physicochemical Measurements *Nicholas A. Katsanos and George Karaiskakis*
Development of High-Speed Countercurrent Chromatography *Yoichiro Ito*
Determination of the Solubility of Gases in Liquids by Gas-Liquid Chromatography *Jon F. Parcher, Monica L. Bell, and Ping J. Lin*
Multiple Detection in Gas Chromatography *Ira S. Krull, Michael E. Swartz, and John N. Driscoll*

Volume 25

Estimation of Physicochemical Properties of Organic Solutes Using HPLC Retention Parameters *Theo L. Hafkenscheid and Eric Tomlinson*
Mobile Phase Optimization in RPLC by an Iterative Regression Design *Leo de Galan and Hugo A. H. Billiet*
Solvent Elimination Techniques for HPLC/FT-IR *Peter R. Griffiths and Christine M. Conroy*
Investigations of Selectivity in RPLC of Polycyclic Aromatic Hydrocarbons *Lane C. Sander and Stephen A. Wise*
Liquid Chromatographic Analysis of the Oxo Acids of Phosphorus *Roswitha S. Ramsey*
HPLC Analysis of Oxypurines and Related Compounds *Katsuyuki Nakano*
HPLC of Glycosphingolipids and Phospholipids *Robert H. McCluer, M. David Ullman, and Firoze B. Jungalwala*

Volume 26

RPLC Retention of Sulfur and Compounds Containing Divalent Sulfur *Hermann J. Möckel*
The Application of Fleuric Devices to Gas Chromatographic Instrumentation *Raymond Annino*
High Performance Hydrophobic Interaction Chromatography *Yoshio Kato*

HPLC for Therapeutic Drug Monitoring and Determination of Toxicity
  Ian D. Watson
Element Selective Plasma Emission Detectors for Gas Chromatography
  A. H. Mohamad and J. A. Caruso
The Use of Retention Data from Capillary GC for Qualitative Analysis: Current
  Aspects *Lars G. Blomberg*
Retention Indices in Reversed-Phase HPLC *Roger M. Smith*
HPLC of Neurotransmitters and Their Metabolites *Emilio Gelpi*

Volume 27

Physicochemical and Analytical Aspects of the Adsorption Phenomena Involved
  in GLC *Victor G. Berezkin*
HPLC in Endocrinology *Richard L. Patience and Elizabeth S. Penny*
Chiral Stationary Phases for the Direct LC Separation of Enantiomers *William H.
  Pirkle and Thomas C. Pochapsky*
The Use of Modified Silica Gels in TLC and HPTLC *Willi Jost and Heinz E.
  Hauck*
Micellar Liquid Chromatography *John G. Dorsey*
Derivatization in Liquid Chromatography *Kazuhiro Imai*
Analytical High-Performance Affinity Chromatography *Georgio Fassina and
  Irwin M. Chaiken*
Characterization of Unsaturated Aliphatic Compounds by GC/Mass Spectrometry
  Lawrence R. Hogge and Jocelyn G. Millar

Volume 28

Theoretical Aspects of Quantitative Affinity Chromatography: An Overview
  Alain Jaulmes and Claire Vidal-Madjar
Column Switching in Gas Chromatography *Donald E. Willis*
The Use and Properties of Mixed Stationary Phases in Gas Chromatography
  Gareth J. Price
On-Line Small-Bore Chromatography for Neurochemical Analysis in the Brain
  William H. Church and Joseph B. Justice, Jr.
The Use of Dynamically Modified Silica in HPLC as an Alternative to Chemically
  Bonded Materials *Per Helboe, Steen Honoré Hansen, and Mogens Thomsen*
Gas Chromatographic Analysis of Plasma Lipids *Arnis Kuksis and John J. Myher*
HPLC of Penicillin Antibiotics *Michel Margosis*

Volume 29

Capillary Electrophoresis *Ross A. Wallingford and Andrew G. Ewing*
Multidimensional Chromatography in Biotechnology *Daniel F. Samain*
High-Performance Immunoaffinity Chromatography *Terry M. Phillips*
Protein Purification by Multidimensional Chromatography *Stephen A. Berkowitz*
Fluorescence Derivitization in High-Performance Liquid Chromatography
  Yosuke Ohkura and Hitoshi Nohta

Volume 30

Mobile and Stationary Phases for Supercritical Fluid Chromatography
  Peter J. Schoenmakers and Louis G. M. Uunk
Polymer-Based Packing Materials for Reversed-Phase Liquid Chromatography
  Nobuo Tanaka and Mikio Araki
Retention Behavior of Large Polycyclic Aromatic Hydrocarbons in Reversed-Phase
  Liquid Chromatography *Kiyokatsu Jinno*
Miniaturization in High-Performance Liquid Chromatography *Masashi Goto,
  Toyohide Takeuchi, and Daido Ishii*
Sources of Errors in the Densitometric Evaluation of Thin-Layer Separations with
  Special Regard to Nonlinear Problems *Viktor A. Pollak*
Electronic Scanning for the Densitometric Analysis of Flat-Bed Separations
  Viktor A. Pollak

Volume 31

Fundamentals of Nonlinear Chromatography: Prediction of Experimental Profiles
  and Band Separation *Anita M. Katti and Georges A. Guiochon*
Problems in Aqueous Size Exclusion Chromatography *Paul L. Dubin*
Chromatography on Thin Layers Impregnated with Organic Stationary Phases
  Jiri Gasparic
Countercurrent Chromatography for the Purification of Peptides *Martha Knight*
Boronate Affinity Chromatography *Ram P. Singhal and S. Shyamali M. DeSilva*
Chromatographic Methods for Determining Carcinogenic Benz(c)-acridine
  Noboru Motohashi, Kunihiro Kamata, and Roger Meyer

Volume 32

Porous Graphitic Carbon in Biomedical Applications *Chang-Kee Lim*
Tryptic Mapping by Reversed Phase Liquid Chromatography *Michael W. Dong*
Determination of Dissolved Gases in Water by Gas Chromatography
  Kevin Robards, Vincent R. Kelly, and Emilios Patsalides
Separation of Polar Lipid Classes into Their Molecular Species Components by
Planar and Column Liquid Chromatography *V. P. Pchelkin and A. G. Vereshchagin*
The Use of Chromatography in Forensic Science *Jack Hubball*
HPLC of Explosives Materials *John B. F. Lloyd*

Volume 33

Planar Chips Technology of Separation Systems: A Developing Perspective in
  Chemical Monitoring *Andreas Manz, D. Jed Harrison, Elizabeth Verpoorte,
  and H. Michael Widmer*
Molecular Biochromatography: An Approach to the Liquid Chromatographic
  Determination of Ligand–Biopolymer Interactions *Irving W. Wainer and
  Terence A. G. Noctor*

Volume 30

Mobile and Stationary Phases for Supercritical Fluid Chromatography
  Peter J. Schoenmakers and Louis G. M. Uunk
Polymer-Based Packing Materials for Reversed-Phase Liquid Chromatography
  Nobuo Tanaka and Mikio Araki
Retention Behavior of Large Polycyclic Aromatic Hydrocarbons in Reversed-Phase Liquid Chromatography *Kiyokatsu Jinno*

Volume 34

High-Performance Capillary Electrophoresis of Human Serum and Plasma Proteins
  Oscar W. Reif, Ralf Lausch, and Ruth Freitag
Analysis of Natural Products by Gas Chromatography/Matrix Isolation/Infrared Spectrometry *W. M. Coleman III and Bert M. Gordon*
Statistical Theories of Peak Overlap in Chromatography *Joe M. Davis*
Capillary Electrophoresis of Carbohydrates *Ziad El Rassi*
Environmental Applications of Supercritical Fluid Chromatography
  Leah J. Mulcahey, Christine L. Rankin, and Mary Ellen P. McNally
HPLC of Homologous Series of Simple Organic Anions and Cations
  Norman E. Hoffman
Uncertainty Structure, Information Theory, and Optimization of Quantitative Analysis in Separation Science *Yuzuru Hayashi and Rieko Matsuda*

Volume 35

Optical Detectors for Capillary Electrophoresis *Edward S. Yeung*
Capillary Electrophoresis Coupled with Mass Spectrometry *Kenneth B. Tomer, Leesa J. Deterding, and Carol E. Parker*
Approaches for the Optimization of Experimental Parameters in Capillary Zone Electrophoresis *Haleem J. Issaq, George M. Janini, King C. Chan, and Ziad El Rassi*
Crawling Out of the Chiral Pool: The Evolution of Pirkle-Type Chiral Stationary Phases *Christopher J. Welch*
Pharmaceutical Analysis by Capillary Electrophoresis *Sam F. Y. Li, Choon Lan Ng, and Chye Peng Ong*
Chromatographic Characterization of Gasolines *Richard E. Pauls*
Reversed-Phase Ion-Pair and Ion-Interaction Chromatography *M. C. Gennaro*
Error Sources in the Determination of Chromatographic Peak Size Ratios
  Veronika R. Meyer

Volume 36

Use of Multivariate Mathematical Methods for the Evaluation of Retention Data Matrices *Tibor Cserháti and Esther Forgács*

Separation of Fullerenes by Liquid Chromatography: Molecular Recognition Mechanism in Liquid Chromatographic Separation   *Kiyokatsu Jinno and Yoshihiro Saito*

Emerging Technologies for Sequencing Antisense Oligonucleotides: Capillary Electrophoresis and Mass Spectrometry   *Aharon S. Cohen, David L. Smisek, and Bing H. Wang*

Capillary Electrophoretic Analysis of Glycoproteins and Glycoprotein-Derived Oligosaccharides   *Robert P. Oda, Benjamin J. Madden, and James P. Landers*

Analysis of Drugs of Abuse in Biological Fluids by Liquid Chromatography   *Steven R. Binder*

Electrochemical Detection of Biomolecules in Liquid Chromatography and Capillary Electrophoresis   *Jian-Ge Chen, Steven J. Woltman, and Steven G. Weber*

The Development and Application of Coupled HPLC-NMR Spectroscopy   *John C. Lindon, Jeremy K. Nicholson, and Ian D. Wilson*

Microdialysis Sampling for Pharmacological Studies: HPLC and CE Analysis   *Susan M. Lunte and Craig E. Lunte*

1
Assessment of Chromatographic Peak Purity

Muhammad A. Sharaf *Perkin-Elmer Corporation, Foster City, California*

I.	INTRODUCTION	2
II.	THE DETECTOR'S VIEW OF THE WORLD	3
III.	SINGLE-CHANNEL TECHNIQUES	5
	A. Derivatives	5
	B. Profile Shape Analysis	7
	C. Curve Fitting	8
	D. Frequency-Domain Techniques	9
IV.	MULTICHANNEL TECHNIQUES	13
	A. Absorbance/Spectral Ratioing	13
	B. Chromatographic Ratioing	15
	C. Spectral Correlations	16
	D. Factor Analysis/Principal Component Analysis	18
	E. Spectral Libraries Matching	19
	F. Other Peak Purity Assessment Techniques	21
V.	OPERATIONAL CONSIDERATIONS OF PEAK PURITY DETECTION STRATEGIES	22
VI.	STATISTICAL CONSIDERATIONS OF PEAK PURITY DETECTION STRATEGIES	23

| VII. | CONCLUDING REMARKS | 26 |
| | REFERENCES | 26 |

I. INTRODUCTION

Long before the term "peak purity" became commonplace in chromatography circles, the techniques that are currently associated with the assessment of peak purity were being developed and applied to a variety of analytical problems. Chromatographers, spectroscopists, and other analytical scientists were hard at work developing and applying many mathematical techniques to their data with the goal of detecting "hard-to-see" shoulders on spectral bands and chromatographic peaks [1,2]. In other scientific areas, astronomers and astrophysicists were (and still are) perfecting the techniques necessary to detect faint and weak signatures in the presence of a large background "noise" [3]. To the delight of chromatographers, when they were ready to assess the purity of their peaks (i.e., to determine whether or not a chromatographic band is made up of more than one coeluting component), there were a multitude of mathematical techniques from which to choose. On the other hand, because many of these techniques were not developed with "peak purity" application in mind, their success was hindered by the compatibility (or lack thereof) of their basic assumptions with the chromatographic application at hand. Different techniques often produce different results when applied to the same chromatograms. Indeed, the same technique may yield two different results when applied to seemingly similar chromatograms.

Although this behavior has caused a fair amount of confusion and frustration to both the practitioners of chromatography and the developers of chromatographic data systems, it is the practitioners who suffer more on the short run as they try to deal with product purity issues and pharmaceutical batch homogeneity as well as a spectrum of regulatory protocols. The purpose of this chapter is to overview the common techniques that may be used to assess the purity of chromatographic peaks. The appropriateness of the assumptions of these mathematical/statistical/chemometric tools will be delineated and their impact on the various application areas will be discussed. Finally, a model will be presented to aid practitioners and developers in assessing their impurity detection strategy from a statistical point of view as well as from an operational point of view. The approach taken here to highlight the statistical and operational implications of the various methods can be applied to the validation of any peak purity detection method under consideration.

II. THE DETECTOR'S VIEW OF THE WORLD

The chromatographic detector serves as the chromatographer's eyes and ears on eluting components. Peak purity assessment techniques can be effectively divided into two categories (see Fig. 1) based on the chromatographic detector—namely single-channel techniques and multichannel techniques. The former are generally more sensitive and computationally cheaper, but relatively information-poor. The latter techniques tend to be computationally and operationally more expensive, but they are information-rich and offer the flexibility of applying a multichannel-based technique as well as a single-channel-based technique where appropriate.

The techniques that have been successfully applied to single-channel chromatographic data include the following:

- Derivatives
- Profile shape analysis
- Curve fitting
- Frequency-domain techniques such as Fourier transforms (FT) and the fast Fourier transform (FFT)

In multichannel detection domain, it is possible to apply any of the above techniques as well as any of the following:

- Spectral ratioing
- Chromatographic ratioing
- Spectral correlations
- Factor analysis/principal component analysis
- Spectral libraries matching
- Other spectral comparisons such as weighted-average wavelength, spectral contrasting, spectral suppression, Kalman filters, and the Biller–Biemann method

Despite their varying degrees of success and appropriateness, all of the above techniques are attractive because they are amenable to a high degree of automation and can be implemented for routine use. Of course, one must not underestimate the ability and potential of the human brain to detect minute amounts of impurities. However, inspection of chromatograms by an expert chromatographer or an expert spectroscopist is a very expensive alternative and cannot be recommended as a routine procedure. In industrial laboratories, there is a need to balance the precision, the accuracy, and the cost of analytical results.

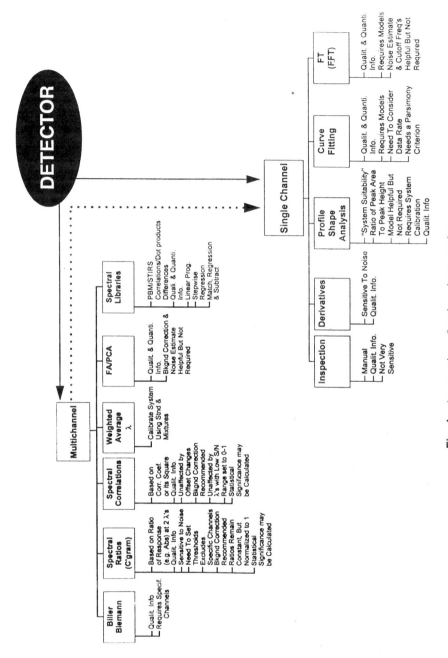

Fig. 1. An overview of peak purity strategies.

III. SINGLE-CHANNEL TECHNIQUES

A. Derivatives

Differentiation of chromatographic signals is based on the assumption that when two or more components coelute, the shape of the resulting profile will have some irregularity such as the appearance of a shoulder on the up- or down-slope of the peak profile. The derivative of the composite profile enhances the detection of such irregularity [4]. This is schematically shown in Fig. 2. Derivative techniques were successfully applied in spectroscopy where they were used to detect a faint spectral transition that appeared as shoulders on major bands. The implementation of this detection strategy in chromatography is quite simple, given the theoretical work and experimental verification of the shape of chromatographic peaks. However, several caveats need to be kept in mind. First, the information obtained from differentiating a chromatographic profile is qualitative in nature. Second, and more importantly, differentiation is a process that is very sensitive to noise, as it will amplify it. The noise amplification can be demonstrated by a very simple example. Consider the subtraction of two data points x_i and y_i, along the profile being differentiated [Eq. (1)]. If x_i and y_i are uncorrelated, the variance of the resultant, d_i, is given, according to error propagation theory, as the sum of the variance of each data point [Eq. (2)]:

$$d_i = x_i - y_i \tag{1}$$

$$\mathrm{Var}(d_i) = \mathrm{Var}(x_i) + \mathrm{Var}(y_i) \tag{2}$$

If the data points are assumed to have the same variance, s^2, the variance in the derivative, d_i, is $2s^2$. If we want to compute the second derivative by subtracting the d_i's, the variance in the resultant ($D_i = d_i - d_{i-1}$) becomes $4s^2$. Although, in practice, derivatives are not necessarily computed as shown above, the outcome remains the same. The effect of noise on the derivatives gets worse as higher-order derivatives are estimated [5]. The interpretation of the noisy derivative profile becomes less straightforward, and detection of shoulders and coeluting impurities becomes more difficult. Derivative techniques are most useful at high signal-to-noise ratios.

In order to build a successful coelution detection system that is based on derivatives, the implementation scheme must "condition" the signal before differentiation, to ensure a high signal-to-noise ratio. Alternatively, a differentiating/smoothing digital filter may be used. Data bunching and box car averaging are two common preprocessing steps that are used to enhance the signal-to-noise ratio. However, the implementation of these two common techniques results in a

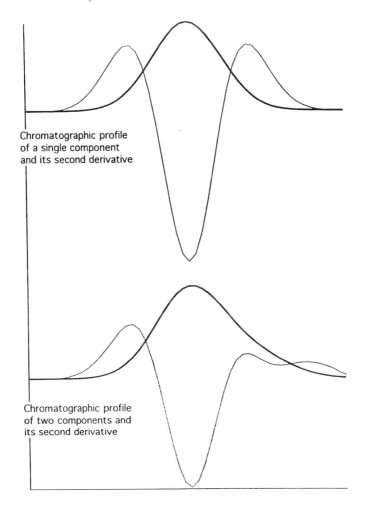

Fig. 2 Chromatographic profiles and their derivatives.

loss of resolution. Another alternative to lessen the effect of random noise is to transform the chromatographic data to the frequency domain and filter out the high-frequency signal component (usually noise components), then back-transform the data to the time domain. The Fourier transform is one way to achieve this frequency-domain filtering of the noise. It must be noted that data bunching, boxcar averaging and frequency-domain filtering are general noise-reduction techniques and can be used for many other purposes, not just for signal conditioning prior to differentiation [6].

B. Profile Shape Analysis

As the name implies, this strategy is based on observing and analyzing the shape of the chromatographic profile/peak [7,10]. Detection of impurities relies on the assumption that the presence of a coeluting component will cause some distortion to the chromatographic profile. This strategy is also the basis for what is known today as the "system suitability" test/validation. In the "system suitability" context, profile shape analysis is used to verify that the chromatographic conditions, as manifested by the shape of certain peaks, meet a set of minimum performance requirements. In the peak-impurity-detection context, profile shape analysis is used to test whether or not the profile is affected by coelutants. Indeed, the ratio of peak area to peak height (a measure of peak width) is one of the parameters used by chromatographers to assess their peaks since the early days of computing integrators [11]. Modern data systems allow a much more extensive analysis of chromatographic profiles with much more convenience than stand-alone computing integrators. Chromatographers can easily obtain important parameters that are related to the presence of impurities. These parameters include the number of inflection points along the profile, the number of local maxima along the profile, peak width at different relative heights, and the symmetry of the profile at different relative heights (e.g., 50% of maximum, 10% of maximum, or 5% of maximum).

The information obtained from profile shape analysis, with respect to the presence of impurity, is qualitative in nature (even though the decision is made using a highly quantitative scheme.) The amount(s) of the coelutant(s) present in the profile is not generally derived directly from the degree of distortion or the amount of deviation from the ideal profile. The implementation of an analysis scheme based on profile shape analysis is relatively simple and, from a computation point of view, inexpensive. Although peak models may be utilized, they are not necessary for implementation. It is necessary, however, to calibrate the separation system using certified standards and to establish a valid statistical sample population of the parameters being monitored. Once this is established, an inference can be made at the appropriate confidence level.

Profile shape analysis methods are highly system dependent. A change in the composition of the mobile phase, column temperature, detection wavelength, or even column vendor may require recalibration, which can be another expensive process. This limitation applies whether the application is for system suitability purposes or for the purpose of detecting impurities.

There are two additional important considerations that must be addressed in any profile shape analysis scheme. The first is the signal-to-noise ratio, and the second is the number of effective sampling points along the chromatographic peak being analyzed. Chromatographic peaks are dynamic systems. The signal-to-noise ratio changes by about a factor of 10 along the profile. When estimating the symmetry figure of merit at 10% and 5% of peak maximum, we are dealing with the noisiest part of the peak. If the signal-to-noise ratio is 30 at peak maximum, it is only 3 at 10% of peak maximum. The precision of a parameter being estimated at such a low signal-to-noise ratio is highly influenced by the number of sampling points. Because the precision, the signal-to-noise ratio, and the number of sampling points are correlated, the symmetry at 10% and 5% need to be considered in view of the signal-to-noise ratio and the number of sampling points or, alternatively, the effective data acquisition rate. Consequently, the 10% and 5% symmetry figures of merit are not reliable indicators of the presence of coeluants at moderate and low signal-to-noise ratio.

C. Curve Fitting

Curve fitting or modeling has been applied successfully to resolve overlapping peaks in chromatography. The basic approach in a modeling strategy is to try to model a chromatographic profile/peak as a single peak or, in general, as the sum of more than one single peak, superimposed on a background signal in the presence of random noise. This general model is shown in Eq. (3):

Profile of interest = Peak A + Peak B + ⋯ + Bkgnd + Random noise   (3)

The left-hand side of Eq. (3) represents the experimental data which, for now, we suspect to be a composite chromatographic profile. Peak A, peak B, . . . are the individual resolved chromatographic peaks assumed to be present under the composite profile on the left-hand side of Eq. (3). These peaks are represented, in the detailed form of Eq. (3), by several parameters (position, width, height, and a decay constant if an exponentially modified Gaussian shape is assumed for them). So for every component that is assumed to be present, there may be as many as four parameters representing it. In addition to the individual component, the background signal is added to account for signal drifts. One may allow the background signal to be a simple offset, a linear function of time, or a nonlinear, typically quadratic, function of time. Finally, the model must include random noise so that we can conveniently explain any "small" differences between the profile of interest and the sum of the peaks and background signal. These deviations are taken as a measure of "lack-of-fit."

Detection of coeluants can be carried out by modeling the profile as a single peak, and the extent of the lack-of-fit is used as an indicator of the presence of coeluants. As the amounts of coeluants increase, and as their resolution from the component of interest increases, the composite profile becomes more and more distorted, and a single peak model begins to show large amount of residuals. When the fit residuals are indistinguishable from random noise, the absence of a coeluant may be inferred, given the appropriate confidence level and the method's detection limit. When the residuals are significantly different from random noise, the single peak model is rejected and the presence of coeluants is inferred. In this mode, the information provided with respect to the coeluant(s) is only qualitative. Curve fitting can also provide quantitative information if all coeluants are included in the model [a full implementation of Eq. (3)].

The estimation of the models represented by Eq. (3) is often treated as a nonlinear optimization problem. Many techniques can be used to estimate the parameters such as Fletcher–Powell, Levenberg–Marquardt, and neural networks [12–15]. The initial guess is critical, as it may lead to divergence if not properly handled. It is often difficult to justify the minimum number of peaks needed to fit the experimental data. A parsimony rule is that the maximum number of peaks allowed in the model should equal the number of crests if there are twice as many inflection points present along the profile.

As mentioned earlier, the number of parameters to be estimated can grow quite rapidly (as many as four for every peak and as many as three for the background.) A profile that is a composite of two peaks may require the estimation of 11 parameters. One can make some justifiable simplifying assumptions such as the peak widths for both peaks are the same, because they elute at about the same time. Also, over the relatively short period of time covering the composite profile of interest, the background may be adequately modeled as a linear function of time. This decreases the number of parameters to be estimated to only nine parameters. The reduction in the number of parameters is important not just from the numerical analysis point of view but also from the point of view of the statistical confidence associated with the model and the estimated parameters. Given a fixed number of effective sampling points along the profile, the fewer the parameters to be estimated, the higher the reliability and the robustness of the estimated parameters. The number of degrees of freedom associated with the residuals are given according to Eq. (4). This is another instance of how chromatographic factors are highly influenced by the data acquisition rate.

$$\nu = N - p - 1 \qquad (4)$$

where ν is the number of degrees of freedom associated with the residuals, N is the effective number of sampling points along the composite profile, and p is the number of parameters being estimated.

D. Frequency-Domain Techniques

A signal system of overlapping chromatographic peaks can be thought of as a linear, time-invariant system where blurring has occurred due to convolution. To resolve a composite chromatographic profile, one needs to carry out a deconvolution operation. In the frequency domain, convolution is a multiplication operation, and deconvolution is a division operation. A chromatographic profile obtained in the time domain may be transformed to the frequency domain, then deconvolved by division in the frequency domain, and finally transformed back to the more familiar time domain [16,17]. The deconvolution steps are summarized in Eqs. (5)–(7):

$$s(t) = c(t)*h(t) \tag{5}$$

where the signal, s, is the true chromatogram, c, which is convolved with a system point spread function, h.

$$S(u) = C(u)H(u) \tag{6}$$

In the frequent-domain representation, the signal, S, is the product of the frequency-domain representation of the true chromatogram, C, and the frequency-domain representation of the point spread function, H.

$$\frac{S(u)}{H(u)} = C(u) \tag{7}$$

The true chromatogram is recovered by division in the frequency domain, then back-transformed to the time domain.

An example of blurring and the recovery of the blurred profile is schematically shown in Fig. 3. While in the frequency domain, another important operation may also be applied to the data, namely eliminating high-frequency noise components. Indeed, the two operations, noise filtering and deconvolution (also called sharpening), are often accomplished in a single-pass transformation.

In the presence of noise (real-life situations), Eqs. (5)–(7) are expressed as

$$s(t) = c(t)*h(t) + noise(t) \tag{8}$$

$$S(u) = C(u)H(u) + NOISE(u) \tag{9}$$

$$\frac{S(u)}{H(u)} = C(u) + \frac{NOISE(u)}{H(u)} \tag{10}$$

An experimental example is shown in Fig. 4 where DNA fragments are resolved. The top trace results from scanning an autoradiography film which was exposed to PAGE (PolyAcrylamide Gel Electrophoresis) gel containing the radioactively labeled DNA fragments following electrophoresis. Deconvolution is carried out in the frequency domain using a Wiener filter [17].

Although simple in principle, the implications of frequency-domain techniques need to be understood in order to interpret the results and the requirements of a successful application. It is necessary to know (or have an estimate of) the blurring function (the point spread function). This is what the chromatography system will do when fed an impulse response. Gaussian and Gaussian-like

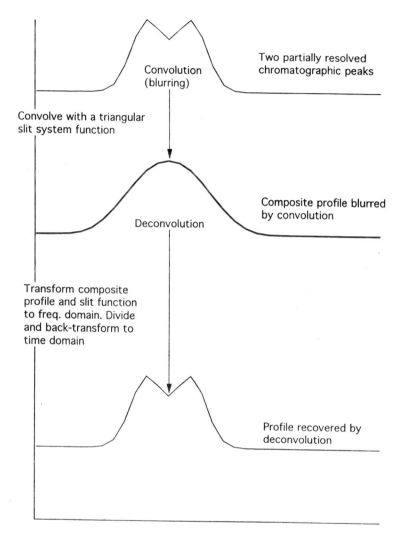

Fig. 3 True profile (top trace) recovered from blurred profile (middle trace) by deconvolution in the frequency domain.

point spread functions have been successfully used in practice. In order to reduce high-frequency noise, you need, again, an estimate of the cutoff frequency. This can be obtained from the signal background levels. The most crucial element in frequency-domain techniques is the division operation (the deconvolution) itself [Eq. (10)]. The frequency transform of the point spread function is likely to have

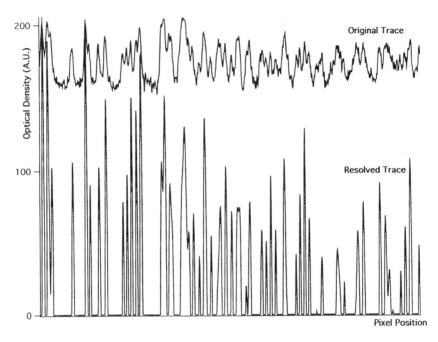

Fig. 4 Electropherogram of DNA fragments (top trace) is resolved by deconvolution in the frequency domain using a Wiener filter.

very low amplitudes (or zero amplitudes) at certain high frequencies (noise frequencies). These frequencies will be amplified, and the transformation to the time domain will reflect noisy components and produce invalid results. One way to deal with this problem is to employ a truncation function that excludes components with low/zero amplitude from participating in the deconvolution. Equation (10) is rewritten as

$$C^*(u) = \frac{S(u)}{H(u)} W(u) \qquad (11)$$

where $C^*(u)$ is an estimate of $C(u)$ and $W(u)$ is a truncation function.

Another challenge to be kept in mind is that chromatographic point spread functions vary as a function of time due to the increased spreading caused by diffusion. More than one point spread function, with progressively increasing peak width, may be necessary to encompass a whole chromatogram. Clearly, a point spread function developed for liquid chromatography could not be expected to work well in capillary gas chromatography where peak widths are generally smaller.

IV. MULTICHANNEL TECHNIQUES

A. Absorbance/Spectral Ratioing

Detecting coeluants based on the monitoring of two detection channels is the simplest of all multichannel detection techniques [18–27]. Despite its simplicity, this technique can be very powerful and highly sensitive. The term "absorbance ratioing" has its roots in the wide application of this technique to liquid chromatography/ultraviolet (LC/UV) data. It is, however, a general approach that can be used in gas chromatography/mass spectroscopy (GC/MS) and LC/fluorescence, as well as other combined chromatography/spectroscopy systems.

The principle involved in spectral rationing is that during the elution of a chromatographic peak of a single component, the spectrum acquired at several points along the chromatographic profile (UV spectrum, MS spectrum, etc.) remains essentially unchanged. Therefore, the ratio of the spectral signals observed in two channels (e.g., the absorbance at two wavelengths in LC/UV) also remains essentially unchanged. Of course, this assumes that certain conditions such as the applicability of Beer's Law (LC/UV) or Henry's Law (GC/MS). These conditions are usually satisfied but must be kept in mind in interpreting the results of certain applications and certain implementation strategies. In practice, the absorbance/spectral ratios are normalized by dividing two ratios, to give an "ideal" indicator of a magnitude of 1; this number is used to indicate the "purity" of the chromatographic peak.

One interpretation of this arbitrary normalization scheme is that for a single component, the result indicates 100% purity. Clearly, if the impurity and the main component have similar or identical spectra, this scheme will fail to detect the presence of the impurity [28].

As described earlier, a coeluants detection strategy based on spectral ratioing is very easy to implement. The information provided, namely deviation from an ideal number of 1, is, in general, qualitative in nature. It must be noted that the effect of random noise in chromatographic detection systems is such that the absorbance ratio indicator is very unlikely to be 1, even for "pure" peaks. The consequence of the ubiquitous noise in chromatographic detection systems is that practitioners of this approach must address the following unavoidable question: How different from 1 could the spectral ratio indicator be before the peak is deemed "impure"? The extent to which this question is addressed can turn a simple ratioing strategy into a serious task of statistical inference. Before we address this issue, let us look at three important features of spectral ratioing.

Computing a ratio involves dividing one number by another. To avoid division by zero (even chemometricians cannot divide by zero) it is necessary to exclude specific channels in spectral ratioing techniques. Spectral ratioing, by its mathematical nature, ignores specific channels and cannot take advantage of the information-rich data. If a suspected coeluant is known to have a response in a specific channel, spectral ratioing may not be the best detection strategy for it. A Biller–Biemann approach (see Section IV.F) may be more sensitive and more appropriate.

Another important feature in spectral ratioing has to do with the effect of random noise. When the noise is channel limited rather than intensity limited, detection channels with low spectral intensity will be influenced by noise much more than channels with high spectral intensity. The spectral ratios of these channels will greatly influence the ratio indicators. This results in a deviation from ideality that can be incorrectly attributed to presence of impurities/coeluants. This behavior is observed in particular at the "edges" of chromatographic peaks and leads to misinterpretation of the so-called ratiograms. To avoid this problem, thresholds values on the spectral intensity need to be set. In other words, only channels with spectral intensities exceeding a certain threshold are considered in the ratioing scheme.

Setting thresholds is a practical solution with two implications. The first is that it introduces an additional parameter (the threshold value) that needs to be controlled and specified as an important experimental parameter when comparing results from different applications and/or different laboratories. The second implication is that by setting a signal threshold, a detection limit (a minimum amount of the analyte) is imposed. Assessing the purity of a peak under these conditions requires a minimum amount to be present, satisfying the threshold requirement. This makes the experimental designs involved in selecting the detection channels, setting the thresholds and setting the detection limits crucial because these factors are very likely to be correlated.

The third important feature in spectral ratioing has to do with background correction. The ratio x/y is not, in general, equal to $(x + c)/(y + c)$ and is not, in general, equal to $(x + a)/(y + b)$. If our target ratio is x/y, our detection strategy needs to be robust enough to handle an offset change such as that represented by $(x + c)/(y + c)$ or a drifting baseline such as that represented by $(x + a)/(y + b)$. To deal with these problems in practice, a background correction needs to be applied to the spectral data before ratioing is implemented. Over the relatively short period of time covered by the chromatographic peak, a linear or a quadratic background correction is often sufficient. This background correction may be the key to successful implementation in gradient elution LC/UV applications.

Now we turn to the unavoidable question of how different from 1 the spectral ratio should be before a peak is deemed "impure"? One approach is to set an appropriate "figure of merit" that is based on knowledge of noise, detection limits, thresholds, and so forth. The chromatographic peak is deemed pure as long as the spectral ratio indicator does not exceed the figure of merit. Such figures of merit have been proposed with values as high as 1.5 [29]. Because such figures of merit are dependent on knowledge of background noise and the choice of channels (wavelengths in LC/UV), they may not transfer well between applications or between laboratories. Indeed, a successful implementation becomes a matter of subjective reasoning and involves a fair amount of art as well as science; it becomes difficult to scientifically justify the inference made based on the detection scheme. The ideal approach is to evaluate the spectral ratios objectively using statistical inference models. This allows us to make our inference with statistical confidence. Such an approach has already been demonstrated with proof and experimental verification [30]. If the absorbance ratio is given as R and the standard deviation of R is denoted by s_R, the relative error (s_R/R) is given by

$$\frac{s_R}{R} = \left(\frac{1}{\text{SNR}_n}\right)\left(1+m^2\right)^{1/2} \tag{12}$$

where $m = \text{SNR}_n/\text{SNR}_d$, and SNR_n and SNR_d are the signal-to-noise ratios of the absorbance in the numerator and denominator, respectively. The above equation shows the dependence of the error in absorbance ratios on both the amplitude of the signal (which is dependent on the concentration) and the wavelengths selected (which is dependent on the experimental design.)

In practice, two spectra are compared. The spectra selected are typically taken on either side of the peak maximum and at a relative height of 50–60% of the height at peak maximum. Theoretically, however, there is no reason why *more* than two spectra cannot be used. It is possible to compute several ratios at several sampling points in order to increase the sensitivity of the detection strategy.

B. Chromatographic Ratioing

Chromatographic ratioing is closely related to spectral ratioing. The rationale is as follows: If two particular channels have enough information to indicate the presence of an impurity based on two (or multiple) sampling points, the information content (and hence the detection sensitivity) can be enhanced by sampling the two channels as long as possible [24,31,32]. The longest sampling period is, of course, equivalent to the baseline width of the chromatographic peak of interest; this detection strategy leads to integrating the channels of interest. Chromatographic ratioing is best implemented as ratioing of the peak areas obtained with each channel. The integration accomplishes a very important goal. It replaces the pairwise ratioing, which is subject to noise effects and thresholding as discussed earlier, with the noise-reducing operation of summing up the individual spectral intensity data in each channel. We end up ratioing the areas of the peaks obtained in the channels of interest.

This technique can be very sensitive even when the spectra are highly correlated. The reason is that the subtle differences in the spectra (as reflected by the selected channels) are accumulated and amplified by the integration process. Table 1 summarizes the results of an example in which trace amount of butyl-

Table 1 Ratios of Peak Areas Monitored at 250 nm, 264 nm, and 274 nm for Pure Toluene and a Trace Amount of Butyl-paraben (0.06% w/w) in Toluene

	$\dfrac{A_{264}}{A_{250}}$	$\dfrac{A_{264}}{A_{274}}$
Pure toluene	1.805 ± 0.004	3.188 ± 0.002
99% Confidence interval	1.782 – 1.828	3.177 – 3.200
Mixture of butyl-paraben in toluene (0.06% w/w)	1.782	3.116

paraben in toluene were determined using high-performance liquid chromatography/ultraviolet (HPLC/UV). These compounds were selected to model two compounds with highly correlated UV spectra coeluting with a chromatographic resolution of less than 2σ. The peak areas of three channels (250 nm, 264 nm, and 274 nm) were monitored. The ratios of the peak areas of pure toluene were first determined in a calibration step. The 99% confidence interval of the ratio was subsequently determined. Detection of a trace amount of butyl-paraben (0.06% w/w) is attempted using the three channels (two ratios; 264 nm/250 nm and 264 nm/274 nm). As shown in Table 1, the presence of the trace amount is inferred using the 264 nm/274 nm ratio, but it is not inferred using the 264 nm/250 nm ratio. This demonstrates the effect of selected channel (the experimental design) shown in Eq. (12).

C. Spectral Correlations

Another way to compare two spectra is to compute their correlation coefficient. If the spectra are considered as two vectors in an n-dimensional space (where n is the number of channels observed), the correlation, r, between the two spectra s_1 and s_2 is given by

$$r = s_1^* \cdot s_2^* \tag{13}$$

where s_i^* is the autoscaled [33] transformation of s_i, and the "·" indicates the dot product of the two vectors (the sum of the pairwise multiplication of spectral intensities at each wavelength). The principle involved here is not unlike that outlined in spectral ratioing. When a single component emerges from a chromatographic column, the spectra acquired during the elution period remain unchanged. By comparing the spectra acquired along the chromatographic profile, it is possible to observe any changes. The correlation coefficient can be used an indicator of these changes [34]. If one or more coeluants are present, the spectra acquired across the profile will differ as the relative amount of each coeluants changes along the profile. If the spectra are different, the correlations between the spectra will differ from 1. Again, one needs to address the question: How different should r be from 1 before the peak is deemed impure? The answer to this question is more readily available because correlation coefficients have been studied much longer than absorbance ratios. The statistical confidence associated with r can be calculated from Fisher's z-transformation [35] which is given as

$$z = \tfrac{1}{2} \ln\left(\frac{1+r}{1-r}\right) \tag{14}$$

Unlike absorbance/spectral ratios, correlation coefficients are not sensitive to offset changes, but they are sensitive to changes in background drifts. Therefore, a background correction is necessary for successful implementation. In general, the information provided by spectral correlation approaches are qualitative in nature. Similar to spectral ratioing, two or ore spectra along the profile are selected and compared. The result is reported in a fashion similar to spectra ratioing, where a "pure" peak is given 100% purity. One may also report the square of the correlation coefficient, as this quantity represents the amount of information or variance that is common to the two spectra being compared.

Spectral ratioing and spectral correlations offer complementary information and should not be viewed as two competing or mutually exclusive techniques. As indicated above, spectral ratioing does not take advantage of specific channels and is affected by noise. By contrast, because they do not employ pairwise division, spectral correlations are more robust at low signal-to-noise ratios and may include specific channels. Also, spectral correlations do not require signal thresholds to be set in the manner that is necessary for spectral ratioing. On the other hand, it must be remembered that correlation coefficients, as indicated by Eq. (13), express a collective property. Consequently, correlation coefficients are not very sensitive to subtle differences in the spectral profile. Spectral ratioing, by contrast, is much more sensitive to subtle changes in the spectral profile as it relies on the response of each individual channel. In applications where the impurity and the main component have highly correlated spectra (such as LC/UV detection of the drugs' metabolites, where the structure of the chromophore remains the same, but some of the side chains are different), ratioing techniques often perform better than correlation techniques because the changes are very subtle. However, in applications where the amount of material is limited and the signal-to-noise ratio of the main component is moderate or low (<20), correlation technique are more robust and are more reliable than ratioing techniques where the effect of low signal-to-noise ratios are much more pronounced.

As with absorbance ratios, the correlations between two or more spectra can be evaluated. The spectra are selected along the chromatographic profile from either side of the peak maximum. The spectra may also be overlayed and presented graphically to the users, with the correlation coefficients (or a suitable transformation) as an indicator of the similarity between two spectra.

D. Factor Analysis/Principal Component Analysis

The application of factor analysis (FA) and principal components analysis (PCA) to chromatography/spectroscopy data is well documented [36,37]. It can be applied in either the spectral domain or, alternatively, in the chromatographic (time) domain. Because FA/PCA relies completely on the acquired data themselves, it is often referred to as a "self-modeling" technique. The FA/PCA techniques have been implemented for peak purity assessment, as well as for peak deconvolution in several variations [38–40].

In 1971, Lawton and Sylvestre [41] demonstrated that mixtures of UV spectra can be used to estimate the spectra of the parent component. The extension of this concept to the estimation of spectra from the mixed spectra acquired along a composite chromatographic profile was reported by Sharaf and Kowalski [42]. With estimates of the parent spectra at hand, one can resolve the composite profiles in at least two different ways. The first uses the parent spectra estimated from FA/PCA and the acquired mixture spectra; multiple linear regression is performed to estimate the relative amounts of the individual components. The second approach takes advantage of the reduced noise in the factor-score space and extracts the quantitative information from the structure of the spectral vectors themselves [43,44]. This structure has been shown to preserve the quantitative information (relative amounts of individual components) as well as the qualitative information (the spectra of the parent components).

Self-modeling deconvolution using FA/PCA offers many advantages. The factor-score space preserves both qualitative information and quantitative information. The method does not require peak models, and it does not require a priori knowledge regarding specific signals. Background correction is unnecessary, and the user does not need to know the number of coeluants in advance. Because of all of these features, self-modeling deconvolution is the closest thing we have to a universal solution, despite some shortcomings. (Of course, adding any of this relevant information to the technique can only improve the results.) A complete application of FA/PCA-based self-modeling will result in estimating the number of significant coeluants, the nature of their spectral profiles, and the individual chromatographic profile of each coeluant. All of these parameters are estimated with bounds, and one can also extract, from the data, an overall analytical figure of merit describing the chromatographic and spectral separation/resolution [45].

One of the most critical aspects in implementing FA/PCA is estimating the number of significant factors or components. In general, factors are extracted from a set of eigenvectors sequentially until the noise level is reached. Malinowski's indicator function and PRESS (PREdiction Sum of Squares) may be used to estimate the significant number of components [36]. Once the number of components is estimated, the parent spectra can be estimated from the factor-score transformation of the experimental data. This is followed by estimating the individual chromatographic profiles.

In the context of peak purity, FA/PCA may be used to estimate the number of significant components within a chromatographic profile. One method is to acquire spectra along the profile and perform eigenvalue analysis on the spectra. If more than one principal components are represent, one can infer the presence of more than one coeluant (impurities.) If, on the other hand, only one significant factor is found to account for the data variance, it may be inferred that no coeluants are present. It must be remembered, however, that the number of significant principal components is highly influenced by the spectral correlations between the coeluant, as well as the chromatographic resolution and the relative amounts of major and minor peaks [42]. In the extreme case where the UV spectrum of a certain drug is identical to that of its metabolite or its degradation products, the presence of the metabolite/degradation products will not be detected. This limitation must be kept in mind when implementing a peak purity detection strategy based on FA/PCA and when interpreting the results of such an implementation. If the spectra of the coeluting components are highly correlated, FA/PCA is not likely to be either sensitive or successful.

E. Spectral Libraries Matching

In the techniques discussed earlier, such as spectral ratioing, spectral correlations, and principal components analysis, one compares the spectra acquired along a chromatographic profile to each other. In spectral libraries matching, one compares the spectra acquired along a chromatographic profile to an external collection of reference spectra (a spectral library). By employing an external reference, it is possible to assess not only the purity of chromatographic peaks but also the performance of the chromatography system. Consider, for example, the case of monitoring a batch of pharmaceutical material where the spectral information are influenced by changes in mobile-phase composition. An external spectral library may be used to control the quality of separation condition, whereas an internal spectral comparison (comparing spectra acquired along the chromatographic profile to each other) will not necessarily reveal these changes. The detection strategy is quite simple. The deviation between the reference library spectrum and the spectra acquired along the chromatographic profile is used to infer the presence of coeluants (impurities). The deviation can also be used to assess the quality of separation and detection components in the chromatography system.

The information obtained by comparing experimental spectra to library spectra is, in general, qualitative in nature. Clearly, this peak purity assessment strategy requires that the components of interest be known a priori. The technique can be used for detection of coeluants as well as for quality control purposes. The matching of library spectra and experimental spectra can be performed in both the forward and reverse modes.

The degree of agreement (or disagreement) between library spectra and experimental spectra can be estimated in a number of ways [46,47]. One of the most common metrics is the dot product between the two spectra. In this approach, the library spectrum and the experimental spectrum are represented by two vectors, and the dot product (a scalar quantity) is computed:

$$s = \sum L_i \cdot E_i \qquad (15)$$

where L_i is the spectral intensity at the ith channel of the library spectrum and E_i is the spectral intensity at the ith channel of the experimental spectrum. Equation (15) has one practical shortcoming, namely it is concentration dependent. In order to eliminate this dependence, both vectors (spectra) are normalized to unit length before the dot product is calculated. The normalized library spectrum, for example, is given as

$$L_i^* = \frac{L_i}{\sqrt{(L_2^1 + L_2^2 + \cdots + L_n^2)}} \qquad (16)$$

where n is the number of channels (e.g., the number of wavelengths in LC/UV).

Once the spectra are normalized to unit length, their dot product becomes a measure of their correlations (association or similarity). The dot product of two vectors, of unit length each, is actually the cosine of the angle between the two vectors. When the vectors are orthogonal (two spectra with no common spectral features), their dot product is 0. When the two vectors are identical (two identical spectra), their dot product is 1. Thus, multiplying the dot product by 100 produces a convenient metric whose limits are 0% agreement (orthogonal spectra) and 100% agreement (two identical spectra.) Other variations (e.g., multiplying by 1000) may also be used.

The dot product metric shares some of the disadvantages of spectral correlation techniques. It is a collective property that may not be sensitive to subtle spectral differences between two spectra. It is not unusual in GC/MS to have the key M/z ratios be the ones with low ion current intensity. In these cases the abundant common signals may obscure the key low-intensity signals. In LC/UV, small but important spectral shoulders may not contribute to the dot product and may not be detected. For these reasons, more sophisticated techniques have been developed for spectral library matching. PBM (Probability Based Matching) and STIRS (Self-Trained Interpretive and Retrieval System), for example, take advantage of the statistical distributions of many key features of the mass spectral information [48]. Other alternative matching methods are based on computing the differences (or the absolute values of the differences) between the library (reference) spectrum and the experimental spectrum. The spectral differences are often manipulated to produce a positive number between 0 (no similarity) and 100 or 1000 (identical spectra).

The use of spectral libraries can be extended to extract quantitative information from unknown mixtures. This was demonstrated by Tunnicliff and Wadsworth [49]. The principle involved in this approach is that once an unknown spectrum is identified by library matching, the relative amounts of its parent components may be estimated from our knowledge of the best library match(es). The application of linear programming and constrained optimization to this problem was demonstrated by Fausett and Weber [50]. Atwater(Fell) used mass spectral libraries to identify unknowns and to suppress the major components, and to match the residuals [51]. Stepwise regression and multiple regression can also be used to evaluate experimental spectra in reference to spectral libraries [52].

Spectral libraries, thus, provide a useful application domain where chromatographic peaks may be assessed on the qualitative and quantitative level. A successful implementation must take advantage of this potential. From a peak purity point of view, library matching offers chromatographers the tools to assess both the identity of coeluants and the relative amounts of each coeluant without having to perform any curve fitting.

F. Other Peak Purity Assessment Techniques

The wide spectrum of chromatographic applications has given rise to numerous variations of the above spectral comparison strategies. One example is the spectral contrasting method where the spectral comparison is based on evaluating the projection of one vector (spectrum) onto the other [53]. Other examples include the weighted-average wavelength computed over the chromatographic profile/peak [54] and the spectral suppression (k-factor) technique [55]. The application of Kalman filters to the analysis of chromatographic peaks was also demonstrated [56,57].

The Biller–Biemann method is one of the earliest techniques applied to chromatographic data acquired with a multichannel hyphenated (in this case GC/MS) application [58]. Its principle and implementation are straightforward. The idea behind the Biller–Biemann approach is that although the chromatographic profile represented by the "total" signal (total ion current in GC/MS or total absorbance in LC/UV) may obscure the detection of coeluants, the profile of the specific detection channel, or the highly selective ones, makes it much easier to detect coeluants. The Biller–Biemann technique, thus, works very well when specific or highly selective channels are present. The information provided by this detection strategy is qualitative and the procedure serves as a good screening tool.

As with other techniques, the implementation of any peak purity detection strategy needs to be evaluated on several operational and statistical principles.

V. OPERATIONAL CONSIDERATIONS OF PEAK PURITY DETECTION STRATEGIES

The ideal peak purity detection implementation must be optimized in terms of cost, sensitivity, and robustness. Cost is critical in the industrial analytical laboratory environment of today. In addition to the price of hardware, cost should also include any human intervention necessary for the peak purity detection operation/system. Any method requiring routine inspection by an experienced chromatographer is, in the opinion of this author, too expensive. The ideal peak purity detection system must be fully automated and must require little, or no, human intervention. It must also be computationally inexpensive. The second important consideration is sensitivity. The ideal peak purity detection system must be able to detect trace amounts of coeluants with high spectral correlations at low chromatographic resolution. The third operational criterion is robustness. The technique must be tolerant of noise and have a large dynamic range. The ideal peak purity detection system must not require frequent system calibration.

Table 2 summarizes the general placement of the above techniques with respect to robustness and sensitivity. Derivative techniques and Fourier transform methods are highly sensitive as well as inexpensive to implement. However, they are not very tolerant of noise. If high levels of noise are suspected, the chromatogram need to be "filtered" and smoothed in a preprocessing step. The other techniques in Table 2 (FA/PCA, library matching, curve fitting of models, multiple regression, ratioing, and correlations) are generally tolerant of noise. Regression, correlations, and ratioing techniques are relatively less expensive than the other three strategies. The sensitivity of these techniques are determined by several factors. Table 3 summarizes the effects of these factors on sensitivity. High

Table 2 General Placements of Peak Purity Detection Strategies in Terms of Sensitivity and Robustness

	Low robustness	High robustness[a]
High sensitivity	Derivatives[b] FT[b]	PCA/FA Library matching Curve fitting Regression[b] Ratioing[b] Correlations[b]
Low sensitivity	⊖	

[a]Sensitivity of these techniques is influenced by the factors shown in Table 3.
[b]Computationally inexpensive.

Table 3 Factors Affecting the Sensitivity of Highly Robust Peak Purity Detection Techniques Shown in Table 2

	High sensitivity	Low sensitivity
High spectral correlations		✓
Low spectral correlations	✓	
Specific channels present	✓	
Specific channels absent		✓
Many channels monitored	✓	
Few channels monitored		✓
Low (good) condition number	✓	
High (bad) condition number		✓
High S/N[a]	✓	
Low S/N		✓
High chromatographic resolution	✓	
Low chromatographic resolution		✓

[a]S/N = signal-to-noise ratio.

sensitivity is attained when the spectral correlation between coeluants is low and when the signal-to-noise ratio is high. The presence of specific channels will increase the sensitivity of techniques based on FA/PCA. In the case of regression techniques, a "good" condition number will increase the sensitivity. As the number of acquisition channels increases, the sensitivity of ratioing techniques and FA/PCA methods increases. An increase in the data acquisition rate improves the sensitivity of modeling and/or curve fitting. When chromatographic resolution increases, the sensitivity of FA/PCA, correlations, and library matching increases.

VI. STATISTICAL CONSIDERATIONS OF PEAK PURITY DETECTION STRATEGIES

The problem of detecting peak impurities is, indeed, a statistical detection problem. This aspect of peak purity assessment is often eclipsed by the immediate challenge of trying to decide what is the "best" technique. However, the statistical aspects of implementing any of the techniques discussed earlier contributes heavily to the reliability of the detection strategy. Table 4 summarizes the response behavior of an ideal peak impurity detection system. As Table 4 indicates, the ideal detection system will give the correct answer 100% of the time. The rate of false positive (indicating the presence of coeluants when there are none) of an ideal peak purity detection system is 0%. Similarly, the rate of false negative (indicating the absence of coeluants when they are present) of an ideal

Table 4 Response Matrix of an Ideal Peak Purity Indicator

Indicator response	Chromatographic peak	
	Pure	Impure
Pure	Correct (100%)	False negative[a] (0%)
Impure	False positive[b] (0%)	Correct (100%)

[a]High spectral correlations and small sampling area increase the rate of false negatives.
[b]Low S/N increases the rate of false positives.

peak purity detection system is also 0%. In practice, the response matrix of a peak purity detection system will be quite different from the ideal response matrix shown in Table 4. We will next demonstrate how a detection strategy based on spectral ratioing can be evaluated in terms of its potential response matrix. The principles involved in the following discussion will apply to other peak purity detection techniques without loss in generalities.

Consider that we wish to assess the purity of a chromatographic peak whose shape can be approximated by the Gaussian profile shown in Figure 5. Also, consider that the peak purity detection strategy being applied is one where two spectra are selected at either side of the peak's crest and at a relative peak height of about 60%. This corresponds to the spectrum acquired at $t_1 = \mu - \sigma$ and the spectrum acquired at $t_2 = \mu + \sigma$. The spectra being compared cover a sampling region of about 68% of the total peak area. The implication of this peak impurity detection strategy is that our sampling scheme (from $t_1 = \mu - \sigma$ to $t_2 = \mu + \sigma$) leaves 32% of the peak area "unguarded." After sampling considerations and before any calculations are performed, there is a 32% chance that an impurity may be present and go undetected. In other words, the rate of a false negative is as high as 32%. In order to reduce this high rate of false negatives, one needs to increase the sampling area. If we sample between $\mu - 2\sigma$ and $\mu + 2\sigma$, our coverage increases from 68% to 95%. However, the spectra acquired at these limits have about 13% of the signal-to-noise ratio at the peak's maximum (Fig. 5). The increase in sampling area is accompanied by a loss in signal-to-noise ratio which, as discussed earlier, will make spectral ratios less reliable and will increase the rate of false positives. Thus, our attempt to reduce the rate of false negatives causes an increase in the rate of false positive.

Minimizing the rate of false positives requires that we maintain an adequately high signal-to-noise ratio. We can maintain a high signal-to-noise ratio

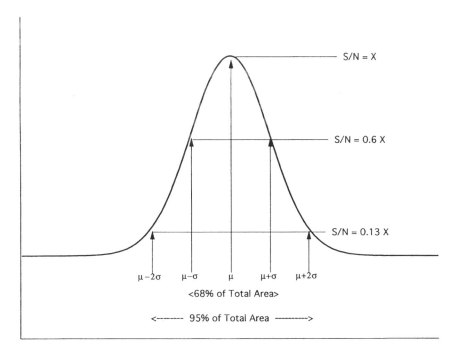

Fig. 5 A hypothetical Gaussian chromatographic peak showing the change in signal-to-noise ratio (S/N) along the profile. The S/N at two standard deviation units away from the mean is about 13% of the S/N at the center of the peak.

by sampling near the center of the chromatographic peak. In doing so, we lose peak coverage and increase the rate of false negatives.

In implementing the above peak detection strategy, we need to balance the rate of false positives and the rate of false negatives in a manner that is compatible with the applications and the risk associated with the false-positive and the false-negative results. Indeed, the results of peak purity assessment must be reported together with the estimated rate of false negatives (or at least the design sampling region) and the estimated rate of false positives (or at least the signal-to-noise ratio of each spectra/channel being analyzed). Without these parameters, it is difficult, if not impossible, to compare results from different chromatography systems. Additionally, in routine applications such as quality control, the response matrix (Table 4) needs to be fully validated in terms of the minimum amounts of coeluants that are necessary to obtain a correct responses near or at the 100% level.

VII. CONCLUDING REMARKS

Peak overlap is *not* a rare event in chromatography [59–62]. Consequently, the need to assess the presence of coeluants will persist. The diversity of chromatographic applications will continue to fuel the development of novel and innovative peak purity assessment methods. It is, however, unlikely that a technique will emerge as the "best" solution to *all* applications. More often than not, it is necessary to combine two or more techniques to satisfy the application at hand.

Practitioners need to be intimately familiar with the strengths and the limitations of the various peak purity assessment methods. This knowledge will help them pair the problem and the technique(s) intelligently. Developers of chromatography systems can help by providing a variety of tools that can be used flexibly in diverse applications. There is one key feature that needs to be addressed by both practitioners and developers when implementing any peak purity assessment technique. This feature is *automation*. Peak purity techniques that rely on graphical displays and/or the interpretation of an expert chromatographer are costly and cannot be used routinely.

When providing a numerical answer that reflects the "purity" of a chromatographic peak, the method must take in consideration the presence of experimental noise. This will become crucial, if not mandatory, as these methods are used in industrial quality control. The use of aribtrary figures of merit to indicate the purity of chromatographic peaks must be avoided. It leads to results that are often unreliable, and the users will not be able to defend their results scientifically.

REFERENCES

1. A. T. Giese and C. S. French, *Appl. Spectrosc.*, *9*: 78 (1955).
2. D. W. Kirmse and A. W. Westerberg, *Anal. Chem.*, *43*: 1035 (1971).
3. J. Skilling and S. F. Gull, in *Maximum Entropy and Bayesian Methods in Inverse Problems*, C. R. Smith and W. T. Grandy, Jr., Eds., Reidel, Dordrecht, 1985, pp. 83–132.
4. A. Grant and P. K. Bhattacharyya, *J Chromatogr.*, *347*: 219 (1985).
5. T. C. O'Haver and T. Begley, *Anal. Chem.*, *53*: 1876 (1981).
6. M. Sharaf, D. Illman, and B. Kowalski, *Chemometrics*, Wiley, New York, 1986, pp. 86–94.
7. J. Glajch, D. Warren, M. Kaiser, and L. B. Rogers, *Anal. Chem.*, *50*: 1962 (1978).
8. J. P. Foley and J. G. Dorsey, *Anal. Chem.*, *55*: 730 (1983).
9. E. Reh, *Trends Anal. Chem.*, *14*: 1 (1995).
10. D. W. Morton and C. L. Young, *J. Chromatogr. Sci.*, *33*: 514 (1995).
11. P. Jones, *Anal. Proc.*, *23*: 261 (1986).
12. W. H. Press, B. P. Flannery, S. A. Teukolsky, and W. T. Vetterling, *Numeri-

cal Recipes in C—The Art of Scientific Computing, Cambridge University Press, Cambridge, 1988, pp. 290–352.
13. A. P. De Weijer, C. B. Lucasius, L. Buydens, G. Kateman, H. M. Heuvel, and H. Mannee, *Anal. Chem.*, *66*: 23 (1994).
14. S. Walsh and D. Diamond, *Talanta*, *42*: 561 (1995).
15. A. Ferry and P. Jacobsson, *Appl. Spectrosc.*, *49*: 273 (1995).
16. E. O. Brigham, *The Fast Fourier Transform and its Applications*, Prentice-Hall, Englewood Cliffs, NJ, 1988, pp. 345–349.
17. W. H. Press, B. P. Flannery, S. A. Teukolsky, and W. T. Vetterling, *Numerical Recipes in C—The Art of Scientific Computing*, Cambridge University Press, Cambridge, 1988, pp. 425–470.
18. J. K. Baker, R. E. Skelton, and C.-Y. Ma, *J. Chromatogr.*, *168*: 417 (1979).
19. P. C. White, *J. Chromatogr.*, *200*: 271 (1980).
20. P. A. Webb, D. Ball, and T. Thornton, *J. Chromatogr. Sci.*, *21*: 447 (1983).
21. A. Fell, H. Scott, R. Gill, and A. Moffat, *J. Chromatogr.*, *282*: 123 (1983).
22. A. Drouen, H. Billiet, and L. De Galan, *Anal. Chem.*, *56*: 971 (1984).
23. H. Cheng and R. Gadde, *J. Chromatogr. Sci.*, *23*: 227 (1985).
24. P. C. White and T. Catterick, *J. Chromatogr.*, *402*: 135 (1987).
25. P. C. White, *Analyst*, *113*: 1625 (1988).
26. P. C. White and A.-M. Harbin, *Analyst*, *114*: 877 (1989).
27. E. L. Inman, M. D. Lantz, and M. M. Strohl, *J. Chromatogr. Sci.*, *28*: 578 (1990).
28. G. Schieffer, *J. Chromatogr.*, *319*: 387 (1985).
29. A. F. Poile and R. D. Conlon, *J. Chromatogr.*, *204*: 149 (1981).
30. M. Sharaf, G. Arroy, and R. Perkins, *J. Chemometrics*, *5*: 291 (1991).
31. R. E. Synovec, E. L. Johnson, T. J. Bahowick, and A. Sulya, *Anal. Chem.*, *62*: 1597 (1990).
32. T. J. Bahowick and R. E. Synovec, *Anal. Chem.*, *64*: 489 (1992).
33. M. Sharaf, D. Illman, and B. Kowalski, *Chemometrics*, Wiley, New York, 1986, pp. 193–194.
34. J. B. Castledine, A. F. Fell, R. Modin, and B. Sellberg, *J. Chromatogr.*, *592*: 27 (1992).
35. J. H. Pollard, *A Handbook of Numerical and Statistical Techniques*, Cambridge University Press, Cambridge, 1977, pp. 196–199.
36. E. R. Malinowski, *Factor Analysis in Chemistry*, 2nd ed., Wiley, New York, 1991.
37. M. Sharaf, D. Illman, and B. Kowalski, *Chemometrics*, Wiley, New York, 1986, pp. 179–292.
38. H. R. Keller, P. Kiechle, F. Erni, D. L. Massart, and J. L. Excoffier, *J. Chromatogr.*, *641*: 1 (1993).
39. Y. Liang and O. M. Kvalheim, *Anal. Chim. Acta*, *292*: 5 (1994).
40. F. C. Sanchez and D. L. Massart, *Anal. Chim. Acta*, *298*: 331 (1994).

41. W. H. Lawton and E. A. Sylvestre, *Technometrics*, *13*: 617 (1971).
42. M. Sharaf and B. Kowalski, *Anal. Chem.*, *53*: 518 (1981).
43. M. Sharaf and B. Kowalski, *Anal. Chem.*, *54*: 1291 (1982).
44. M. Sharaf, *Anal. Chem.*, *58*: 3084 (1986).
45. M. Sharaf, *2nd Hidden Peak Symposium on Computer-Enhanced Analytical Spectroscopy, Snowbird*, UT, June 1–3, 1988.
46. H.-J. Sievert and A. Drouen, in *Diode Array Detection in HPLC*, L. Huber and S. George, Eds., Marcel Dekker, Inc., New York, 1993, pp. 51–126.
47. S. Stein and D. Scott, *J. Am. Soc. Mass Spectrom.*, *5*: 859 (1994).
48. F. W. McLafferty, S. Y. Loh, and D. B. Stauffer, in *Computer-Enhanced Analytical Spectroscopy*, H. Meuzelaar, Ed., Plenum, New York, 1990, pp. 163–181.
49. D. Tunnicliff and D. Wadsworth, *Anal. Chem.*, *37*: 1082 (1965).
50. D. Fausett and J. Weber, *Anal. Chem.*, *50*: 722 (1978).
51. B. Atwater(Fell), R. Venkataraghavan, and F. McLafferty, *Anal. Chem.*, *51*: 1945 (1979).
52. M. Sharaf, *Pittsburgh Conference on Analytical Chemistry and Applied Spectroscopy*, New Orleans, LA, Feb. 22–26, 1988.
53. M. Gorenstein, J. Li, J. Antwerp, and D. Chapman, *LC/GC*, *12*: 768 (1994).
54. T. Alfredson, T. Sheehan, T. Lenert, S. Aamodt, and L. Correia, *J. Chromatogr.*, *385*: 213 (1987).
55. E. Owino, B. Clark, and A. Fell, *J. Chromatogr. Sci.*, *29*: 298 (1991).
56. T. Barker and S. D. Brown, *Anal. Chim. Acta*, *225*: 53 (1989).
57. P. Wentzell, S. Hughes, and S. Vanslyke, *Anal. Chim. Acta*, *307*: 459 (1995).
58. J. Biller and K. Biemann, *Anal. Lett.*, *7*: 252 (1974).
59. D. Rosenthal, *Anal. Chem.*, *54*: 63 (1982).
60. J. M. Davis and J. C. Giddings, *Anal. Chem.*, *55*: 418 (1983).
61. J. M. Davis, *Anal. Chem.*, *66*: 735 (1994).
62. M. C. Pietrogrande, F. Dondi, A. Felinger, and J. M. Davis, *Chemometrics Intell. Lab. Sys.*, *28*: 239 (1995).

2
Fluorescence Detectors in HPLC

Maria Brak Smalley* and Linda B. McGown *P. M. Gross Chemical Laboratory, Duke University, Durham, North Carolina*

I.	INTRODUCTION	30
II.	AN OVERVIEW OF FLUORESCENCE DETECTION IN HPLC	30
III.	STEADY-STATE FLUORESCENCE INTENSITY AND SPECTRAL DETECTION	31
	A. Filter Channel Selection of Emission Wavelength	31
	B. Monochromator Selection of Emission Wavelength	32
	C. Rapid Scanning and Array Detection	32
	D. Excitation–Emission Matrix Detection	34
	E. Synchronous Fluorescence Spectral Detection	37
IV.	INNOVATIONS IN DETECTOR TECHNOLOGY AND METHODOLOGY	37
	A. Selective Modulation	37
	B. Beta-Induced Fluorescence Detection	38
	C. Sequentially Excited Fluorescence Detection	39
	D. Two-Photon-Excited Fluorescence Detection	40
	E. Supersonic Jet Laser Fluorometric Detection	41
	F. Fluorescence-Detected Circular Dichroism Detection	41

**Current affiliation*: Johnson & Johnson Clinical Diagnostics, Rochester, New York

	G. On-Column Fluorescence Detection	42
	H. Indirect Detection	44
	I. Laser-Induced Fluorescence Detection	45
	J. Fluorescence Detection Combined with Other Detectors	46
V.	DYNAMIC-STATE FLUORESCENCE DETECTION	46
	A. Time-Domain Resolution and Lifetime Detection	50
	B. Frequency-Domain Lifetime Detection	59
VI.	CONCLUSIONS	67
	REFERENCES	68

I. INTRODUCTION

The advent of high-performance liquid chromatography (HPLC) and its application to the analysis of complex samples are important developments in the field of chemical analysis. The ability of HPLC to separate sample concomitants is reaching new levels due to the large variety of solid-phase packing materials and particle sizes that are available. Even more impressive are the increasing sensitivity and selectivity of the HPLC detectors used for analyte identification and quantification. Comprehensive reviews of various HPLC detection techniques may be found in the literature (e.g., Refs. 1–3). The focus of this chapter is on HPLC detectors based on measurement of molecular fluorescence, which are among the most selective and sensitive types of HPLC detectors. Particular emphasis is placed on advances that have been made in the past decade, although unique approaches prior to this time are also included. This is not intended as a comprehensive review of fluorescence detection, several of which have been provided elsewhere (e.g., Refs. 4–10).

II. AN OVERVIEW OF FLUORESCENCE DETECTION IN HPLC

Upon absorption of light of appropriate wavelength, a molecule is excited to a higher electronic state. Relaxation back to the more stable ground state may occur through a variety of competing processes, both radiative and nonradiative. Fluorescence occurs when molecules in singlet excited states deexcite through photon emission. It is important to note that the extent to which the radiative and nonradiative processes occur or compete may be affected by the environment of the fluorophore as well as its intrinsic properties.

Fluorescence detection techniques may be applied to intrinsically fluorescent, fluorescent-derivatized, and fluorescent-labeled compounds. Fluorescence detectors are designed to monitor emission through some manner of wavelength selection. Monochromators may be used in the single or multiwavelength modes to acquire intensity or spectral information, respectively. Total fluorescence intensity information may also be obtained for particular wavelength ranges using filter combinations. Other techniques are based on detection of the fluorescence lifetime of the analyte (i.e., the average time a molecule exists in the excited state before returning to the ground state) in the time or frequency domain.

High-performance liquid chromatography fluorescence detectors typically have four main components: an excitation source, a flow cell, a wavelength selector (monochromator or optical filters), and an emission detector. As a requirement, excitation sources only need to electronically excite a molecule into an excited state. This is generally achieved through the use of light sources, which include lasers (continuous wave or pulsed) and lamps (continuum, line, or both). However β particles have also been used as a fluorescence excitation source [11–14]. Emission can be detected at individual wavelengths by photomultiplier tube (PMT) detectors or over a range of wavelengths by array detectors. The data may be collected in a variety of formats: fluorescence intensity measured at one or several discrete excitation–emission wavelength pairs; fluorescence excitation, emission, or synchronous spectra; total luminescence spectra (excitation–emission matrices, or EEMs); or fluorescence lifetime at one or more wavelengths. The detected signal may arise directly from the analyte, from the derivatized analyte, or from a background fluorophore as in indirect detection.

III. STEADY-STATE FLUORESCENCE INTENSITY AND SPECTRAL DETECTION

Steady-state fluorescence detection techniques provide information regarding the intensity of the emission, which is related to the concentration of the analyte, and the spectral characteristics of the emission, which are related to the identity of the analyte. Spectral differences provide a basis for resolution of overlapping peaks, enhancement of selectivity, and reduction of background signals. Fluorescence detection may provide intensity at one or more discrete pairs of excitation–emission wavelengths using filter or monochromator wavelength selection, or it may be used to generate one of several spectral formats, such as an emission spectrum, a synchronous spectrum, or an excitation–emission matrix, by means of scanning monochromators or array detectors.

A. Filter Channel Selection of Emission Wavelength

Measurements of a few different emission wavelength windows provide more information than single-channel detection without the decreases in intensity that occur with monochromator wavelength selection. Winefordner and co-workers [15] simultaneously measured fluorescence intensity at four different filter channel PMTs, each monitoring a different emission wavelength window. They deconvoluted multicomponent peaks by using a least-squares multiple-regression technique. Others have used commercially available instruments as HPLC detectors. In one instance, a commercially available spectrofluorometer, designed to measure the native fluorescence of biogenic amines, was used with lamp excitation and filter channel emission for determination of catecholamine, serotinin, and metabolites [16].

Some HPLC fluorescence detectors are tailored to particular applications. For example, Johnson et al. [17] demonstrated the use of a filter fluorometer that they designed as an HPLC detector for the determination of polynuclear aromatics, aflatoxins, and LSD (N,N-diethyl-D-lysergamide).

B. Monochromator Selection of Emission Wavelength

Organ and co-workers used three pairs of monochromator selected excitation–emission wavelengths for fluorescence detection [18]. They combined wavelength programming with solvent programming to improve component resolution. A single excitation–emission wavelength pair was also used for detection purposes in the HPLC separation of major phospholipid classes [19]. The novelty of this technique is that phospholipid fluorescence was achieved after the postcolumn formation of mixed micelles with 1,6-diphenyl-1,3,5-hexatriene. The investigators used ultraviolet (UV) absorption detection at 205 nm in conjunction with fluorescence detection.

A programmable, dual-wavelength controller for choosing optimum excitation–emission wavelengths pairs to be used during elution has been described [20]. The controller adjusts the monochromator settings on a commercially available fluorescence spectrophotometer that uses a lamp as the excitation source. Also, using a commercially available instrument, Marriott et al. investigated the procedures needed for the optimization of programmable fluorescence detectors. Additional programming options were made possible by using a computer to control the instrument, instead of its own software [21].

C. Rapid Scanning and Array Detection

Rapid scanning of emission spectra during chromatographic elution using commercially available fluorescence instrumentation as the detector has been described. In one study [22], emission spectra were rapidly scanned at multiple excitation wavelengths during peak elution to determine peak purity. If the spectra at the different excitation wavelengths were the same across the peak, then homogeneity (i.e., the presence of a single compound in the peak) was indicated because the emission spectrum of a compound is generally independent of the excitation wavelength used. If different emission spectra were obtained at the different excitation wavelengths, then peak heterogeneity (i.e., coeluting compounds) was indicated (Fig. 1). In another study, the emission spectrum was scanned in the stopped-flow mode at a single excitation wavelength [23]. The flow cell was of a square geometry, reducing the amount of stray light produced.

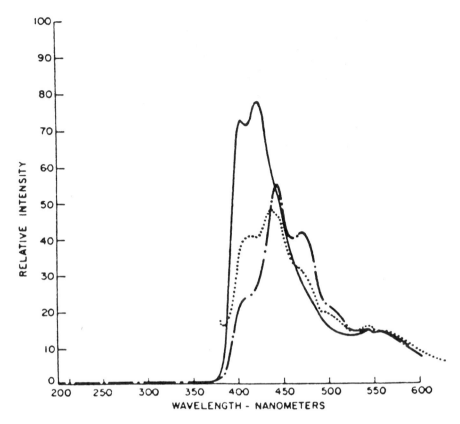

Fig. 1 Emission spectra recorded at excitation wavelengths of 285 nm (—), 350 nm (•••), and 250 nm (–•–) of a chromatographic peak containing a mixture of polycyclic aromatic hydrocarbons. (Reprinted with permission from Ref. 22, copyright 1973 American Chemical Society.)

Photodiode array detectors have been used for continuous monitoring of emission spectra during chromatographic elution. The spectra are compared to those of standard compounds directly [24,25] or after spectral subtraction [26]. In the latter, an intensified linear diode array detector was used with xenon arc lamp excitation for the capillary liquid-chromatographic determination of high-molecular-weight neutral aromatics extracted from fossil fuel material. Jadamec et al. used an optical multichannel analyzer controlled, silicon-intensified target, electronic imager as a multichannel fluorescence detector to monitor the emission spectra of petroleum-derived polyaromatics during chromatographic elution [27]. An example of their data is shown in Fig. 2, which displays the spectral response-time dependency for consecutive spectra of fluorene.

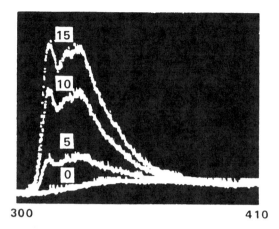

Fig. 2 Multiple-exposure photograph of the CRT screen, at indicated elution time intervals, of the fluorene peak (real-time monitoring). Each exposure represents two separate OMA scans and displays. (Reprinted with permission from Ref. 27, copyright 1977 American Chemical Society.)

D. Excitation–Emission Matrix Detection

Hershberger and co-workers [28] collected EEMs with a videofluorometer and resolved the contributing components of heterogeneous peaks by analyzing the pixel chromatograms of interest. An EEM collected on the fly during elution of benzo[a]pyrene (BaP) is shown in Fig. 3. Warner and co-workers have described two types of deconvolution algorithms, eigenanalysis and ratio analysis, for analysis of multicomponent EEMs collected with a video fluorometer [29,30]. A simplified method for displaying all of the excitation–emission information obtained on the fly was also developed [29]. Excitation spectra or emission spectra may be displayed separately as a function of time, as shown in Fig. 4. In addition to excitation and emission information, total fluorescence intensity and retention-time information is also acquired. This is truly a multidimensional approach.

A second-generation videofluorometer has been described [31], in which a nitrogen-laser-pumped dye laser is the excitation source and a quarter-meter polychromator/microchannel plate-intensified diode array is the detector (Fig. 5). Advantages over the first-generation videofluorometer include increased sensitivity and selectivity. Coupling the microchannel plate image intensifier to the linear photodiode array results in an individual channel sensitivity approaching that of a PMT. Selectivity was improved by rapid tuning of the dye laser to the optimum excitation wavelength of each eluting peak by a scanner-driven, holographic reflecting grating.

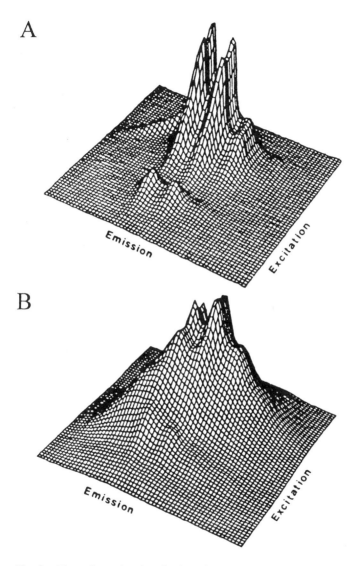

Fig. 3 Three-dimensional projection of EEMs, corresponding to the chromatographic peak of B*a*P, from identical retention lines of the chromatograms of (A) a B*a*P standard and (B) a shale oil sample. (Reprinted with permission from Ref. 28, copyright 1981 American Chemical Society.)

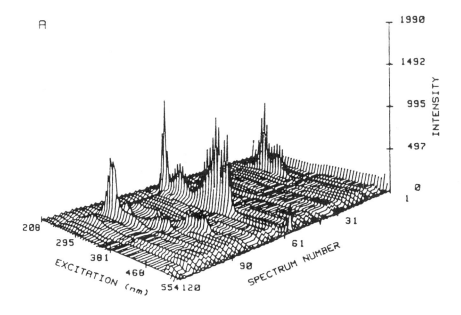
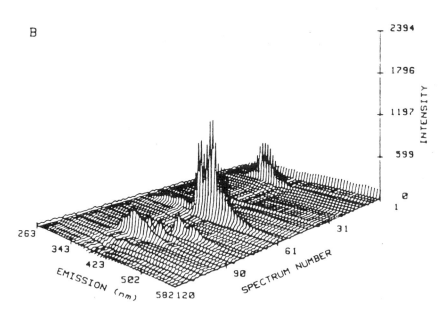

Fig. 4 Time–luminescence matrices derived from the chromatography of a mixture of 18 polynuclear aromatic compounds: (A) time-excitation matrix (TEXM); (B) time–emission matrix (TEMM). (Reprinted with permission from Ref. 29.)

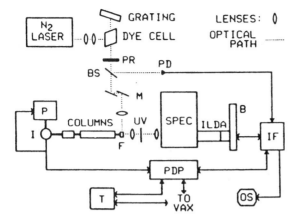

Fig. 5 Block diagram of the laser videofluorometer–HPLC system: PR = dye laser partial reflector; BS = beam splitter; PD = monitor photodiode; M = beam steering mirrors; P = HPLC pumps, mixer, and solvent programmer; I = HPLC injector; F = HPLC flow cell; UV = ultraviolet cutoff filter; SPEC = spectrograph; ILDA = intensified and cooled linear photodiode array; B = Reticon evaluation board; IF = computer interface; PDP = PDP 11/04 computer; T = CRT terminal; OS = oscilloscope. (Reprinted with permission from Ref. 31, copyright 1986 American Chemical Scoiety.)

E. Synchronous Fluorescence Spectral Detection

Kerkhoff and Winfordner used a rapid-scanning fluorimeter (scan rate = 200 nm s^{-1}) to collect a constant-energy synchronous fluorescence spectrum of the components in a mixture of polycyclic aromatic hydrocarbon (PAH) compounds as they were eluted during HPLC [32]. In the collection of synchronous spectra, the excitation and emission monochromators are synchronously scanned at a constant energy difference, maintaining a constant frequency difference of $\Delta\nu$. The resulting spectral peaks correspond to the absorption–fluorescence transitions of the PAHs with an overall vibrational energy loss equal to the $\Delta\nu$ value chosen; in this case, 4800 cm^{-1}. Relative to traditional emission spectra, synchronous spectra provide simpler spectra with narrower spectral bandwidths for improved resolution between different components, as well as reduction of Rayleigh and Raman scattering interference [33,34].

IV. INNOVATIONS IN DETECTOR TECHNOLOGY AND METHODOLOGY

A. Selective Modulation

O'Haver and Parks used selective modulation to resolve chromatographically overlapping PAH pairs that also have overlapping excitation spectra, which would hamper identification of the components by traditional spectral methods [35]. In the selective modulation approach, the sample is excited by excitation wavelengths corresponding to the excitation spectral maxima of the overlapping PAH compounds, and the wavelengths are modulated such that the excitation maximum of each compound is modulated at a different frequency. The resulting emission spectrum of each compound will be modulated at the same frequency as the respective excitation maximum. A lock-in amplifier is used to independently monitor the emission of each of the components.

B. Beta-Induced Fluorescence Detection

The novel use of β⁻ decay radiation of ^{63}Ni and ^{149}Pm as an excitation source for fluorescence detection in HPLC [11–14] has been shown to be more stable than traditional excitation sources, which tend to exhibit intensity fluctuations. Initial studies of β-induced fluorescence (BIF) were performed using the β⁻ decay from a nickel wire that was positioned in the eluent flow (Fig. 6) [11]. A PMT was used to detect the fluorescent scintillations from material passing through the flow cell. The scintillation counts under each peak were integrated and used to identify and quantify the analyte. In later studies, ^{147}Pm was used as the source, as its β⁻ decay energy is greater than that of ^{63}Ni. In this case, the ^{147}Pm source

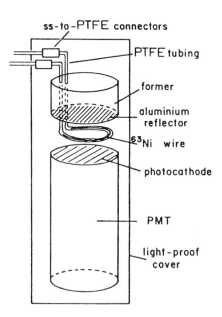

Fig. 6 Schematic diagram of β-induced fluorescence flow cell. Electronic connections to the PMT are not shown. (Reprinted with permission from Ref. 11.)

was introduced through a hole in the side of the flow cell using a positioning screw, so that various flow-cell volumes could be used (Fig. 7). Fluorescence was monitored by a PMT as the eluent passed over the source. Because PMTs cannot detect β particles, monitoring the fluorescence at 90° from the excitation source is unnecessary [12]. This technique can also be used as an indirect detection method for nonfluorescent analytes (see below) by monitoring the quenching of a mobile-phase fluorophore in their presence [14].

C. Sequentially Excited Fluorescence Detection

A vast majority of molecular fluorescence detection techniques monitor the transition from the lowest singlet excited state to the ground state. However, fluorescence from higher excited states may also be detected. In one such case, sequentially excited fluorescence, a molecule is excited to the lowest singlet excited state, which exhibits a symmetry different from the ground state, by ab-

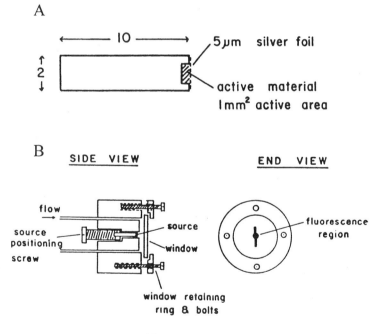

Fig. 7 (A) Structure of the ^{147}Pm source used in the prototype BIF detector system. The cylindrical body is of stainless steel and the dimensions are in millimeters. The source is available from the Radiochemical Centre, code PHC.32. (B) Schematic detail of construction of prototype BIF flow cell. The body of the cell was of stainless steel and the window of Spectrosil A. (Reprinted with permission from Ref. 12.)

sorption of a photon. A second photon is then absorbed by the excited molecule, causing promotion to a higher excited state which has the same symmetry as the ground state. As transitions between states with the same symmetry are generally forbidden, fluorescence from the higher excited state back to the ground state requires interconversion to an intermediate state of different symmetry. Because fluorescence is occurring from a state other than the first excited state, different spectral information about the molecule may be obtained. In their studies [36], Huff et al. used a tunable laser source (nitrogen-pumped dye laser or argon ion laser) to induce sequential resonant excitation. Fluorescence from this highly excited molecular state was monitored by a PMT using a saturated $CuSO_4$ solution and bandpass filters for wavelength selection. The large blue shift that occurs between the excitation and emission wavelengths allows for rejection of stray radiation. The authors suggest that although their work uses only a single laser excitation source, the greatest analytical selectivity would result from using two independently tunable excitation sources.

D. Two-Photon-Excited Fluorescence Detection

Molecules can be excited to higher quantum states through simultaneous absorption of two photons of the same energy. These molecular two-photon states are as common as one-photon states but follow different selection rules (e.g., fluorescence from two-photon excited states is generally forbidden). Different absorption spectra are produced from the absorption of two photons versus the absorption of a single photon because different electronic states are excited, thereby offering possibilities for improved selectivity. Polarization selection is also possible because the two-photon excitation process is dependent on individual polarization conditions of the photons. Two-photon excitation is enhanced if the energy of the individual photons corresponds to one of the electronic states of the molecule, thereby providing a unique type of wavelength dependence [37].

As in sequentially excited fluorescence, a third electronic state is required in two-photon excitation, making this a more restrictive process than normal fluorescence. It is necessary to use a high-power laser for the excitation source because the strength of the two-photon absorption is small and the probability of two-photon excitation increases with increasing photon density. Emission is monitored by a filter channel PMT. Background in two-photon excited fluorometry is almost always negligible because visible photons are absorbed (e.g., from an argon ion or copper vapor laser) and UV photons are emitted. Moreover, the signal increases quadratically with laser power for the two-photon process, whereas the background signal exhibits a linear relations with incident power. Laser beam and power stability are critical to achieve reproducible results because two-photon signals are nonlinearly dependent on laser power density [37,38]. Two-photon excitation has been compared with single-photon excitation for fluorescence detection in HPLC [39].

E. Supersonic Jet Laser Fluorometric Detection

Although fluorescence detectors are sensitive, they are not particularly selective when analytes have poorly resolved spectra. However, spectral resolution may be increased if samples are supercooled before spectral acquisition, which reduces spectral band broadening [40]. Imasaka and co-workers described the application of a supersonic jet laser fluorometric detector to the HPLC of anthracene derivatives [41]. In their instrumental design (Fig. 8), the column effluent is introduced into a continuous-flow supersonic jet nozzle. The analyte molecules are cooled to several Kelvin, and, upon exciting the nozzle, are excited by an excimer-laser-pumped dye laser. The analyte's fluorescence signal is passed through a monochromator for wavelength selection and detected by a PMT. The resulting signal is measured with a boxcar integrator. This technique was able to be resolved three chloroanthracene derivatives (Fig. 9) which could not be resolved by conventional fluorescence detection.

F. Fluorescence-Detected Circular Dichroism Detection

In fluorescence-detected circular dichroism (FDCD), the optical activity of ground-state molecules is measured through detection of the fluorescence emission that is excited alternately by left-handed and right-handed circularly polar-

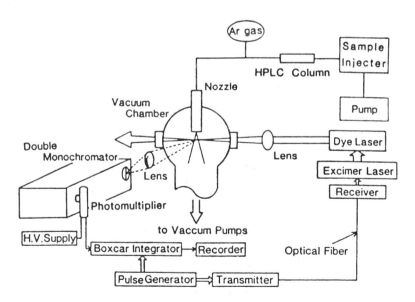

Fig. 8 Supersonic jet spectrometer combined with HPLC. (Reprinted with permission from Ref. 41, copyright 1987 American Chemical Society.)

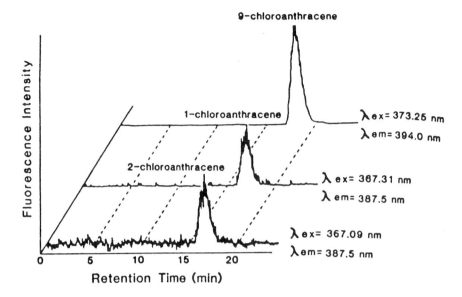

Fig. 9 Chromatograms for a mixture of chloroanthracene derivatives by supersonic jet/fluorometric detector. (Reprinted with permission from Ref. 41, copyright 1987 American Chemical Society.)

ized light [42]. FDCD offers higher sensitivity and detectability than absorption circular dichroism (CD) due to lower background signals and direct dependence of fluorescence emission on the intensity of the excitation beam. Moreover, FDCD is more selective because nonfluorescent chirophores do not contribute directly to the FDCD signal (indirect effects can be corrected through measurements of total absorbance and CD).

Increased selectivity may be achieved by taking advantage of an analyte's intrinsic fluorescence and its optical activity. Synovec and Yeung described a FDCD detector for HPLC (Fig. 10) that uses He–Cd laser excitation at 325 nm and filter channel PMT detection [43]. A Pockel's cell in the excitation beam was used to generate, alternately, right-handed and left-handed circularly polarized light. A lock-in amplifier was used for phase-sensitive detection of the resulting FDCD emission signal. The technique was applied to the determination of optically active (−)riboflavin in a synthetic mixture with optically inactive 4-methylumbelliferone, which did not contribute to the FDCD signal of the riboflavin.

G. On-Column Fluorescence Detection

Although analyte fluorescence is generally measured in solution, measuring the fluorescence of a molecule in the adsorbed state has its advantages. In general, the more rigid a molecule, the greater its fluorescence intensity. For example,

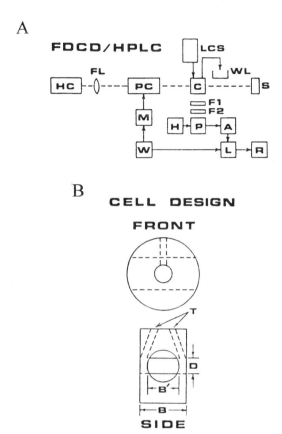

Fig. 10 (A) Fluorescence-detected circular dichroism–HPLC system. HC = He–Cd laser, 8 mW, FL = 50 cm focal length quartz lens, PC = electrooptic modulator (Pockel's cell), M = modulation driver, W = waveform generator, LCS = liquid-chromatographic system, WL = waste liquid, C = detection cell (detailed in B), S = beam stop, F1 = 4-65 Corning filter, F2 = 0-52 Corning filter, P = photomultiplier tube, H = high-voltage power supply, A = AC amplifier, L = lock-in amplifier, R = chart recorder. (B) Chromatographic detection cell. B = 1.8 cm total path length, B' = 1.2 cm observed path length, D = 1.0 mm i.d. quartz tubing, T = chromatography tubing, inlet, and outlet. (Reprinted with permission from Ref. 43.)

Panalaks and Scott [44] packed a detector flow cell with the solid-phase Lichrosorb Si-60 so that the aflatoxins of interest would be detected in the adsorbed state, resulting in increased sensitivity and decreased detection limits.

Analyte fluorescence has been monitored on-column and used for theoretical applications, including investigations of elution nonequilibrium [45] and elimination of extracolumn sources of variance so that variance due only to on-column processes could be determined more accurately [46]. In both applications, two identical, mobile detectors were used to monitor a single solute zone during elution. Excitation at 325 nm was provided by a He–Cd laser and directed to the column by an optical fiber. The resulting fluorescence, transmitted through a second optical fiber, was passed through a monochromator and detected by a PMT. In the first case [45], a single solute zone was detected on-column before the end frit and then again immediately after exiting the column (Fig. 11). This allowed for the direct experimental measurement of changes in the zone profile upon column exiting. In the second case [46], a single solute zone was detected at two points on the column. The first detector was placed near the column inlet; the second was placed near the column end. Any observed difference between the signals at the two detectors could be attributed to column effects only, because extracolumn effects will be "seen" by both detectors, whereas column effects are "seen" only by the detector near the column outlet.

H. Indirect Detection

Any HPLC column and any fluorescence detector may be used in the indirect fluorometric detection of analytes. A mobile-phase additive provides a baseline fluorescence signal that will increase or decrease upon the elution of an analyte.

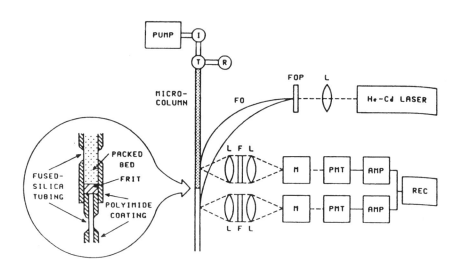

Fig. 11 Schematic diagram of liquid chromatographic system with sequential fluorescence detection. I = injection valve, T = splitting tee, R = restricting capillary, FOP = fiber-optic positioner, FO = fiber optic, L = lens, F = filter, M = monochromator, PMT = photomultiplier tube, AMP = amplifier/current-to-voltage converter, REC = chart recorder. (Reprinted with permission from Ref. 45.)

If the analyte fluoresces with a quantum efficiency similar to, or greater than, the additive, then a positive peak is observed. However, if the analyte is weakly or nonfluorescent, or if the analyte absorbs at the excitation or emission wavelength of the additive, a negative peak is observed [47]. Thus, this approach to indirect detection extends the applicability of fluorescence detection to include nonfluorescent analytes without the need for precolumn or postcolumn chemical derivatization of the analyte.

One of the more popular mobile-phase additives is aniline, which has been used not only for indirect detection [48] but also as a means for studying the mechanism of indirect fluorescence [47]. Other examples of indirect detection differ only in the combination of column and detector type, the mobile-phase additive, and the analytes. Takeuchi and co-workers applied indirect fluorometric detection to both reversed-phase and ion chromatography in micro-HPLC [49]. Chrysene was used as the visualization reagant for the determination of hydrocarbons in the former, whereas salicylate was the mobile-phase additive for the determination of inorganic anions in the latter. Other examples of indirect detection of organic and inorganic analytes may be found in a review of the discussions at the International Symposium on Detection in HPLC and FIA (flow injection analysis [50]).

Another indirect detection method takes advantage of a photoreduction reaction between a mobile-phase additive and the eluting analytes of interest [51]. In this detector, which was applied to the determination of cardiac glycosides and saccharides, the mobile-phase additive anthraquinone-2,6-disulphonate absorbs UV radiation from excitation source and reacts with the analytes to produce a fluorescent product (dihydroxyanthracene-2,6-disulfonate). Emission was monitored by a commercially available fluorometer using either a monochromator-selected wavelength or a filter channel. This approach is particularly useful for the determination of alcohols, aldehydes, ethers, or saccharides that cannot be determined by traditional HPLC methods due to low UV–visible molar absorptivities.

I. Laser-Induced Fluorescence Detection

A detailed discussion of laser-based detection techniques in liquid chromatography, with special attention given to laser-induced fluorescence (LIF) detection, may be found in Ref. 52. Laser-induced fluorescence detection in HPLC is simply the detection of the fluorescence signal that results from laser excitation of an eluting fluorophore. Although the fluorescence intensity resulting from laser excitation is often greater than that resulting from lamp excitation due to the concentration of monochromatic light in a narrow laser beam that can be focused on the small volume of the flow cell, sensitivity and selectivity are not necessarily improved [1]. This is due to increased background signals from scattered light, mobile-phase impurities, and luminescence from the flow-cell walls. Moreover,

Table 1 Examples of LIF Detectors for HPLC

Laser	Flow cell	λ Selector	Detector	Analyte(s)	Ref.	Fig.
Argon ion 488 nm	Sub-µL flow-through cuvette based on sheath flow principle	520-nm and 625-nm cutoff filters	Microscope; PMT	Porphyrins (test compounds)	53	12
Argon ion 488 nm	Quartz capillary tube; optical fiber positioned in flow cell collects emission; others refer to as "Yeung type"	Monochromator and interference filter	PMT	Adriamycin and daunorubicin in human urine	54	13
Krypton ion UV mode 351 and 356 nm	1. Free-falling jet produced by small bore capillary and syringe needle 2. Fused silica capillary tube, polyimide coating removed for window	Monochromator and filters	PMT	Fluoranthene; (dimethylamino)–naphthalene–sulfonylhydrazone derivatives of carbonyl compounds	55	14
Argon ion 351.1 and 363.8 nm	Quartz capillary tube; optical fiber carries excitation directly into solution in cell; 17 µL effective volume	Monochromator and interference filter	Commercial fluorescence spectrometer (PMT)	Dansyl-alanine and fluorescamine derivatives of dopamine and norepinephrine	56	15
Semiconductor 780 nm	Quartz; 1 cm path length; 18 µL effective volume	Monochromator and filter	PMT	Human serum protein labeled with indocyanine green	57	16

Laser	Flow cell	λ Selector	Detector	Analyte(s)	Ref.	Fig.
Helium–cadmium 325 nm	Rectangular quartz; 100 nL effective volume	Filters	PMT; single-electron pulse mode	Dansyl-amino acids	58	—
Argon ion 334–363 nm[a] and 457.9 nm[b]	"Yeung type"	450-nm[a] and 490-nm[b] interference filter	PMT	OPA[a] and NDA[b] derivatized amino acids	59	—
Nitrogen-laser-pumped dye laser 453.9 nm	"Yeung type" and windowless solution droplet type	Bandpass filter	PMT; boxcar integrator	NDA-derivatized amino acids	60	17
Diode 635, 650, and 670 nm	Quartz; external dimensions = 4 × 4 × 10 mm; optical fiber-positioned in flow cell collects emission	Various filter combinations	PMT and photon counter	Aluminum phthalocyanine photosensitizers in biological samples	61	—

inherent instabilities in laser intensity, as compared to traditional sources, contribute to the fluctuations in the background signal, further limiting detectability. Therefore, these limitations must be considered and minimized during detector design.

Various lasers have been used as the excitation source. As in other fluorescence detectors, LIF detectors use filters, monochromators, or a combination of the two for emission wavelength selection and PMTs for detection. The exciting light has been focused on the face of the flow cell or introduced into the flow cell using optical fibers. Likewise, emission has been monitored directly from a flow-cell face, 90° to the excitation, or collected by an optical fiber contained in the flow cell. Because the manners of excitation introduction and emission collection differ, so do the configurations of the flow cells. Various combinations of LIF detector components, as described in the literature, are listed in Table 1. As indicated in the table, several of the LIF systems and flow cells are shown in Figs. 12–17. Some unique approaches to LIF detection are described in other sections of this chapter.

J. Fluorescence Detection Combined with Other Detectors

Fluorescence has been used in conjunction with other techniques to offer multiple detector approaches to component identification. Das and Thomas [62] used four techniques, requiring multiple sample injections, to identify the components of a heterogeneous peak: UV absorption detection, standard addition, differential fluorescence, and peak height ratios at two excitation–emission wavelength pairs. The results of each technique were compared to those for standards. Other multidimensional approaches have been described by Yeung and co-workers. One approach entails collecting UV absorption and one- and two-photon fluorescence spectra simultaneously [63]. A second approach combines UV absorbance, fluorescence intensity, and refractive index measurements to gain more information about a complex mixture [64]. Schmidt and Scott used a trifunctional detector cell (Fig. 18) to monitor column effluent by UV absorption, fluorescence intensity, and electrochemical response [65]. A trifunctional detector allowing measurements of fluorescence, absorbance, and refractive index difference signals was used for the determination of DNA, proteins, and salts in mixtures [66].

V. DYNAMIC-STATE FLUORESCENCE DETECTION

Traditional steady-state approaches based on measurements of intensity at one or more wavelengths have several limitations. Background signals will contribute to the total signal and ultimately limit detectability. For closely eluting compounds, chromatographic resolution may require long run times for isocratic elution conditions or shorter run times with long reequilibration times under gradient elution conditions. Detection at only one or a few wavelengths may be

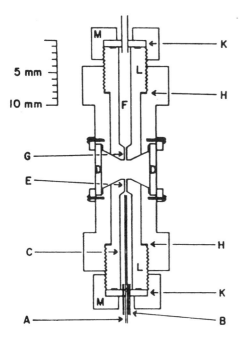

Fig. 12 Schematic diagram of the submicroliter fluorescence flow-through cuvette. A = effluent entry tube, B = sheath entry tubes, C = sheath inlet channel, D = 8 mm diameter quartz windows, E = 500 μm diameter inlet alignment bore, F = exit channel, G = 500 μm diameter exit alignment bore, H = Teflon O-ring, K = inlet or outlet tube holder, L = inlet or outlet probe, and M = stainless-steel nut. (Reprinted with permission from Ref. 53, copyright 1979 American Chemical Society.)

inadequate to resolve chromatographically overlapping peaks, whereas the use of array detection to collect an entire spectrum can decrease sensitivity as a result of dispersion. Furthermore, the techniques may not be readily able to detect the presence of minor components or matrix effects.

This section discusses the use of dynamic fluorescence measurements to improve signal-to-noise by time-resolved detection and to detect fluorescence lifetime as an independent dimension for improved spectral resolution, identification, and determination of overlapping peaks. The discussion of lifetime detection includes both time-domain and frequency-domain techniques. As fluorescence lifetimes are generally concentration independent, they serve as a powerful complement to concentration-dependent intensity and spectral measurements. Isocratic elution is generally necessary because fluorescence lifetimes are sensitive to the solvent polarity changes that occur in gradient elution programming. On

(a)

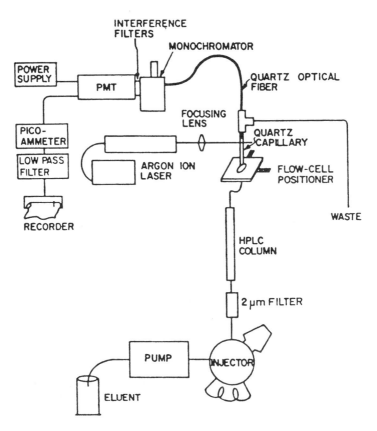

Fig. 13 (a) Apparatus used for the laser fluorometric HPLC detection of adriamycin and daunorubicin. (b) Close-up of fiber-optic flow cell. A = 1/16 in. stainless-steel tee, B = 1/16 in. stainless-steel nut, C = graphite ferrule, D = optical fiber (0.99 mm diameter core, 1.5 mm o.d.), E = 1/16 in. stainless-steel tubing, F = quartz capillary tube (1.05 mm i.d., 2 mm o.d.), G = focused laser beam. (Reprinted with permission from Ref. 54.)

the other hand, the additional information from lifetime detection can provide better resolution of overlapping peaks, thereby circumventing the need for the superior chromatographic resolution of gradient elution.

A. Time-Domain Resolution and Lifetime Detection

In the time domain, a sample is excited with a pulse of light at a given excitation wavelength, λ_{ex}. If the sample contains a single fluorescent species and the dura-

(b)

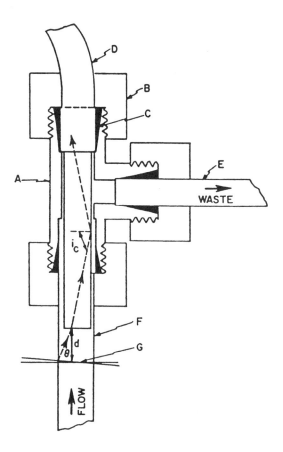

tion of the pulse is negligible, then the time-dependent fluorescence emission, $F(t)$, at a given emission wavelength, λ_{em}, will exhibit the monoexponential decay associated with first-order kinetic processes:

$$F(t) = \left(\frac{F_0}{\tau}\right) e^{-t/\tau} \tag{1}$$

The initial fluorescence intensity, F_0, can be expressed as

$$F_0 = K I_0 \Phi_F \varepsilon b c \tag{2}$$

where K is a constant containing instrumental response factors, I_0 is the intensity of the exciting light, Φ_F is the fluorescence quantum yield of the compound, ε is the molar absorptivity of the compound, b is the cell path length, and c is the

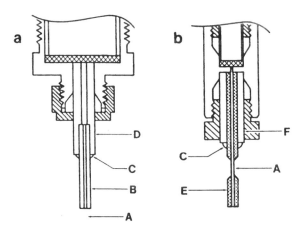

Fig. 14 Schematic diagram of detector cell arrangements: (a) free-falling jet, (b) microbore column quartz capillary. A = laser focusing point, B = syringe needle SGE (17 mm, 0.12 mm i.d., 0.5 mm o.d.), C = cyanoacrylate adhesive, D = 1/16 in. stainless-steel tubing (15 mm, 0.2 mm i.d.), E = fused silica capillary (60 mm, 0.20 mm i.d., 0.30 mm o.d.), F = 1/16 in. stainless-steel tubing (17 mm, 0.35 mm i.d.). (Reprinted with permission from Ref. 55, copyright 1982 American Chemical Society.)

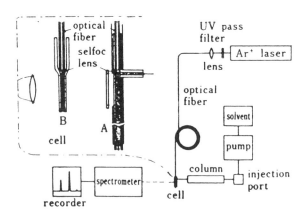

Fig. 15 Schematic of the laser fluorescence spectrometric system with an optical fiber showing detail of microliter size cells. (A) Flowing liquid sample cell, combined with HPLC. (B) Static liquid sample cell. (Reprinted with permission from Ref. 56.)

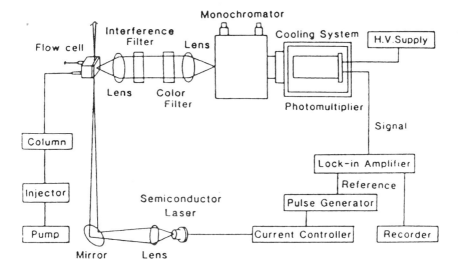

Fig. 16 Block diagram of experimental apparatus for laser-induced fluorescence detection. (Reprinted with permission from Ref. 57, copyright 1986 American Chemical Society.)

concentration of the compound. The fluorescence lifetime, τ, is the mean excited-state lifetime [i.e., the time at which the intensity equals $1/e$ of the initial intensity ($F = F_0/e$ when $t = \tau$)].

High-intensity, pulsed laser excitation has been used to improve detection limits for PAHs in HPLC systems by several orders of magnitude [67,68]. The fluorescence signals of the analytes in the chromatographic peaks are measured after a given time interval following the excitation pulse. The result is an enhancement of the analyte signal relative to shorter-lived background signals, such as those due to fluorescent impurities in the solvent, Raman scattering by the solvent, scattered light from the flow cell, and the excitation pulse itself. The improvement in signal-to-noise by the use of time resolution has been demonstrated for samples extracted from airborne particulates [67]. The instrument used in this study is shown in Fig. 19 [69]. The investigators used a subnanosecond dye laser for excitation, which provided time resolution in the nanosecond range. Their detector system required the collection of a chromatogram at each delay time (Fig. 20). Using this same instrumental system, Ishibashi and coworkers later applied their time-resolved fluorescence detection technique to the determination of protein in human serum, using 1-anilino-8-naphthalene sulfonate (ANS) and N-(1-anilinonaphthyl-4)-maleimide (ANM) as the fluorescent probes [69]. The delay time used for a particular experiment will depend on the decay characteristics of both the background signal and the fluorescent analytes;

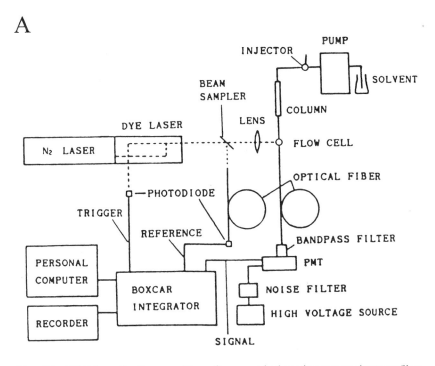

Fig. 17 (A) Schematic diagram of laser fluorometric detection system that uses fiber-optic-based flow cells. (B) Detail of system's fiber-optic-based flow cells. (a) "Yeung-type" flow cell: 1 = optical fiber (25 or 50 m polymer clad silica, 400 μm core diameter, 21 MHz km^{-1} cutoff frequency), 2 = quartz capillary tube (0.52 mm i.d., 0.65 mm o.d.), cell volume ≈ 0.4 μL. (b) Windowless solution droplet cell: 1 = optical fiber (same as above), 2 = quartz tube (0.8 mm i.d., 1.5 mm o.d.), 3 = stainless-steel tube (0.1 mm i.d., 1.6 mm o.d.), cell volume ≈ 4 μL. (Reprinted with permission from Ref. 60.)

a long delay maximizes the reduction of the background signal but also reduces the intensity of the analytes. This results in a curve for S/N (signal-to-noise ratio) versus delay time that rises to a maximum value as the background signal decreases, and then decreases as the analyte signal continues to decay [67].

The effect of delay time on chromatograms was studied for a mixture of 16 PAHs and for lake water extracts using fluorescence-intensity chromatograms which were measured using various delay times (0–40 ns) following the excitation pulse from a nitrogen laser-pumped dye laser [68]. Again, a chromatogram was required for each different delay time. Moreover, the lifetime resolution in this study was limited by the pulse width of 10 ns, which was sufficient only to distinguish between groups of shorter-lived and longer-lived signals but could not accomplish resolution of individual components within the groups, nor were actual lifetimes of individual components determined.

B

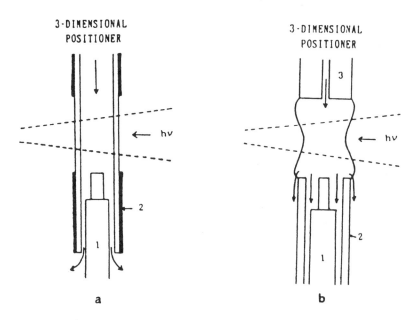

Fig. 17 Continued

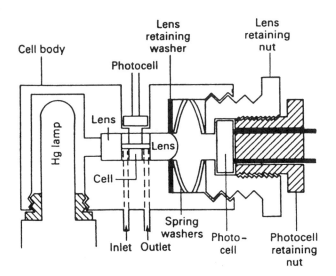

Fig. 18 Schematic diagram of the multifunctional detector combining UV absorption at 254 nm, fluorescence by excitation at 254 nm, and electrical conductivity. (Reprinted with permission from Ref. 65.)

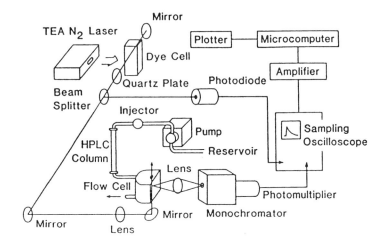

Fig. 19 Schematic diagram of the time-resolved fluorometric detection experimental apparatus. (Reprinted with permission from Ref. 69.)

In addition to improving the S/N ratio through fluorescence lifetime-based background reduction, time-domain techniques have been used to provide information about actual lifetimes of the eluting components [70]. In this study, measurements at multiple delay times were achieved in a single chromatogram by continuous monitoring of the signal resulting from pulsed excitation at a repetition rate of 20–30 Hz as the compounds were eluted. The chromatogram for a given delay time could then be constructed from the appropriate point following each pulse in the original chromatogram in order to obtain decay curves for the various chromatographic peaks, from which the lifetimes of the various peaks could be determined. However, the signal-to-noise ratio was not adequate for resolution of coeluting peaks or recovery of the individual lifetimes of coeluting analytes from multiexponential curves.

In a different lifetime-detection scheme, the PMT anode current was split in half and one of the halves was delayed by 10 ns before being sent to its channel of a two-channel oscilloscope (Fig. 21) [71]. Two intensity chromatograms were thus obtained representing two different delay times that are 10 ns apart. A ratio of the two chromatograms provides a "ratiogram" in which each lifetime is identified by a unique, concentration-independent ratio (Fig. 22). A changing ratio across a peak is indicative of coelution. Indication and resolution of severely overlapped peaks was problematic and could be solved only sometimes by the use of detection at a second wavelength.

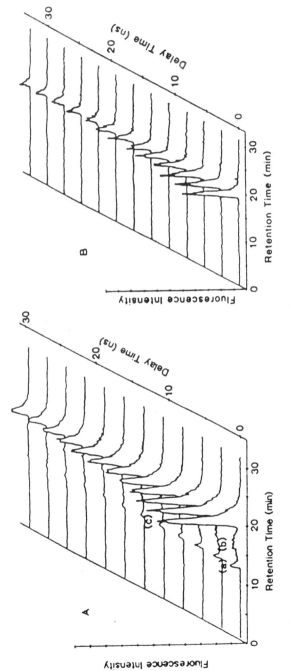

Fig. 20 Three-dimensional chromatograms for human serum. (A) The ANS preadded elution method: (a) α-globulin; (b) γ-globulin; (c) albumin. (B) The ANS prelabeling method; the single peak is assigned to albumin, whereas α- and γ-globulins are not observed because of their small binding constants with ANS. (Reprinted with permission from Ref. 69.)

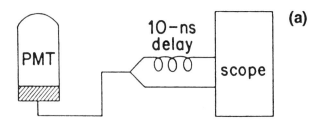

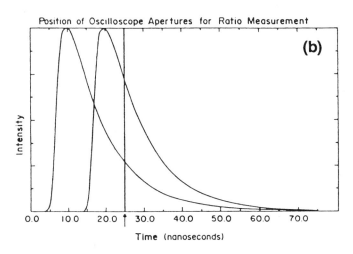

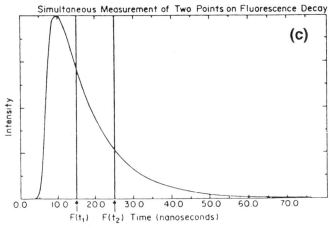

Fig. 21 (a) Simplified diagram illustrating how the signal from the photomultiplier is divided, with half being given a 10 ns delay. (b) Signals seen by the two-channel sampling oscilloscope. (c) Effective monitoring of the fluorescence decay at two points in time by poising the sampling apertures at the point shown in B. (Reprinted with permission from Ref. 71, copyright 1987 American Chemical Society.)

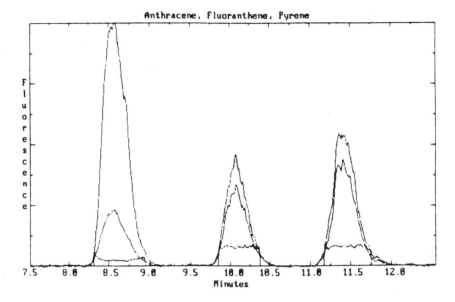

Fig. 22 Channel 1 and channel 2 chromatograms with ratiogram for three PAH standards. From left to right: anthracene, fluoranthene, and pyrene. (Reprinted with permission from Ref. 71, copyright 1987 American Chemical Society.)

B. Frequency-Domain Lifetime Detection

In the frequency-domain experiment [72], the sample is excited with a continuous beam of light of wavelength λ that is intensity modulated at a high (MHz–GHz) frequency, producing a time-dependent intensity signal $E(t)$ that is comprised of a sinusoidal (AC) component superimposed on a time-independent (DC) intensity component:

$$E(t) = A(1 + m_{ex} \sin \omega t) \qquad (3)$$

where A is the steady-state (DC) excitation intensity, $m_{ex} \sin \omega t$ is the time-dependent (AC) component, m_{ex} is the modulation depth (i.e., the ratio of the AC amplitude to the DC intensity), and ω is the angular modulation frequency ($\omega = 2\pi f$, where f is the applied linear modulation frequency in hertz).

Due to the finite lifetime of the excited state of a fluorescent compound, the resulting fluorescence emission of the compound, $F(t)$, will be phased shifted by an angle ϕ and demodulated by a factor m relative to the exciting light:

$$F(t) = A'[1 + m_{ex} m \sin(\omega t - \phi)] \qquad (4)$$

where A' is the steady-state (DC) emission intensity. The fluorescence lifetime can be calculated independently from ϕ,

$$\tau_p = \left(\frac{1}{\omega}\right)\tan\phi \tag{5}$$

and from m,

$$\tau_m = \left(\frac{1}{\omega}\right)\left[\left(\frac{1}{m^2}\right) - 1\right]^{1/2} \tag{6}$$

The excitation phase and modulation (m_{ex}) are calibrated by the use of a reference solution, which can be either a scattering solution ($\tau = 0$ ns) or a fluorophore with a known fluorescence lifetime under the conditions of the experiment.

For a solution containing a single fluorescence lifetime species (i.e., one that would exhibit monoexponential fluorescence decay following pulsed excitation in the time-domain experiment), the observed phase and modulation lifetimes will be equal ($\tau_p \equiv \tau_m$). For a heterogeneous system of multiple components with different fluorescence lifetimes, the observed τ_m will be greater than the observed τ_p. This is because a phase shift gives heavier weight to shorter-lived fluorescence components, whereas demodulation gives heavier weight to longer-lived components. Therefore, $\tau_m > \tau_p$ indicates ground-state heterogeneity in the frequency-domain experiment, just as multiexponential decay indicates heterogeneity in a pulsed experiment.

In its earliest application to HPLC, phase–modulation lifetime detection was limited by available instrumentation, which could only collect data at one modulation frequency per chromatogram and therefore required multiple injections of the sample [73–76]. The resulting chromatograms had to be overlaid perfectly and the data combined at each point along the chromatographic peaks in order to determine fluorescence lifetimes and intensity at each of the points. This technique was applied to an 11-component PAH mixture [76]. Phase and modulation data, along with fluorescence intensity, were collected on-the-fly at 1 s intervals for the mixture at each of five modulation frequencies (5, 10, 15, 25, and 40 MHz), thereby generating five individual chromatograms. The fluorescence-intensity chromatogram at a modulation frequency of 10 MHz is shown in Fig. 23 and the observed phase and modulation lifetime chromatogram obtained at the same frequency is shown in Fig. 24. In order to resolve the peaks of the first six components, which coeluted in pairs, the phase–modulation chromatograms at the five different modulation frequencies were combined in a nonlinear least-squares analysis. Results are shown in Fig. 25. Peaks of components 3 and 4 were only partially resolved, due to the similarities in their lifetimes. The last five components to be eluted were all chromatographically resolved, as indicated by unchanging lifetime across the peak of each component.

Unlike previous lifetime detection in both the time and frequency domains,

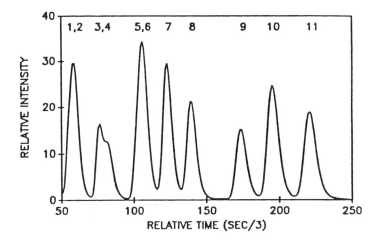

Fig. 23 Steady-state (DC) chromatogram of an 11-component PAH mixture detected at 10 MHz. Numbers above peaks refer to PAH components. (Reprinted with permission from Ref. 76, copyright 1990 American Chemical Society.)

detection by a recently introduced, commercially available multiharmonic Fourier transform phase–modulation fluorometer (MHF, Fig. 26) is sufficiently fast and information-rich to provide on-the-fly lifetime determination, indication of chromatographic peak overlap, and resolution of coeluting peaks, all in a single chromatographic run. This is made possible by the simultaneous collection of fluorescence intensity and multifrequency phase–modulation data at subsecond intervals during elution. The intensity of the exciting light, provided by a continuous-wave laser, is electrooptically modulated by a Pockel's cell at as many as 75 frequencies simultaneously, in the MHz range. The resulting emission is modulated at the same multiple frequencies and is detected using the cross-correlation technique [77]. The data are then digitized and Fourier transformed into the resultant phase and modulation information used to determine fluorescence lifetime.

The first use of the MHF for on-the-fly fluorescence lifetime detection has been described and demonstrated for reverse-phase HPLC of one- and two-component systems of PAHs [78,79]. Fluorescence lifetime and intensity information were collected simultaneously at 0.096 s intervals during chromatographic peak elution. The acquired data were then averaged over 0.96 s intervals (or 10 signal samplings) for presentation purposes. Because fluorescence lifetime of a compound is concentration independent over a wide dynamic range, the lifetime should remain constant across a chromatographic peak of a resolved compound. This is shown in Fig. 27, which illustrates the excellent precision of on-the-fly

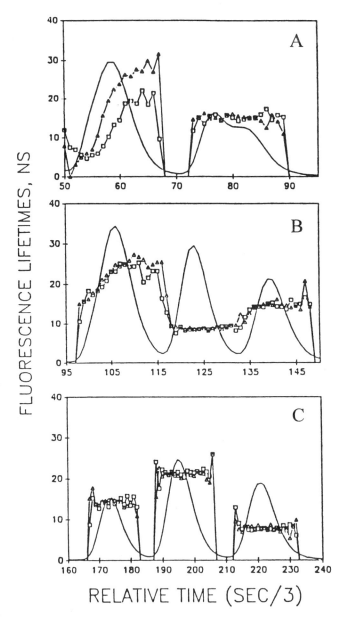

Fig. 24 Fluorescence lifetime chromatogram of the 11-component PAH mixture detected at 10 MHz, with an expanded time axis relative to Fig. 23: (A) peaks 1 through 4, expanded time axis; (B) peaks 5 through 9, expanded time axis; (C) peaks 9 through 11, expanded time axis. Legend: (—) steady-state (DC) intensity, (Δ) τ_m, (□) τ_p. (Reprinted with permission from Ref. 76, copyright 1990 American Chemical Society.)

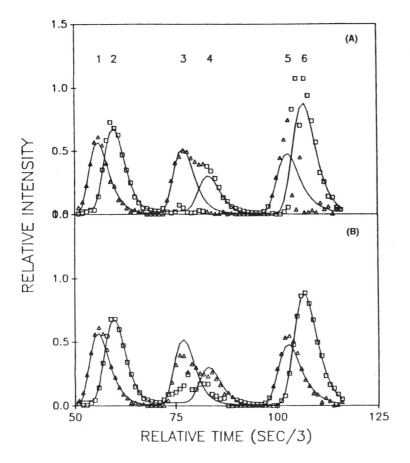

Fig. 25 Results for lifetime resolution of overlapping peaks of components 1 through 6 in the 11-component mixture: (A) lifetimes of the components not fixed during analysis; (B) lifetimes fixed during analysis to values found from the lifetime chromatogram in Fig. 24. (Reprinted with permission from Ref. 76, copyright 1990 American Chemical Society.)

MHF detection, which extends well into the low concentrations at the peak peripheries.

The sensitivity of lifetime detection to minor signal contributions facilitates resolution of chromatographically overlapping components when the signals of the components are far from equal. For example, the resolution of overlapping peaks of benzo[*b*]fluoranthene (B*b*F) and benzo[*k*]fluoranthene (B*k*F) is shown in Figs. 28 and 29, for which the injection amount of B*k*F was varied while that of B*b*F remained constant, and timed injections were used to accomplish coelu-

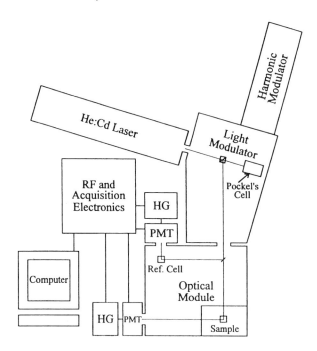

Fig. 26 Schematic of the multiharmonic Fourier transform fluorescence lifetime instrument (MHF): HG = harmonic comb generator, RF = radio frequency. The radio-frequency generator, which feeds the base modulation and cross-correlation frequencies to the harmonic modulator and HGs, respectively, is not shown. (Courtesy of SLM Instruments, Inc.)

tion. Although the steady-state chromatogram for plot A appears to be a severely tailing peak of a single component, the observed lifetime changes across the eluting peak (Fig. 28). Construction of the contributions of the individual components indeed shows coelution of a minor component, in this case BkF (Fig. 29). In addition to indicating coelution, lifetime detection can also indicate photophysical interactions among the coeluting components, which might otherwise lead to inaccurate resolution [79]. Most recently, MHF lifetime detection has been applied to complex PAH mixtures and real wastewater samples with encouraging and exciting results [80].

The concentration independence of fluorescence lifetime precludes the use of traditional methods for determining detection limits. Therefore, a new method has been proposed and applied to on-the-fly detection [79]. Limits of lifetime de-

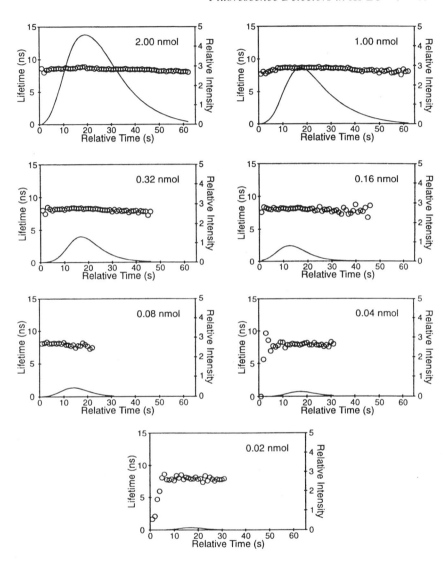

Fig. 27 Fluorescence intensity (—) and lifetime (○) recovered from raw MHF data collected on-the-fly for various injection amounts of BkF. (Reprinted with permission from Ref. 78, copyright 1993 American Chemical Society.)

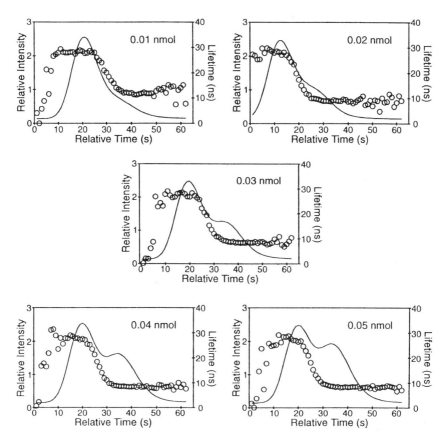

Fig. 28 Fluorescence intensity (—) and lifetime (O) recovered from raw MHF data collected on-the-fly for overlapping BbF and BkF peaks. The injection amount of BbF was 0.20 nmol in all cases, and the injection amount of BkF was varied as indicated on each plot. (Reprinted with permission from Ref. 78, copyright 1993 American Chemical Society.)

tection were reported to be 6.3 pmol (1.6 ng) for BbF and 0.31 pmol (79 pg) for BkF in mixtures of the two. Limits of lifetime resolution for MHF detection were also reported in the same article. Using simple, two-component mixtures, it was found that if the lifetime ratio of two coeluting components is less than 1.2, the lifetime analysis is unable to indicate the presence of more than one lifetime component, thereby making component resolution impossible.

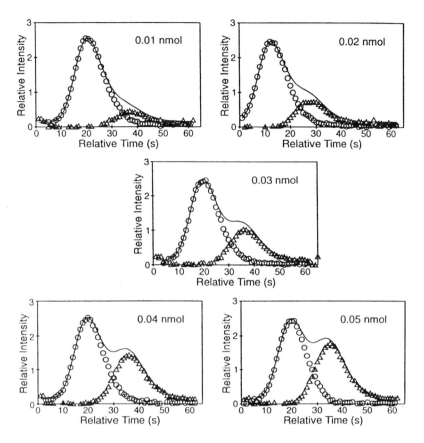

Fig. 29 Results for lifetime resolution of overlapping BbF (O) and BkF (Δ). The lifetime data were fit to a two-component model, with the lifetimes fixed to values recovered from separate injections of BbF and BkF. (—) represents the fluorescence intensity chromatogram for the overlapping components, collected on-the-fly. (Reprinted with permission from Ref. 78, copyright 1993 American Chemical Society.)

VI. CONCLUSIONS

New technological advances show promise for chromatographic detectors with increased sensitivity and selectivity, including such frontiers as single-molecule detection, indirect detection of nonfluorescent analytes, and dynamic lifetime detection. Such advances, combined with multidimensional detection approaches, reduce the need to achieve complete, physical separation of compounds. This can reduce analysis times as well as the need for robust sample clean-up. In addition, complex systems may be more accurately and easily analyzed by multidimensional detection approaches that account for matrix effects and interference.

REFERENCES

1. E. S. Yeung and R. E. Synovec, *Anal. Chem.*, 58: 1237A (1986).
2. C. A. Bruckner, M. D. Foster, L. R. Lima, R. E. Synovec, R. J. Berman, C. N. Renn, and E. L. Johnson, *Anal. Chem.*, 66: 1R (1994).
3. I. S. Krull, in *Chromatography and Separation Chemistry: Advances and Developments*, S. Ahuja, Ed., ACS Syposium Series 297, American Chemical Society, Washington, DC, 1986, Chap. 9.
4. R. Weinberger, *Lab. Prac.*, 36: 65 (1987).
5. P. Froehlich, *BioChromatography*, 2: 144 (1987).
6. D. C. Shelly and I. M. Warner, in *Liquid Chromatography Detectors*, T. M. Vickrey, Ed., Marcell Dekker, Inc., New York, 1983, Chap. 3.
7. A. Hulshoff and H. Lingeman, in *Molecular Luminescence Spectroscopy*, S. G. Schulman, Ed., ACS Monograph 77, John Wiley & Sons, New York, 1985, Chap. 7.
8. M. J. Sepaniak and C. N. Kettler, in *Detectors for Liquid Chromatography*, E. S. Yeung, Ed., ACS Monograph 89, John Wiley & Sons, New York, 1986, Chap. 5.
9. H. Lingeman, C. Gooijer, N. H. Velthorst, and U. A. T. Brinkman, in *Analytical Applications of Spectroscopy II*, A. M. C. Davies and C. S. Creaser, Eds., The Royal Society of Chemistry, Cambridge, 1991, pp. 189–212.
10. H. Lingeman, W. J. M. Underberg, A. Takadate, and A. Hulshoff, *J. Liq. Chromatogr.*, 8: 789 (1985).
11. D. J. Malcolme-Lawes, P. Warwick, and L. A. Gifford, *J. Chromatogr.*, 176: 157 (1979).
12. D. J. Malcolme-Lawes and P. Warwick, *J. Chromatogr.*, 200: 47 (1980).
13. D. J. Malcolme-Lawes, S. Massey, and P. Warwick, *J. Chem. Soc., Faraday Trans. 2*, 77: 1795 (1981).
14. D. J. Malcolme-Lawes, S. Massey, and P. Warwick, *J. Chem. Soc., Faraday Trans. 2*, 77: 1807 (1981).
15. K. Tanabe, M. Glick, B. Smith, E. Voigtman, and J. D. Winefordner, *Anal. Chem.*, 59: 1125 (1987).
16. T. Honma, *Anal. Lett.*, 19: 417 (1986).
17. E. Johnson, A. Abu-Shumays, and S. R. Abbott, *J. Chromatogr.*, 134: 107 (1977).
18. K. Ogan, E. Katz, and W. Slavin, *Anal. Chem.*, 51: 1315 (1979).
19. W. Bernhard, M. Linck, H. Creutzburg, A. D. Postle, A. Arning, I. Martin-Carrera, K.-Fr. Sewing, *Anal. Biochem.*, 220: 172 (1994).

20. G. K. C. Low, G. E. Batley, and B. S. Moore, *Anal. Instrum. (N.Y.)*, *17*: 339 (1988).
21. P. J. Marriott, P. D. Carpenter, P. H. Brady, M. J. McCormick, A. J. Griffiths, T. S. G. Hatvani, and S. G. Rasdell, *J. Liq. Chromatogr.*, *16*: 3229 (1993).
22. E. D. Pellizzari and C. M. Sparacino, *Anal. Chem.*, *45*: 378 (1973).
23. W. Slavin, A. T. Rhys Williams, and R. F. Adams, *J. Chromatogr.*, *134*: 121 (1977).
24. J. Wegrzyn, G. Patonay, M. Ford, and I. Warner, *Anal. Chem.*, *62*: 1754 (1990).
25. R. J. van de Neese, C. Gooijer, G. P. Hoornweg, U. A. T. Brinkman, N. H. Velthorst, and S. J. van der Bent, *Anal. Lett.*, *23*: 1235 (1990).
26. J. C. Gluckman, D. C. Shelly, and M. V. Novotny, *Anal. Chem.*, *57*: 1546 (1985).
27. J. R. Jadamec, W. A. Saner, and Y. Talmi, *Anal. Chem.*, *49*: 1316 (1977).
28. L. W. Hershberger, J. B. Callis, and G. D. Christian, *Anal. Chem.*, *53*: 971 (1981).
29. M. P. Fogarty, D. C. Shelly, and I. M. Warner, *HRC CC, J. High Resolut. Chromatogr. Chromatogr. Commun.*, *4*: 561 (1981).
30. D. C. Shelly, M. P. Fogarty, and I. M. Warner, *HRC CC, J. High Resolut. Chromatogr. Chromatogr. Commun.*, *4*: 616 (1981).
31. D. B. Skoropinski, J. B. Callis, J. D. S. Danielson, and G. D. Christian, *Anal. Chem.*, *58*: 2831 (1986).
32. M. J. Kerkohoff and J. D. Winefordner, *Anal. Chim. Acta*, *175*: 257 (1985).
33. J. B. F. Lloyd, *Nature (London) Phys. Sci.*, *231*: 64 (1971).
34. T. Vo-Dinh, *Anal. Chem.*, *50*: 396 (1978).
35. T. C. O'Haver and W. M. Parks, *Anal. Chem.*, *46*: 1886 (1974).
36. P. B. Huff, B. J. Tromberg, and M. J. Sepaniak, *Anal. Chem.*, *54*: 946 (1982).
37. W. D. Pfeffer and E. S. Yeung, *Anal. Chem.*, *58*: 2103 (1986).
38. M. J. Sepaniak and E. S. Yeung, *Anal. Chem.*, *49*: 1554 (1977).
39. R. J. van de Neese, A. J. G. Mank, G. P. Hoornweg, C. Gooijer, U. A. T. Brinkman, and N. H. Velthorst, *Anal. Chem.*, *63*: 2685 (1991).
40. J. M. Hayes and G. J. Small, *Anal. Chem.*, *55*: 565A (1983).
41. T. Imasaka, N. Yamaga, and N. Ishibashi, *Anal. Chem.*, *59*: 419 (1987).
42. D. H. Turner, I. Tinoco, and M. F. Maestre, *J. Am. Chem. Soc.*, *96*: 4340 (1974).
43. R. E. Synovec and E. S. Yeung, *J. Chromatogr.*, *368*: 85 (1986).
44. T. Panalaks and P. M. Scott, *J. Assoc. Off. Anal. Chem.*, *60*: 583 (1977).
45. C. E. Evans and V. L. McGuffin, *J. Liq. Chromatogr.*, *11*: 1907 (1988).
46. C. E. Evans and V. L. McGuffin, *Anal. Chem.*, *60*: 573 (1988).
47. E. L. Torres, F. van Geel, and J. D. Winefordner, *Anal. Lett.*, *16*: 1207 (1983).

48. S. Y. Su, A. Jurgensen, D. Bolton, and J. D. Winefordner, *Anal. Lett.*, *14*: 1 (1981).
49. T. Takeuchi, M. Murayama, and D. Ishii, *Chromatographia*, *25*: 1072 (1988).
50. G. Christian, I. Krull, and J. Tyson, *Anal. Chem.*, *62*: 455A (1990).
51. M. S. Gandelman, J. W. Birks, U. A. T. Brinkman, and R. W. Frei, *J. Chromatogr.*, *282*: 193 (1983).
52. C. M. B. Van Den Beld and H. Lingeman, in *Luminescence Techniques in Chemical and Biochemical Analysis*, W. R. G. Baeyens, D. De Keukeliere, and K. Korkidis, Eds., Marcel Dekker, Inc., New York, 1991, Chap. 9.
53. L. W. Hershberger, J. B. Callis, and G. D. Christian, *Anal. Chem.*, *51*: 1444 (1979).
54. M. J. Sepaniak and E. S. Yeung, *J. Chromatogr.*, *190*: 377 (1980).
55. S. Folestad, L. Johnson, B. Josefsson, and B. Galle, *Anal. Chem.*, *54*: 925 (1982).
56. H. Todoriki and A. Y. Hirakawa, *Chem. Pharm. Bull.*, *32*: 193 (1984).
57. K. Sauda, T. Imaska, and N. Ishibashi, *Anal. Chem.*, *58*: 2649 (1986).
58. A. P. Lobazov, V. A. Mostovnikov, S. V. Nechaev, B. G. Belenkii, J. J. Kever, E. M. Korolyova, and V. G. Maltsev, *J. Chromatogr.*, *365*: 321 (1986).
59. M. C. Roach and M. D. Harmony, *Anal. Chem.*, *59*: 411 (1987).
60. K. Tsunoda, A. Nomura, J. Yamada, and S. Nishi, *Anal. Chim. Acta*, *229*: 3 (1990).
61. A. J. G. Mank, C. Gooijer, H. Lingeman, N. H. Velthorst, and U. A. T. Brinkman, *Anal. Chim. Acta*, *290*: 103 (1994).
62. B. S. Das and G. H. Thomas, *Anal. Chem.*, *50*: 967 (1978).
63. E. S. Yeung and M. J. Sepaniak, *Anal. Chem.*, *52*: 1465A (1980).
64. S. A. Wilson and E. S. Yeung, *Anal. Chem.*, *57*: 2611 (1985).
65. G. J. Schmidt and R. P. W. Scott, *Analyst*, *110*: 757 (1985).
66. A. A. Evstrapov, B. P. Kuz'min, K. L. Matisen, and N. A. Perevezentseva, *Opt.-Mekh. Prom-st.*, *8*: 23 (1986).
67. T. Imasaka, K. Ishibashi, and N. Ishibashi, *Anal. Chim. Acta*, *142*: 1 (1982).
68. N. Furuta and A. Otsuki, *Anal. Chem.*, *55*: 2407 (1983).
69. Y. Kawabata, K. Sauda T. Imasaka, and N. Ishibashi, *Anal. Chim. Acta*, *208*: 255 (1988).
70. K. Ishibashi, T. Imasaka, and N. Ishibashi, *Anal. Chim. Acta*, *173*: 165 (1985).
71. D. J. Desilets, P. T. Kissinger, and F. E. Lytle, *Anal. Chem.*, *59*: 1830 (1987).
72. J. R. Lakowicz, *Principles of Fluorescence Spectroscopy*, Plenum, New York, 1983.
73. W. T. Cobb and L. B. McGown, *Appl. Spectrosc.*, *41*: 1275 (1987).

74. W. T. Cobb, K. Nithipatikom, and L. B. McGown, in *Progress in Analytical Luminescence*, D. Eastwood and L. J. Cline Love, Eds., ASTM, Philadelphia, PA, 1988, pp. 12–25.
75. W. T. Cobb and L. B. McGown, *Appl. Spectrosc.*, *43*: 1363 (1989).
76. W. T. Cobb and L. B. McGown, *Anal. Chem.*, *62*: 186 (1990).
77. R. D. Spencer and G. Weber, *Ann. N.Y. Acad. Sci.*, *158*: 361 (1969).
78. M. B. Smalley, J. M. Shaver, and L. B. McGown, *Anal. Chem.*, *65*: 3466 (1993).
79. M. B. Smalley and L. B. McGown, *Anal. Chem.*, *67*: 1371 (1995).
80. M. B. Smalley, Dissertation, Duke University, 1995.

3A
Carbon-Based Packing Materials for Liquid Chromatography
Structure, Performance, and Retention Mechanisms

John H. Knox *University of Edinburgh, Edinburgh, United Kingdom*
Paul Ross *Hypersil, Runcorn, Cheshire, United Kingdom*

I.	INTRODUCTION	74
II.	DEVELOPMENT AND PRODUCTION OF CARBON-BASED PACKING MATERIALS FOR HPLC	75
III.	STRUCTURE OF PGC/HYPERCARB®	78
IV.	PERFORMANCE OF PGC/HYPERCARB®	79
V.	BASIS OF RETENTION BY GRAPHITE IN LC AS SEEN IN 1988	81
VI.	RETENTION STUDIES POST-1988	85
VII.	EXISTING THEORIES OF RETENTION IN REVERSED-PHASE CHROMATOGRAPHY AND THEIR APPLICATION TO GRAPHITE	93
VIII.	CORRELATION OF RETENTION ON GRAPHITE WITH MEASURED OR CALCULATED PHYSICAL PROPERTIES	99
	A. Single-Parameter Approach	104
	B. Factor Analysis Approach	113
IX.	CONCLUSIONS	116
	REFERENCES	116

I. INTRODUCTION

An earlier review by Knox and Kaur [1] discussed the role of carbon in liquid chromatography as it appeared in 1988. Since then a number of important developments and advances have occurred which we now cover, at the same time briefly reviewing the background. For fuller accounts of work on carbon prior to 1988, the reader is referred to the above review and to that of Knox et al. [2] published in 1983.

Carbon was frequently used in the early days of chromatography [3], but the active carbons then available showed strongly nonlinear adsorption isotherms which made them unsuitable for elution chromatography as now practiced. With the development of thin-layer chromatography, they were displaced by silica gels. With the emergence of high-performance liquid chromatography (HPLC) in the 1970s, major improvements were made in materials for liquid chromatography, the most important being the development of the bonded silica gels, typified by the alkyl silicas [4]. These have been supplemented by a wide variety of specialized bonded materials and polymeric supports which evolved from the early polystyrene-based ion exchangers [5]. Berek and Novák [6] have reviewed the characteristics of the main classes of support materials used for HPLC.

Since they were first described in the early 1970s, a number of problems have surfaced with the bonded silica gels. The chief of these are as follows: (1) the difficulty experienced by manufacturers in maintaining reproducibility, which is partly due to uncontrollable variability of the starting materials used in the manufacture of silica gel itself and partly to that of the reagents used in the subsequent bonding procedures; (2) the very significant differences in retentive properties of supposedly equivalent bonded silicas made by different manufacturers (e.g., the octadecyl silica gels); (3) the limited hydrolytic stability of silica-based packing materials, which restricts the pH of eluents for long-term use to the range 2–8; (4) the unavoidable presence of unreacted silanol groups; and (5) the variable content of metallic impurities. Problems 4 and 5 are thought to be the major causes of variability noted in problems 1 and 2.

There have, therefore, been many attempts to develop packings with improved characteristics. Carbon in the form of graphite is one of the most attractive, as it should be free from the disadvantages peculiar to silica gels. The reason is simple. Pure graphite is a crystalline material made up of sheets containing huge numbers of hexagonally arranged carbon atoms linked by the same conjugated 1.5-order bonds which are present in any large polynuclear aromatic hydrocarbon. There are, in principle, no adventitious functional groups on the surface because the aromatic C atoms have all valencies satisfied within the graphitic sheets. The carbon atoms in any sheet are coplanar and about 1.4 Å apart. The individual sheets of carbon atoms are about 3.4 Å apart and are held together by London dispersion interactions (i.e., instantaneous-dipole–induced-dipole interactions between the carbon atoms in adjacent sheets). Perfect graphite is therefore an intrinsically reproducible material with a completely uniform surface, free from any functional groups. Its properties should not, and indeed do not, depend on its method of manufacture, except those, such as specific surface area and pore volume, which depend on particle geometry. Furthermore, graphite is one of the most unreactive substances known and can withstand any eluant which will not attack the HPLC equipment itself. Kiselev and Yashin [7] have pointed out that graphite should be the ideal chromatographic material.

Problems, however, may arise when one asks for a material of high specific surface area, as this implies that the graphite crystallites will have to be very small, having dimensions of the order of 1–10 nm. Surface defects must therefore be present at the edges of such crystallites and the ideality of bulk graphite is to some extent lost. However, it should still be possible to provide high-surface-area graphites with minimal adventitious surface groups. As an example, a square array of graphitic carbon atoms with a side of 10 nm will contain about 40,000 C atoms, of which about 160 will be at the edge of the sheet. The edge atoms will thus comprise about 0.4% of the total C atoms. Such edge atoms will have unsatisfied valencies which are likely to be occupied by –OH, =O, and –COOH groups, and possibly N-containing groups. If the crystallites are about 30 layers thick, then their specific surface area will be of the order of 300 m^2/g, a value typical for an HPLC material.

II. DEVELOPMENT AND PRODUCTION OF CARBON-BASED PACKING MATERIALS FOR HPLC

As pointed out in an earlier review [1], the ideal HPLC packing material needs to have a combination of properties:

(a) Availability in the form of small spherical particles of mean size from 3–10 μm, with at least 95% of the particles falling within a two-fold size range

(b) Adequate porosity, typically more than 50%, to allow good mass transfer within the particles

(c) Adequate specific surface area, typically between 30 and 300 m^2/g, to provide a reasonable chromatographic capacity

(d) Adequate mechanical strength to withstand pressure gradients up to 5000 bar/m typically used in column packing

(e) Good surface homogeneity, which is essential for good peak symmetry, coupled with an absence of highly adsorptive sites (even if these are rare)

(f) Good chemical stability
(g) Ease of reproducible manufacture on the 1–100 kg scale

Notwithstanding the problems noted above, silica gels currently used in HPLC largely meet these requirements. By contrast, until quite recently, there have been no carbons which remotely do so. In spite of the bewildering variety of carbons available in bulk form (activated charcoals, carbon blacks, glassy carbons, pyrocarbons, and industrial graphites), none of them are suitable for HPLC. In spite of numerous attempts to develop suitable chromatographic carbons, few have been successful, and all that have been are based on graphite. Unmodified graphitized carbon black (GCB) was widely used by Di Corcia and Liberti for gas chromatography [8] and is marketed by Supelco Ltd under the trade name Carbopack. This material is unfortunately too fragile for HPLC, although the LC separations demonstrated by Ciccioli et al. in 1981 [9] showed excellent peak symmetry and chromatographic performance, considering the relatively low pressures and large particles used. A few years previously, in 1976, Colin et al. [10–12] had attempted to improve the strength of GCB by depositing pyrolytic carbon on its surface. Although they succeeded in producing a stronger material which could be used in HPLC, its chromatographic properties were mediocre and its manufacture difficult. Attempts to make porous carbons by reduction of polytetrafluoroethylene by Plzák, Dousek, and co-workers [13–15] were unsuccessful, as were those of Unger starting with active cokes [16]. However, in 1979 Knox and Gilbert patented a novel method of making a robust porous carbon [17], whose method of manufacture could readily be scaled up. A high-porosity HPLC silica gel was impregnated with a phenol-formaldehyde mixture. This was polymerized within the pores of the silica gel to give phenol-formaldehyde resin. The resin was carbonized at 1000°C in nitrogen, and the silica finally dissolved out with alkali. Because pyrolysis of phenol-formaldehyde resin is known to form glassy carbon, Knox and Gilbert called this material "porous glassy carbon" (PGC). The material proved to be microporous, and its LC performance poor. However, by heating it above 2000°C, the micropores closed and the material became graphitized [18]. Its chromatographic performance was then similar to that of Colin et al.'s material. X-ray powder diffraction showed it to be a two-dimensional graphite, similar to GCB, but with a more robust structure [19]. This material was called "porous graphitized carbon." The original acronym was conveniently still applicable and is now used exclusively for the graphitized material made by the method of Knox and Gilbert. Initially, PGC was manufactured by the Wolfson Liquid Chromatography Unit in the Chemistry Department at the University of Edinburgh. Subsequently manufacture was taken over by Shandon HPLC, now Hypersil (a division of Life Sciences International) using the method of Knox and Gilbert. The material is now marketed under the trade name Hypercarb®. In the years between the original development of PGC, its launch on the market in 1988, and the present time, the chromatographic performance of PGC has been greatly improved and modern Hypercarb compares favorably with the bonded silica gel in terms of peak sharpness and symmetry [20].

Soon after Knox and Gilbert took out their patent, Novák and Berek [21] disclosed an almost identical method for manufacturing porous graphite. Its performance has been reported by Skutchanová et al. [22] In 1991, two Japanese groups [23,24] disclosed porous graphites made by entirely different procedures. In the method of Obayashi, Ozawa, and Kawase of the Tonen Corporation [23], a roughly 50/50 mixture of low-molecular-weight pitch (MW ~ 300) and a polymerizable monomer, such as styrene and/or divibylbenzene, along with a suitable initiator, is suspended in water and the monomer polymerized. The beads are then separated and heated in stages to 1100°C, and finally to around 2800°C. The typical material so produced has a specific pore volume of about 0.4 ml/g, with 80% of the pore volume contained in pores with diameters from 200 to 500 Å.

In the method disclosed by Ichikawa, Yokoyama, Kawai, Moriya, Komiya, and Kato of the Nippon Carbon Company and the Tosoh Corporation, [24], typically, equal amounts of carbon black (colloid particle diameter about 300 Å) and a phenolic resin are dissolved in methanol, made into a slurry, and then spray-dried to give roughly spherical particles in the range 3–100 μm. The polymerization of the phenolic resin is completed at around 140°C, and the particles heated at a controlled rate in nitrogen to 1000°C and then to 2800°C. The final materials have specific surface areas ranging from 20 to 120 m^2/g, and specific pore volumes from 0.3 to 1.0 mL/g. This procedure is analogous to that originally used by Colin et al. [10].

Table 1 compares the physical and chromatographic characteristics typical of the three materials: PGC, Tonen carbon, and Nippon/Tosoh carbon.

In confirmation of the view that graphite, however produced, should show the same chromatographic characteristics, it is encouraging to note that in spite of widely different methods of production and the somewhat different physical properties of the three porous graphites, the absolute and relative k'-values for the test phenols are similar. Only PGC/Hypercarb, however, is widely available commercially, and as a result, virtually all applications which have been published since 1988 have used either the original PGC manufactured in the University of Edinburgh by Knox and Kaur or Hypercarb manufactured by Shandon. The original PGC had a mean particle size of 7 μm. Hypercarb is now available in 5 μm and 7 μm grades. Since the previous review was written in 1988 [1], there have been significant improvements in the performance of PGC/Hypercarb, and advances in our understanding of the basis of retention by graphite in HPLC. There has also been a considerable expansion of applications of PGC/Hypercarb. These are reviewed separately in Chapter 3B of this volume.

Table 1 Comparison of Physical and Chromatographic Properties of Porous Graphites

	PV[a]	SA[b]	PD[c]	Eluant[d]	Phenol	k'-values[e] Para-cresol	3,5-xylenol
PGC	0.85	120	350	95%	0.25 (0.31)	0.8 (1.00)	1.8 (2.2)
Tonen [23]	0.44		~300	95%	0.5 (0.45)	1.1 (1.00)	1.7 (1.5$_5$)
Nippon/Tosoh [24]	0.95	24	355	100%	0.3 (0.33)	0.9 (1.00)	2.1 (2.3)

[a] Pore volume in mL/g.
[b] Surface area in m^2/g.
[c] Mean pore diameter in Å.
[d] Percentage methanol in eluant, remainder water.
[e] Bracketed values relative to k' for p-cresol.

III. STRUCTURE OF PGC/HYPERCARB®

High-resolution transmission electron microscopy [19] indicates that PGC consists of intertwined ribbons of carbon, where each ribbon consists of some 30 discrete layers or sheets about 3.4 Å apart. This structure has striking similarities to that of graphitized carbon black, except that GCB appears to be made up of independent roughly hexagonal particles whereas PGC is composed of interconnected ribbons. It is this ribbon form which accounts for the high mechanical strength of PGC compared to that of GCB. Both structures are typical of the so-called "turbostratic carbons" first defined by Warren [26]. The layer spacing seen in the electron micrographs is the same as that in graphite, and this is confirmed by x-ray powder diffraction, although the latter shows that PGC and GCB are two-dimensional rather than three-dimensional graphites. In true three-dimensional graphite, the sheets are precisely registered with the carbon atoms in one sheet above and below the centers of the six carbon atom rings in the adjacent layers. This arrangement is known as the Bernal structure [27]. X-ray powder diffraction shows sharp (h, k, l) reflections where h, k, and l can all simultaneously be non zero. By contrast, turbostratic graphites give only the $(0, 0, l)$, and $(h, k, 0)$ reflections, and these are relatively broad. This means that although there is a well-defined (l)-spacing between layers, and well-defined (h, k)-spacings of the atoms within the layers, there is no registration of one layer relative to those below and above. The crystallographic spacings in PGC are typical of turbostratic two-dimensional graphites. The layer spacing is slightly larger than in true graphite (3.44 Å compared with 3.354 Å), whereas the spacing of the atoms within the sheets is slightly less (1.40 Å compared to 1.42 Å). From the width of the reflection peaks it was deduced [19] that the crystallite sizes in PGC were of the order of 100 Å across. Interestingly, the material disclosed in the Nippon/Tosoh patent [24], based upon GCB, has very similar characteristics: layer spacing 3.41–3.45Å, and crystallite thickness 50–75 Å.

In chromatography we are primarily concerned with the surface of a packing material (its internal surface if porous) rather than the underlying atomic/molecular structure, and therefore it makes little difference whether the material has two-dimensional or three-dimensional order. Because the surface layers in two- and three-dimensional graphites are almost identical, we expect their adsorptive properties to be essentially the same. Indeed, many of the studies of adsorption by graphite have, in fact, been carried out with GCB, not true three-dimensional graphite [28].

By virtue of its method of production, PGC/Hypercarb is a "template material" in which the original template is a mesoporous silica gel. The pore size and to some extent the pore volume can be controlled by selecting an appropriate pore size and porosity for the silica-gel template. In the original manufacture of PGC, Gilbert et al. [18] chose a silica-gel template with a high porosity of 1.2 ml/g. Because silica has a density of 2.2 g/cm^3, its fractional porosity was 72%. In the production of PGC, the pore volume was filled with phenol-formaldehyde resin whose density is about 1.2 g/cm^3. After carbonization, the final carbon has approximately 50% of the weight of the original polymer [29], and because graphite has a density of around 2.2 g/cm^3, the final graphite occupied only 27% of the original pore space of the silica, or 20% of the original volume of the silica particle. The final PGC, therefore, had a porosity of about 80%. This was too high for adequate strength for HPLC, and the early batches could not be packed into HPLC columns at typical packing pressures (e.g., 400 bars for a 100 mm long column). To improve the mechanical strength, an additional impregnation with phenol-formaldehyde followed by polymerization and carbonization was introduced for the commercial product. The porosity of Hypercarb as now manufactured [20] is about 70% and its specific surface area about 120 m^2/g. PGC/Hypercarb is almost as robust as silica gel and can be packed without damage into 100 mm columns using a packing pressure of 500 bars. Hypercarb is fully compatible with current HPLC procedures and meets all of the desiderata listed above for an HPLC packing material.

IV. PERFORMANCE OF PGC/ HYPERCARB®

As noted earlier, the initial HPLC tests with PGC [18] were somewhat disappointing, the quality of the chromatograms being very similar to that demonstrated by Colin et al. [10–12] in which significantly retained peaks were tailed, indicating a nonlinear adsorption isotherm, and therefore heterogeneity of adsorption sites. Early problems were put down to the difficulty of obtaining reli-

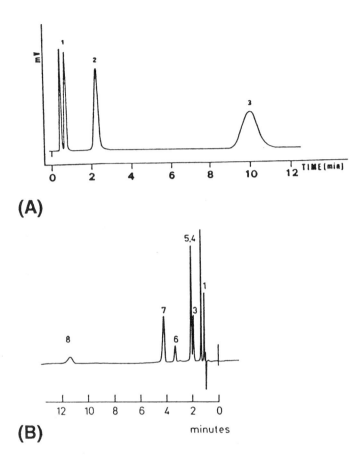

Fig. 1 Liquid chromatogram on graphite. (A) On GCB from Ciccioli et al. [9]. Solutes: 1—phenylacetamide; 2—phenacetin; 3—caffeine. (B) On PGC from Knox et al. [19]. Solutes: 1—benzene; 2—toluene; 3—m-xylene; 4—p-xylene; 5—o-xylene; 6—135 trimethylbenzene; 7—124 trimethylbenzene; 8—1245 tetramethylbenzene. (C) On Hypercarb from Hypersil [20]. Solutes: 1—acetone; 2—phenol; 3—anisole; 4—p-cresol; 5—phenetole; 6—3,5-xylenol.

able graphitization of the porous glassy carbon. The crucial breakthrough came in 1986 when the first high-quality material, batch PGC94, was produced [19]. Figure 1 reproduces comparative chromatograms, on the GCB of Ciccioli et al. [9], on the 1986 PGC sample and on a modern sample of high-performance Hypercarb.

These chromatograms demonstrate that excellent peak symmetry can now

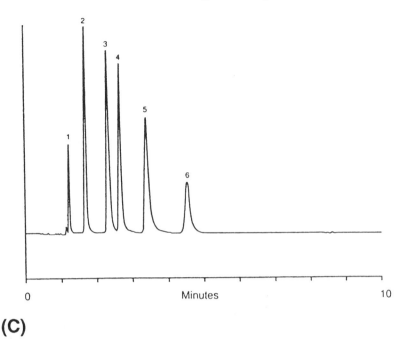

(C)

be obtained from graphitic carbons. This has resulted from an intensive development program in which particular attention was paid to conditions of graphitization, to achieving good particle uniformity and to improving column packing technology. As shown by Fig. 2, the plate efficiency of Hypercarb is now comparable with that of bonded silica gels, with a minimum HETP below three particle diameters.

V. BASIS OF RETENTION BY GRAPHITE IN LC AS SEEN IN 1988

By 1988 several major features of retention by graphite in HPLC had been established. First, it was shown by Dias [30], using gas chromatography, that the heats of adsorption of small molecules, particularly those of common HPLC solvents, closely followed the areas which they occupied on the surface when adsorbed. The heats of adsorption were determined from the dependence of $\log k'$ on temperature using the van't Hoff equation. Figure 3 replots Dias' original data.

It may be noted that points for polar molecules, such as methanol, acetonitrile, and ethyl acetate, lie on the same line as those for nonpolar molecules such as hexane, benzene, and cyclopentane. Furthermore, the point for benzene lies on the same line showing that, at this level of approximation, there is no indica-

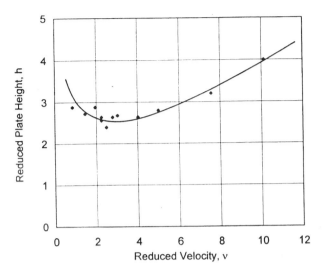

Fig. 2 Reduced plate height, reduced velocity plot for 5 μm Hypercarb [20]. Note: Reduced velocity of 5 is equivalent to a linear flow rate of 1 mm/sec, and a volume flow rate in a standard 4.6 mm bore column of 1 mL/min.

tion of a significant contribution to the heat of adsorption from π–electron exchange. It may be deduced from these results that the major interactions of molecules with graphite, when adsorbed from the gas phase, are through nonspecific dispersive interactions. These arise from the interaction of instantaneous dipoles present in the graphite and the analyte molecules. The instantaneous dipoles arise from the nonuniform distribution of the electrons in any atom at any particular instant, even when the long time distribution is uniform [31]. The dispersion forces are those which account for the nonideal properties of the noble gases. The interaction energies of adsorption from the gas phase are, of course much greater (about 10 times as large) than those for adsorption from solution. It is therefore likely that secondary effects which are not revealed by heat of adsorption measurements from the gas phase will emerge when we come to study adsorption from solution; that is, when we come to study retention in liquid chromatography.

A second important factor which is noted from gas adsorption studies was that heats of adsorption of nonpolar molecules by graphite [28,32] were 20–50% higher than their heats of vaporization from their liquid state [33]. Table 2 presents some examples.

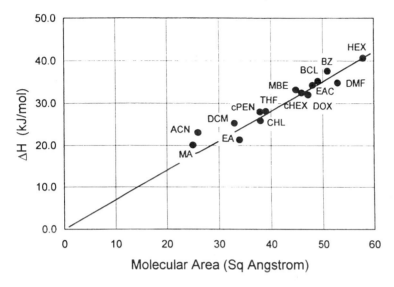

Fig. 3 Heats of adsorption of various organic solvents on to PGC from the gas phase determined by Dias [30] as a function of their adsorbed surface areas. Identification: MA—Methanol; ACN—acetonitrile; EA—ethanol; DCM—dichloromethane; CHL—chloroform; cPEN—cyclopentane; THF—tetrahydrofuran; MBE—methyl t-butyl ether; cHEX—cyclohexane; DOX—dioxan; EAC—ethyl acetate; BCL—butyl chloride; BZ—benzene; DMF—dimethylformamide; HEX—hexane.

Table 2 Heats of Adsorption onto Graphite from the Gas Phase

Substance	ΔH_{ads} (kJ/mol)	ΔH_{vap} (kJ/mol)	$\Delta H_{ads}/\Delta H_{vap}$	Ref.
Hexane	47	32	1.47	28
Benzene	43	34	1.26	28
Neon	3.1	1.86	1.67	32, 33
Argon	8.9	6.5	1.37	32, 33
Krypton	11.7	9.0	1.30	32, 33
Xenon	15.5	12.6	1.23	32, 33

At first sight, it is surprising that heats of adsorption onto graphite are so much greater than heats of condensation, because a molecule in the liquid state is completely surrounded by other similar molecules, whereas when adsorbed on graphite, it is contacted on one side only. However, detailed calculations of the dispersive interactions of the noble gases on to graphite have provided close agreement between theory and experiment [32,34], indicating that no other sources of interaction are needed to explain the results. The high values for heats of adsorption may be attributed to the very small amplitude of the thermal vibration of the carbon atoms in the surface of graphite which allows an adsorbate molecule to be closer to the graphite surface than it would be to neighboring molecules in the liquid state and to the high atomic density of carbon atoms (density of graphite = 2.2 g/mL, density of a typical organic liquid = 0.9 g/mL). The important points to note are (1) that the dispersive interactions of adsorbate molecules with graphite are very strong and (2) that even with polar molecules, such as HCl, dispersive interactions are far more important than dipole–induced-dipole interactions [35].

A third notable feature concerning liquid chromatography with graphite is the absence of a universal eluotropic series. In this regard, graphite is in marked contrast to the oxide adsorbents. Although it is always possible to arrange solvents in an order of eluting power for a particular functional group, for example, $-CH_2-$ or $-CH_3$ [11,16], the order of solvent strength, as shown by Kaur [36], is different for different functional groups. Nevertheless, certain solvents are generally stronger than others, and water is a particularly weak solvent for hydrophobic species. Thus, Kaur [36] shows that for ethylbenzene, phenol, aniline, and nitrobenzene, methanol is weak, whereas dimethylformamide, chloroform, and dichloromethane are strong. For homolog selectivity (i.e., CH_2 addition to aliphatic groups or CH_3 substitution into a benzene ring), methanol and acetonitrile are weak, whereas dioxane and chloroform are strong. Along with the lack of any universal eluotropic series of solvents, the range of eluotropic strength with graphite (using the Snyder scale [37]) for any particular compound or functional group is very small, typically being around 0.2 units (omitting water) compared to about 1 unit for oxide adsorbents such as silica gel and alumina. This major difference is readily explained using Snyder's own theory, according to which, adsorption of an analyte from a solution is thought of as a process in which a molecule of analyte replaces one or more molecules of eluant at the adsorbent surface. With suitable assumptions applicable to oxide adsorbents, the eluotropic strength of a solvent is related to its heat of adsorption per unit area. For such adsorbents, these are strongly dependent on the ability of solvent molecules to form hydrogen bonds with the adsorbent surface groups (e.g., $-Si-OH$ groups on silica gel) and therefore cover a wide range as one moves from non-H-bonding solvents such as hydrocarbons to alcohols and acids, and ultimately water, which form strong H bonds. The H-bonding interactions are indeed so dominant that differences in the solution interaction energies of the analyte with solvent, and solvent with solvent can be neglected. With graphite, by contrast, H-bond formation is impossible, and the dominant contribution to the heat of adsorption and chromatographic selectivity found with oxide sorbents is now absent. Solution interactions can no longer be ignored, and indeed they now dominate selectivity considerations. Because the heats of adsorption of solvents per unit area from the gas phase are, in fact, very similar, graphite, relatively speaking, should be unselective with regard to functionality.

A fourth feature of retention by graphite, which became clear around 1986, is its unusual selectivity with respect to structural isomers [1]. This was ascribed to the favored adsorption of those isomers which can most easily be accommodated to the flat surface of graphite, thus giving higher dispersive interaction energies for such compounds. These results paralleled those found in gas chromatography [7,8].

From all these observations, it was concluded that the primary interactions of molecules with graphite were dispersive and that graphite was unspecific toward functionality. It was, therefore, argued that graphite should behave as a strong reversed-phase packing material similar to a hydrocarbon-bonded silica gel or a hydrocarbon polymer such as polystyrene. Indeed, it was argued that graphite could be the perfect reversed-phase material. Initial experiments broadly confirmed this [18,19,36]. The elution order was generally similar to that for an alkyl-bonded silica gel, but stronger eluants containing lower proportions of water were normally required.

This view has had to be radically altered in the light of more recent work.

VI. RETENTION STUDIES POST-1988

Kaliszan and co-workers [38,39] were the first to show that graphite did not behave as a "perfect" reversed-phase material. Using heptane as solvent, they studied the retention by PGC of a range of benzene derivatives having a wide spread of polarity. By contrast to what might be expected from a reversed-phase packing, more polar solutes were more strongly retained than less polar solutes. Log k' showed a poor correlation with the estimated energy of overlap between the π electrons of the graphite surface and those of the adsorbate but showed a reasonable correlation with Kaliszan's polarity parameter Δ [35]. The correlation indicated that "a localised polar segment of the molecule was responsible for retention rather than the π-orbitals of the aromatic ring." The same authors also noted the similarity between the retentive behavor of graphite and the metal palladium. This was in accord with the well-known fact that graphite is an electrical conductor, albeit a two-dimensional conductor, there being conduction within each graphitic layer but not between layers.

Lim and co-workers [40,41] reached similar conclusions as a result of their work on the separation of pertechnetate and perrhennate anions on PGC with a predominantly aqueous eluant, separations totally unexplained on the basis of dispersive interactions alone. There appeared to be strong attractive interaction between the charged centers in the analyte and the graphitic surface leading to the observed retention from an aqueous eluant. The remarkable separations of Lim et al. suggest that PGC may well be an important stationary phase for the separation of ionized solutes by a mechanism other than ion exchange. Indeed, Lim et al. showed that positively and negatively charged ions could be separated in the same chromatogram.

Comprehensive studies on the retention by PGC of simple aliphatic and aromatic compounds have been carried out by Möckel and co-workers [42], by Tanaka and co-workers [43], and by Kříž and co-workers [44].

Möckel et al. [42] compared PGC with a C_{18}-bonded silica gel using a range of homologous alkane derivatives, with methanol as eluant. They obtained good linear plots of log k' against carbon number, n, according to Equ. (1):

$$\log_{10} k' = a + bn \tag{1}$$

The gradients, b, of the plots for the different homologous series $CH_3(CH_2)_{(n-1)}X$ were more or less the same irrespective of X and gave a measure the selectivity of the materials with respect to addition of the $-CH_2-$ group. For Inertsil C_{18}, b [$=\log(k'_{n+1}/k'_n)$] had a value of 0.08, and for PGC a value of 0.26. Evidently, PGC shows much greater discrimination for the methylene group than does the C_{18} phase.

Values of the intercepts, a, for various substituent groups measure the retentive power of the material for a particular functional group, X. They are listed in columns 2 and 3 of Table 3. In general, the intercepts for PGC are lower (i.e., more negative) than those for Inertsil C_{18}, indicating that with pure methanol, PGC has a lower retentive power than Inertsil C_{18}. If, however, the differences in a relative to the value for the alkanes (X = H) are computed, one notes that with Inertsil C_{18} (column 4), substitution of $-H$ by a functional group (except $-I$) reduces retention quite considerably. This is what one expects for replacement of $-H$ by a more polar group. Substitution increases the energy of solvation more than it increases the dispersive interaction energy with the C_{18} stationary phase. Only with phenyl is retention increased, but this is, of course, expected, as phenyl counts effectively as addition of a C_6 hydrocarbon moeity. If, more realistically, we consider the replacement of the hexyl group by phenyl (data line labelled "phenyl less 6 CH_2"), retention is decreased, in line with the greater polarizability of the phenyl group. Turning to PGC (column 5) the situation is very different. On the whole, substitution of $-H$ by a hydrophilic group either increases retention or has a neutral effect. Replacement of hexyl by phenyl decreases retention only slightly. The expected decrease in retention arising from

Table 3 Retention Parameters for Homologous Alkanes and Substituted Alkanes, According to Möckel et al. [42]; Eluant Methanol

Group X	Inertsil C_{18}	PGC	Inertsil C_{18}	PGC	PGC Less Inertsil C_{18}
	Note 1		Note 2		Note 3
	a Values		$a_X - a_H$		
H	−0.60	−2.56	0.00	0.00	0.00
CH_2 [b of Eq. (1)]	0.08	0.25			0.17
Phenyl–	−0.46	−1.22	0.14	1.34	1.20
Phenyl less 6 CH_2 (Note 4)	−0.95	−2.71	−0.35	−0.15	0.19
Cl–	−0.73	−2.30	−0.13	0.26	0.39
Br–	−0.69	−2.14	−0.09	0.42	0.50
I–	−0.59	−1.91	0.02	0.65	0.64
CN–	−1.25	−2.51	−0.64	0.05	0.70
$COCH_3$	−1.16	−2.62	−0.56	−0.06	0.50
OH–	−1.14	−2.59	−0.54	−0.03	0.50

Notes:
1. Values of a from Eq. (1).
2. Value of a for X less value of a for H.
3. Difference between values in columns 4 and 5.
4. Values in line labeled "phenyl less 6 CH_2" are the increments obtained when –$(CH_2)_5CH_3$ is substituted by phenyl.

increased solubility when –H is substituted by a more polar group is substantially compensated by an increased affinity for graphite. The adsorption energy for polar groups onto graphite must therefore contain terms additional to those for purely dispersive interactions.

Thus, when PGC is compared to Inertsil ODS (last column in Table 3), all substitutions, relatively speaking, have a positive effect on retention. Graphite has a greater affinity for polar or polarizable functional groups than Inertsil C_{18}. In a number of cases (halogen substitution) the additional affinity of graphite for the substituent group exceeds its additional affinity for methanol; in others (–CN, –$COCH_3$, –OH) the two effects more or less cancel (column 5 in Table 3).

Tanaka et al. [43] have also published an extensive comparison of the selectivity of carbon- and silica-based packings using homologous alkanes and simple derivatives thereof, as part of a wider study of polymeric and other HPLC stationary phases. Reference 43 is unique, being the only article which compares two different carbons, namely Hypercarb (Shandon) and Carbonex (Tonen). The two graphites are made in totally different ways, yet exhibit almost identical properties, particularly their respective selectivities. This is the first, and so far only, experimental confirmation that graphites developed for LC, however produced, show nearly identical chromatographic properties.

Tanaka et al. [43] analyzed their data in terms of Equ. (1), and their results using methanol/water 80/20 v/v are given in Table 4. They complement those of Möckel et al. [42] and show good agreement for the alkanols. Again, with ODS silica, polar substitutuents reduce retention. With PGC, substitution by $-COOCH_3$ and $-CHO$ increase retention, implying that their additional affinity for the solvent is more than compensated by their additional affinity for graphite. As Möckel et al. found, when PGC is compared with any of the bonded silica gels, the effect of substitution, relatively speaking, produces a positive change (last three columns). This is most pronounced when PGC is compared with C_{18} silica gel. Tanaka's pyrenylethyl-silica (PYE-silica) behaved in an intermediate way.

Tanaka et al. [43] also found that over a wide range of methanol/water compositions (0–80% methanol), graphite shows a higher selectivity for homologs than either C_{18} or PYE silica. That b values are nearly independent of the substituentas are also shown by Möckel et al. [42]. Their dependences on methanol content is shown in Fig. 4. The final points for pure methanol are taken from Mockel et al. [42], Gilbert et al. [18], and Kříž et al. [44] who used alkyl benzenes. The points for pure water agree with the extrapolated values of derived by Cocquart [45] using alkyl benzenes.

For both C_{18} silica and PGC, b falls as the methanol content increases, but the decline is steeper with C_{18} silica. Broadly, the decline in b can be attributed to

Table 4 Retention Parameters for Homologous Alkanes and Substituted Alkanes According to Tanaka et al. [43]; Eluant Methanol/Water 80/20 V/V)

Group X	a Values			Relative to H $a_X - a_H$			Difference of values relative to H		
	C18	PYE	PGC	C_{18}	PYE	PGC	PYE–C_{18}	PGC–C_{18}	PGC–PYE
H	−0.39	−0.84	−1.96	0.00	0.00	0.00	0.00	0.00	0.00
CH_2 [b of Eq. (1)]	0.18	0.15	0.31				−0.03	0.13	0.16
Phenyl	−0.03	−0.36	−0.38	0.36	0.48	1.58	0.12	1.22	1.10
Phenyl less 6 CH_2	−1.05	−1.26	−2.24	−0.66	−0.42	−0.28	0.24	0.38	0.14
$COOCH_3$	−0.80	−0.71	−1.52	−0.41	0.13	0.44	0.55	0.85	0.41
CHO	−1.09	−1.19	−1.62	−0.70	−0.35	0.34	0.35	1.04	0.69
OH	−1.33	−1.45	−2.28	−0.94	−0.61	−0.34	0.33	0.60	0.27

Note: Values in line labeled "phenyl less 6 CH_2" gives the increments obtained when $-(CH_2)_5CH_3$ is substituted by phenyl.

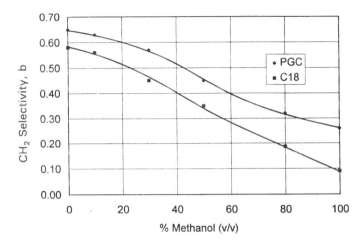

Fig. 4 Dependence of methylene selectivity, b, on composition for PGC and C_{18} silica gels. (Data from Refs. 18, 42, 43, and 45.)

the better solubilization of the CH_2 group in methanol than in water, a view confirmed by the fact that b is close to zero with hexane as eluant [38,39]. What is at first difficult to explain is why the values at the same eluant composition differ so much for C_{18} silica and PGC, and why the decline is more rapid with ODS silica than with PGC. In both cases, the analyte has to displace solvent molecules from the surface of the support material to form the final close association with the stationary phase where there is probably no solvent between the face of the analyte and the stationary phase, although the exposed surface of the analyte molecule must still be solvated. Tanaka et al. were surprised and commented that "the solvent-like flexibility of a C18 stationary phase is expected to give such an interaction to reduce the contact between the C–H surface of a solute and water, [while] the rigid carbon surface would not be able to completely surround the alkyl chain of the solute." The explanation may lie with the phenomenon already referred to, namely the much greater dispersive interactions of an adsorbed species with graphite than with a liquid. Because of this, any differences in adsorptive strength will be amplified so that graphite should be more selective with respect to any displacement of one substance by another (e.g., of methanol by analyte) than a quasi-liquid phase such as a bonded silica gel. The differences in b values are therefore not unexpected. However, there is no such explanation for the much steeper decline of b with methanol concentration for ODS silica.

Likewise, there is no simple explanation for the very different effects of substitution when using PGC and ODS silica. These indicate a positive interaction between graphite and the polar or polarizable functional groups, an interaction which is not present with an ODS phase, although we note that there is evidence for it with the pyrenylethyl (PYE) phase.

Tanaka et al. [43] also plotted log k' for the various stationary phases against log P, where P is the octanol–water partition ratio. For C_{18} and methanol/water 60/40 v/v the correlation was very good. With pyrenylethyl silica (PYE) and nitrophenyl silica (NPE), log k'_w values for polar compounds lay somewhat above the line representing the dependence of log k' on log P for the alkylbenzenes. This tendency was greatly accentuated with graphite. Substances like Φ–$COCH_3$, Φ–NO_2, Φ–$COOCH_3$, and Φ–$N(CH_3)_2$ (where $\Phi = C_6H_5$), exhibiting k' values about 15 times higher than expected on the basis of their log P, whereas Φ–NH_2, Φ–OH, and Φ–CH_3 gave k'_w values about 4 times higher.

Coquart and Hennion [45–47], as part of their study of the use of Hypercarb as an adsorbent for removing contaminants from water samples, again correlated log k' and log P for a number of stationary phases including PGC. The correlations with C_{18} and C_8 silicas were again very good; with PRP-1 the data points were more scattered, and with PGC, data points for polar compounds lay well above the line for the alkyl benzenes or alkanes. The measurements of the gradients, $d(\log k')/d(\log P)$, of the lines through the points for the alkyl benzenes, along with those of Tanaka, are listed in Table 5. Where there is overlap, there is good agreement.

These gradients apparently reflect three factors:

1. Eluant polarity: Methanol is less polar than water and more similar to octanol so that the gradient $d(\log k')/d(\log P)$ should decrease as the methanol content increases. This is demonstrated by the figures for C_{18}–silica, PRP-1, and PGC.

2. The degree to which the solute is surrounded by the stationary phase: For a bonded stationary phase this should be somewhat less than for liquid octanol, so that the partitioning into the bonded phase is likely to be less with a bonded phase than with octanol, and the gradient less. Thus, even with pure water, the gradient $d(\log k')/d(\log P)$ for C_{18}–silica is less than unity.

3. The strength of the dispersive interactions of the solute with the stationary phase: As the stationary phase becomes more polar (hydrophilic), it becomes more like water and less like octanol, so that with the same degree of solvation, $d(\log k')/d(\log P)$, should be less. This is observed in the series C_{18}–silica, PYE, NPE. PGC stands out as the most hydrophobic stationary phase and shows the highest values of $d(\log k')/d(\log P)$ at all methanol concentrations. As Tanaka has pointed out, one might have expected the values for PGC to be less than for a C_{18} phase, as only one side of the molecule can be in contact with the graphite surface; however, the strength of the dispersive interactions, as shown by the heats of adsorp-

Table 5 Values of $d(\log k')/d(\log P)$ for Homologous Series Using Various Stationary Phases and Methanol/Water Compositions

Methanol/water⇒	0/100	50/50	60/40	80/20
Stationary phase		$d(\log k')/d(\log P)$		
C_{18}	0.79		0.43	
C_8	0.79			
PRP-1	1.31	0.73		0.48
PYE			0.37	
NPE			0.29	
PGC	1.31		0.55	0.57
Reference	45	45	43	45

tion from the gas phase, are very large. The great strength of the dispersive interactions more than compensates for the lesser solvation. Thus, $d(\log k')/d(\log P)$ values are larger with PGC than with C_{18} phases. PRP-1 lies somewhere between C_{18}-silica and graphite. However, with pure water, both PRP-1 and PGC show the $d(\log k')/d(\log P)$ value of 1.31. These must reflect the very strong dispersive interactions of graphite and the probability that PRP-1 can more or less embrace the entire molecule of adsorbate when water is the eluant.

Although there are similarities between PRP-1 and PGC with regard to their $d(\log k'/d(\log P)$ values for homologs, we still have to remember that a large difference remains: With polar analytes PRP-1 behaves very similarly to a reversed-phase material, whereas graphite shows additional retention.

Another important feature which Tanaka et al. [43] confirmed, was the high stereoselectivity of graphite especially for cis- and trans-isomers of di-substituted cyclohexanes. The more planar the isomer, the more it was retained. Although a similar selectivity was seen with C_{18}-silica and PYE, the selectivity was less. The flat surface of graphite was held to be responsible, allowing stronger dispersive interactions with those molecules which could accommodate themselves better to the flat surface.

The study by Kříž and co-workers [44] compared the retention of some 42 aromatic hydrocarbons on Hypersil ODS, PGC, silica, and alumina. Methanol or methanol/water was used as the eluant for the first three and hexane for the latter two. Their main conclusion confirmed previous work [2] that the b values for PGC were respectively 0.22 for the addition of CH_2 groups and 0.46 for the substitution of CH_3 groups in the ortho position in a benzene ring. With ODS Hypersil, *b*-values for both CH_2 addition and CH_3 substitution were 0.17, whereas for phenyl silica they were 0.10. These values are in agreement with those of Möckel et al. [42]. The work of Kříž et al. [44] further emphasizes the unique stereoselectivity of graphite. Its very different selectivity for CH_2 addition and CH_3 substitution mirrors that of the bare adsorbents, silica and alumina.

Forgács co-workers [48–57] have more recently carried out extensive studies of the retention of various classes of compounds, including barbiturates, amines, and phenols by PGC. They found similar results (discussed in more detail below). Contrary to what had been expected for a reversed-phase packing material, the various substances in most of these groups were eluted not in order of their hydrophobicity but in the order of their polarity. The more polar or hydrophilic compounds eluted later. Thus, the increased affinity of a solute for water within these series was more than balanced by the increased affinity for graphite.

Further key evidence on the mechanism of retention by graphite has been provided by Coquart and Hennion [45–47]. As a part of a study of the use of Hypercarb as an adsorbent for removing contaminants from water samples, they determined the effective k' values for various pollutants in pure water, which they termed k'_w. As did Forgács et al., they determined k' for each analyte as a function of the volume fraction of methanol in the eluant, *C*. Generally, plots were linear according to Eq. (2), so that log k'_w could be determined by extrapolation if direct determination was not possible:

$$\log k' = \log k'_w + mC \qquad (2)$$

Among other surprising discoveries, they found that the k'_w values of mono-, di- and trihydric phenols increased with the number of OH groups in the molecule, whereas with ODS silica gel and PRP-1, they decreased as one would have expected. Similar results were obtained for the carboxy-benzenes. Their results are evidently similar to those of Forgács et al. and are considered in more detail below.

To summarize, the retention by graphite from aqueous/organic eluants is apparently determined by a balance of two factors: (1) hydrophobicity, which is primarily a solution effect tending to drive analytes out of solution and (2) the interaction of polarizable or polarized functional groups in the analyte with graphite, which are additional to the normal dispersive interactions. The second effect is particularly strong when the stereochemistry of the analyte molecules forces the polar group to be close to the graphite surface. We shall call this the "polar retention effect by graphite (PREG)." It appears to be an effect which is additional to the normal hydrophobic and dispersive effects found with conventional reversed-phase materials.

The big question now facing us is to quantify the interactions between graphite and polar groups responsible for PREG and to identify their origin. Regrettably we are still some way from being able to do either of these things.

VII. EXISTING THEORIES OF RETENTION IN REVERSED-PHASE CHROMATOGRAPHY AND THEIR APPLICATION TO GRAPHITE

From what has been stated earlier, there seem to be four main factors which influence retention by graphite in LC.

1. Eluant–analyte interactions (dispersive, dipole–dipole and hydrogen bonding interactions) which occur in the eluant. These discourage retention.
2. Hydrophobic eluant–analyte repulsions (arising from resistance to the disruption of the structure of hydrogen-bonded solvents by non-hydrogen-bonding analytes). These occur between a hydrophilic eluant and any nonpolar segments of the analyte. These encourage retention.
3. Dispersive interactions of the London type between the graphite surface and the analyte. These are largely balanced by similar interactions between the graphite surface and the eluant which is displaced by the analyte. Their net effect may either encourage or discourage retention, but they may have an important effect on selectivity.
4. Charge-induced interactions of the analyte with the graphite which promote the retention of polar molecules (PREG). These interactions are compensated to a greater or lesser extent by polar interactions of the analyte with the eluant (type 1 above). These charge-induced interactions are strongest when the polar groups of the analyte are forced into direct contact with the graphite surface by the stereochemistry of the analyte molecule. In such cases, the additional interactions resulting from substitution of –H by a polar group can then be so strong that they more than compensate for the increased analyte–solvent interactions. When stereochemistry does not force direct contact of the polar group with the surface, the effect is less strong but still significant.

The overall effect of these competing interactions is that increasing the hydrophobicity of the analyte, say by adding alkyl groups into a molecule, always increases retention, as expected in a typical reversed-phase mode. However, increasing the polarity of the analyte by adding groups which can either donate or accept electrons or can polarize the graphite surface may also increase retention, particularly if these groups are constrained to be in close contact with the graphite surface.

It appears that the balance between types 1 and 2 is reasonably well represented by the hydrophobicity parameter log P and that this parameter is fairly well correlated with log k'_w on, say, ODS silica gel or PRP resin. What we need is a theory for contribution 4, namely PREG; before this can be achieved, we need some way of quantifying the effect.

We first examine how far existing theories for retention in liquid chromatography can explain retention by graphite. The main theories are those of Horváth et al. [58], and of Martire and Boehm [59] for reversed-phase chromatography, and of Snyder [37] for adsorption chromatography.

The solvophobic theory of Horváth et al. [58] is based on the analysis of the equilibrium between an isolated solvated molecule of analyte A, an isolated solvated ligand L, and their solvated combination AL. A is in the eluent phase, whereas the underlined species L and AL are in the stationary phase. The distribution equilibrium is represented by

$$\underline{L} + A \Leftrightarrow \underline{LA} \tag{3}$$

The analyte molecule is imagined to consist of a hydrophobic part and a hydrophilic part, whereas the surface of the ligand is hydrophobic. The eluent is imagined to consist of a mixture of hydrophilic and hydrophobic components (although both "components" may be combined in the same eluent molecule). The hydrophobic parts of the analyte and ligand are solvated by the hydrophobic component in the eluent, whereas the hydrophilic parts of the analyte are solvated by the hydrophilic component of the eluent. Horváth al. focus attention on the decrease in the area of contact with the solvent when the ligand, L, and analyte, A, combine to form the associate LA. It is noted that the new contact surface is made by joining the hydrophobic side of the analyte with the hydrophobic surface of the ligand with the elimination of some of the hydrophobic component of the eluent. The association process is analyzed with the help of a thermodynamic bridge, as illustrated by Fig. 5, where the subscript "solv" indicates a solvated species, and the subscript "gas" indicates a species in the gas phase.

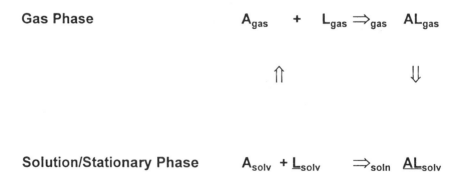

Fig. 5 Thermodynamic bridge illustrating transfer of analyte, A, from solution to association with ligand, L, via evaporation into the gas phase. Underlined species are in the stationary phase.

According to the theory, the overall process of transfer represented by the lower process labeled $\Rightarrow_{soln}$ in Fig. 5 can be achieved by desolvating A and L by transfer into the gas phase (process ⇑), linking them together in the gas phase (process $\Rightarrow_{gas}$), and then resolvating AL (process ⇓). The free-energy change for the linking process in the gas phase is not discussed in detail, but it seems to be assumed that the bonding between A and L is entirely by London dispersion interactions. (Horváth et al. call these Van der Waals interactions, but, strictly speaking, van der Waals interactions include all interactions which lead to nonideality of gases: they include dipole–dipole, dipole–induced dipole, and as well as London dispersion interactions.) The processes represented by ⇑ and ⇓ are split into two parts for each substance. Taking the solvation of AL (process ⇓), a cavity is first of all created in the solvent of appropriate size to take the molecule AL. AL is then placed inside the cavity and solvative bonds are formed between AL and the solvent molecules at the wall of the cavity. The free energy to form the cavity is mainly determined by the surface tension of the solvent and the wall area of the cavity, although a complex correction for the cavity size, reorganization energy and entropy, is included; that is,

$$\Delta F_{cavity} = CA_{cavity}\gamma \tag{4}$$

where A_{cavity} is the surface area of the cavity, γ is the surface tension of the solvent, and C is the correction factor.

The free-energy change when the molecule is introduced into the cavity includes two contributions; the first comes from London dispersion interactions, and the second from dipole–dipole and dipole–induced–dipole interactions between the molecule and the solvent cage. The contributions to the overall free energy change for the desolvation of S and L (process ⇑) are calculated in the same way but have opposite signs.

Although the theory is complex mathematically, it is conceptually simple and has long been regarded as the basis for understanding reversed-phase chromatography. When applied to chromatography on graphite, the ligand would be a small part of the graphite surface; the desolvation of the ligand (⇑) would be the process of removing solvent from the surface forming bare graphite and a corresponding surface of free solvent, whereas desolvation of A would be represented as it is in the original theory. The formation of the AL in the gas phase ($\Rightarrow_{gas}$) would correspond to adsorbing the hydrophobic side of A on to this small area of graphite in the gas phase. The solvation of AL (⇓) would correspond to wetting the element of graphite surface which now had A adsorbed onto it. The crucial assumption of the whole theory is that the bond formed between A and L is the result of dispersive interactions. We have already seen that this assumption is incorrect when applied to graphite because there are additional interactions resulting in PREG. These are not allowed for in the hydrophobic theory. Accordingly, the hydrophobic theory gives no explanation of PREG.

The unified retention theory developed by Martire and Boehm [59] starts from a totally different standpoint, namely that of a lattice containing atomic-sized cubic cells. A strand of ligand (say, a C_{18} moeity) will occupy a number of such cells which are connected together; this number is denoted by r_{ligand}. Likewise, solvent molecules (the theory as presented allows for two eluant species 1 and 2) occupy r_1 and r_2 cells, whereas an analyte molecule occupies $r_{analyte}$ cells. Various energy terms are introduced to allow for bond bending in the hydrocarbon chain, but the crucial energy terms are those for the interactions between segments of the various species. These are called χ values. The treatment is statistical and provides a rigorous treatment of both the entropy and enthalpy changes for the basic equilibrium $A + \underline{L} \Leftrightarrow \underline{AL}$. It enables various properties of the system to be predicted, such as the degree of solvation of the bonded hydrocarbon chain and the differential partitioning of solvent components into the hydrocarbon layer. The theory is conceptually elegant and could readily be adapted to adsorption onto graphite. Unfortunately, it gives no guidance as to how to calculate the key energetic parameters, namely the interaction energies between segments (i.e., the χ values), so it likewise gives no lead on the interpretation of PREG.

The theory of Snyder [37] was originally developed to explain and quantify the idea of an eluotropic series for oxide adsorbents. The theory can be applied to graphite if the specific assumptions made for oxide adsorbents are revised.

Snyder regards the process of adsorption of the analyte A as a displacement process of the form

$$A_{solv} + n\underline{S}_{ads} = \underline{A}_{ads} + nS_{solv} \tag{5}$$

where A is an analyte molecule and S a solvent molecule. The analyte displaces n solvent molecules from the surface. The free-energy change, ΔF, can then be written

$$\Delta F = nF(S_{solv}) + F(\underline{A}_{ads}) - nF(\underline{S}_{ads}) - F(A_{solv}) \tag{6}$$

The free energies, F, are taken with respect to the isolated (i.e., unsolvated) species in the gas phase. The free energies of solvated species are evidently similar to those used by Horváth et al. in their theory [58]. The free energies of adsorbed species would contain a free-energy contribution for attachment of one face of the molecule to the adsorbate, and a second contribution for solvation of the side of the molecule in contact with the solvent. The second part of this contribution will cancel part of the total free energy of the fully solvated species, so we only need to consider the free energy of solvation of those parts of A and S which become desolvated when A and S are adsorbed. When Snyder applied this model to oxide adsorbents, he assumed that the solvation free energies, $nF(S_{solv})$ and $F(A_{solv})$ cancelled. This was the crucial assumption of the theory and it was justified only because the major interactions of oxide adsorbents with typical solvents, $F(\underline{S}_{ads})$, are dominated by the possibilty of hydrogen-bond formation between the –OH groups in the adsorbents and atoms such as –O–,–N–, –Cl, and so forth in the solvents. Accordingly, solvents could be effectively organized in the order of their hydrogen-bonding power or, more generally, their polarity. The theory enabled a semiquantitative measure of the eluting strength of solvents to be made on the basis of their chromatographic performance averaged over a large number of analytes. With these assumptions, Snyder derived the following experession for log k':

$$\log k' = \log(V_s/V_m) + \beta(S° A_s \varepsilon°) \qquad (7)$$

where V_s/V_m is the ratio of the volume of solvent in an adsorbed monolayer to the volume in the mobile phase, β is a surface activity coefficient (equal to 1 or less), $S°$ is a dimensionless free energy of adsorption of solute ($S° = \Delta G°/2.303RT$) onto the surface from a standard eluant (pentane), A_s is the contact area of the analyte in units of 8.5 Å^2, and $\varepsilon°$ is the eluotropic strength parameter of the solvent. $\varepsilon°$ is the dimensionless free energy of adsorption of the solvent per unit area of 8.5 Å^2. The theory gives a semiquantitative explanation of the dependence of retention on alumina and silica gel on the solvent used. Weak solvents, such as hydrocarbons, have $\varepsilon°$ values near zero, whereas strong solvents such as methanol have $\varepsilon°$ values around unity. Although the theory supposedly treats free-energy changes, it seems to apply primarily to enthalpy changes and there is no discussion of entropy changes.

As soon as we apply this theory to a nonpolar adsorbent such as an ODS silica gel, a polymeric phase, or graphite, we lose the unifying concept that solvents can be organized in order to their hydrogen-bonding power to these adsorbents, as they do not form H bonds with either solvent or analyte. It is then no longer appropriate to cancel the solvation enthalpies of A and nS. Indeed, they now become the key to retention selectivity. It is thus no surprise that there is no single eluotropic series for nonpolar adsorbents and that if one tries to apply the Snyder equation, the range of $\varepsilon°$ values is very small (apart from the value of water). Regrettably, once again, the Snyder theory gives no guidance as to how the unusual retentive properties of graphite can be explained.

The three theories with very different starting points all emphasise the importance of a thorough understanding of the forces which bind both the analyte and the solvent to the graphite surface. As already noted, such binding involves not only London–dispersion interactions, but additional interactions, so far unidentified, which are responsible for PREG. Regrettably, we have little understanding of the nature and magnitude of these additional interactions. All three theories emphasize the importance of changes in solvation enthalpy when the new contact surface is created between the analyte and graphite. The theories of Horváth et al. and of Mantine et al. include discussion of entropy changes which must also be important.

Overall, the three theories give a surprisingly unified picture of the process of displacement of solvent by analyte at the surface of graphite which may be represented as in Fig. 6, where the hydrophobic surfaces of the analyte and eluant molecules are shaded. The solvent molecules which are adjacent to surfaces are preferentially orientated, depending on whether their contact surface is hydrophilic or hydrophobic. When the new contact surface is formed between analyte and graphite, the solvent molecules which are preferentially orientated at the original hydrophobic surfaces are displaced and join the bulk solvent in random orientation. The diagram can obviously be redrawn for a two-component eluant comprising hydrophilic and hydrophobic components. The hydrophobic component will then be preferentially adsorbed at the hydrophobic surfaces, and vice versa.

This section may be concluded by summarizing the key aspects of the three theories:

(1) The Horváth theory [58] emphasizes surface interactions of analyte with solvent but assumes that the interactions of analyte with stationary phase involve only dispersive interactions.

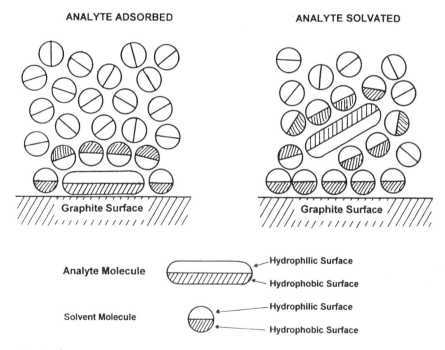

Fig. 6 Schematic representing equilibrium between analyte in adsorbed and solvated states. Shaded portions represent hydrophobic surfaces; unshaded portions represent hydrophilic surfaces; unshaded solvent molecules are randomly orientated.

(2) The theory of Martire and Boehm [59] essentially deals with the entropy of the displacement process, leaving the energy terms to be ascribed independently.

(3) The Snyder theory [37], in its original form, recognizes the displacement process shown in Fig. 6 but concentrates primarily on hydrogen-bonding interactions with an oxide stationary phase.

None of these theories address the central question of how the polar retention effect on graphite, PREG, arises.

VIII. CORRELATION OF RETENTION ON GRAPHITE WITH MEASURED OR CALCULATED PHYSICAL PROPERTIES

Any attempt to explain the peculiar retentive properties of graphite requires definition of the precise property one is trying to explain. Broadly, there are two approaches. The first tries to explain $\log k'$ and its dependence on experimental parameters in terms of independent physical properties of the analyte. It is thereby hoped that one can identify one or more key physical parameters which can correlate with retention. The second tries to separate the polar retentive effect, PREG, from the remaining contributions to $\log k'$. This requires an assessment of the contribution to $\log k'$ from the remaining, mainly hydrophobic, interactions, so that these can in some way be factored out. Nearly all attempts to explain the retentive behavior of graphite have used the first approach and have attempted to correlate $\log k'$ with various physico-chemical parameters. We believe that the approach most likely to succeed will, in fact, be the second one, where the polar retentive effect is first quantified.

Nevertheless, without a good theoretical lead, the best hope of explaining PREG must lie in an examination of correlations between retention and the values of physical parameters which are related to the electronic properties of selected analytes. There are a large number of such physical parameters, some empirical and some theoretical. There are two basic approaches to identifying appropriate physical properties, the *single-parameter* approach and the *factor analysis* approach. In the former, correlations are sought with a single, hopefully dominant, physical parameter, whereas in the latter, a weighted group of parameters is sought which gives an optimum correlation of the experimental data with independent physical parameters.

One group of parameters comprises those which are obtained by quantum mechanical modeling (see Ref. 35 for an introductory treatment). At its simplest level, this involves molecular orbital calculations on analyte molecules, which take account of the valency electrons but do not involve adjacent molecules or surfaces. A number of parameters are evaluated by these calculations. Of most immediate interest is the evaluation of the (excess) electronic charges on the individual atoms of analyte molecules. These excess charges are both positive and negative, with the total being zero for a neutral molecule. A number of parameters can then be derived. Kalizsan's "submolecular polarity parameter" Δ [35], for example, is defined as the sum of the largest excess negative and the largest excess positive charge in the molecule and so, in a sense, represents an overall charge profile. The "local dipole index" [35] is obtained by evaluating the charge difference over the atoms at the ends of each bond in the molecule and then adding them together. Another parameter used by Cocquart [45] is the "excess charge" C_n, which is defined by Eq. (8):

$$C_n = \tfrac{1}{2} \sum |q_i| \qquad (8)$$

where the q_i are the excess charges on the individual atoms. There seems to be little theoretical basis for using the sum of charges as a descriptor because forces between charges and the associated energies depend on the products of charges. A more likely descriptor would seem to be the sum of squares of the charges because these could reflect the interactions of the excess charge on each atom with corresponding features in some kind of induced-charge footprint in the graphite surface. Such a parameter would be given by

$$S = \sum (q_i)^2 \qquad (9)$$

However, there is a little more theoretical foundation for this modified descriptor than for C_n, as the following discussion shows.

Graphite, as was noted by Kaliszan et al. [38,39], has properties analogous to a metal. It is a two-dimensional conductor, the graphitic layers being metallic through the presence of an extended π-electron cloud. The theory of the interaction of a charged body with a conductor is well known [60]. The interaction energy can be calculated using a simple model. As shown in the upper diagram of Fig. 7, the lines of force entering any conductor must do so at right angles to the surface. From this, it may be readily shown that the energy of interaction of a charged body with a conductor is equal to its energy of interaction with its imaginary charge image (of opposite sign) reflected in the surface. The lower diagram in Fig. 7 shows how this works for a molecule with three charge centers.

The interaction energies, U_{ij}, of each atom in the analyte molecule with each atom in its image, the two separated by a distance r_{ij}, can be represented by the terms in the matrix shown in Table 6. Each energy term is given by

$$U_{ij} = \left(\frac{1}{4\pi\varepsilon_0}\right)\left(\frac{q_i q_j}{r_{ij}}\right) \qquad (10)$$

The total energy of interaction, U, is given by the sum of all nine energy terms in the matrix of Table 6; that is,

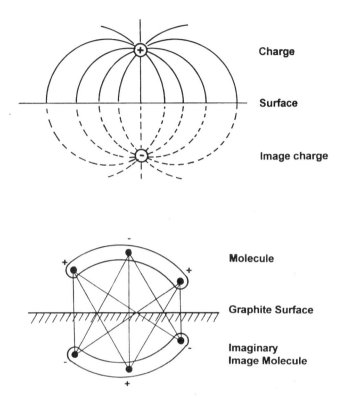

Fig. 7 Image model for interaction of a charged species with a conductor. *Upper:* Lines of force between charged particle and its imaginary image in the surface. *Lower:* Interaction of a three charge body with its oppositely charged image; lines represent interactions which have to be taken into account in determining total energy of interaction; see also Table 6.

Table 6 Matrix of Interaction Energy Terms for the Interaction Energy of a Charged Analyte With Its Image in the Graphite Surface

Image atom ⇒				1	2	3
Analyte atom	Coordinates of analyte atom			Interatom energies		
1	x_1	y_1	z_1	U_{11}	U_{12}	U_{13}
2	x_2	y_2	z_2	U_{21}	U_{22}	U_{23}
3	x_3	y_3	z_3	U_{31}	U_{32}	U_{33}

$$U = \frac{1}{4\pi\varepsilon_0} \sum \left(\frac{q_i q_j}{r_{ij}} \right) \tag{11}$$

The diagonal elements of the matrix represent the interaction energies of each atom in the analyte with the atom in its image which is directly below. Their sum, denoted by U', is given by

$$U' = \frac{1}{4\pi\varepsilon_0} \sum \left(\frac{q_i^2}{r_{ij}} \right) \tag{12}$$

U' is closely approximated by $U'' = S/4\pi\varepsilon_0 r$ where it is assumed that all r_{ii} are the same and equal to r (as they would be for a planar molecule parallel to the surface). The overall interaction energy U has to include all nine terms in the matrix and will approach U' only if the distances between the atoms in the molecule and its image, r_{ij}, are much smaller than the interatomic distances within the molecule itself. This, in fact, can never be the case. So, although S or C_n may give a general assessment of the overall charge separation in a molecule, they cannot represent the interaction energy of a charged body with a conductor. The full calculation of U must be made.

This calculation of U is readily carried out using a spreadsheet. However, a major assumption which is still made in the simple model is that the molecule is adsorbed from a vacuum, for which the relative dielectric constant is unity. When the molecule is adsorbed from a liquid, allowance should be made for the fact that lines of force emerging above the horizontal (see Fig. 7) from the adsorbate pass through a medium of high dielectric constant and so are effectively eliminated, whereas those emerging from below the horizontal are more or less unaffected by the solvent. A proper allowance for the effect of solvent is difficult, and, as a first approach to testing the relevance of this model as an explanation of PREG, we assume that Eqs. (10) applies without modification. Any effect of dielectric shielding will presumably reduce the interaction energy.

Other parameters which can be calculated by quantum mechanical methods [35] include the energies of the highest occupied molecular orbital, E_{homo} and the energy of the lowest unoccupied molecular orbital E_{lumo} of the analyte. These are expected to relate to the electron donor and acceptor capabilities of the analyte. They may well be relevant because graphite, being a conductor, can certainly act as an ready donor or acceptor of electrons. Other parameters such as the total binding energy E_T and various connectivity and structural parameters relate primarily to the size of a molecule and are likely to be correlated with its overall capacity to take part in dispersive interactions. This last group is unlikely to correlate with PREG.

More detailed molecular orbital calculations which take account of the substrate itself can, of course, be carried out using supercomputers [61]. Such calculations may in the end be the only way of providing a quantitative account of retention of polar species from say methanol/water eluents by graphite, but they are clearly not going to be useful for predictive purposes in the short term. To date, there have been no full molecular modeling studies of retention on graphite although there have been such studies of the interactions of molecules with alkyl-bonded silica gels [62]. The application of such methods to graphite could be highly informative. The calculations should be simpler with graphite than for bonded silicas, where a vast amount of computing effort is required simply to model the silica and its bonded alkyl groups before going on to consider the addition of solvent and analyte. However, we should be cautious: Unless the quantum mechanical calculations can correctly take account of all interactions which are relevant, they cannot give a correct answer.

A second important group of empirical parameters are those which are related to σ, the parameter originally devised by Hammet [63] in order to quantify, and hence predict, the effects of benzene-ring substituents on the rate and equilibrium constants of groups of reactions (e.g., hydrolysis of esters) which involve a "reactive centre which is influenced by a substituent." σ for any substituent was determined from

$$\sigma_X = \log K_{a(X)} - \log K_{a(H)} \tag{13}$$

where $K_{a(H)}$ and $K_{a(X)}$ are respectively the ionization constants of benzoic acid and benzoic acid substituted with X. When the substituent X is an electron-withdrawing group, σ values are positive, and when it is an electron-donating group, σ values are negative. The total range of values is from about -1 to $+1$. The value of σ for any substituent, X, depends on whether it is in the ortho, meta, or para position. The "Hammett equation" for any rate or equilibrium constant, k_X, can then expressed by

$$\log k_X = \rho \sigma_X + \log k_H \tag{14}$$

where k_H and k_X are the rate or equilibrium constants for the test molecule and the molecules with the substituent X, respectively, The constant, ρ, is characteristic of the process or equilibrium under consideration.

This basic framework has been extended over the years [64]. For example, it was found that when the ionization constants of phenols or amines were used as the standard in Eq. (13) rather than benzoic acids, the values of σ were significantly different and higher, ranging from around -0.2 to 1.4. This was attributed to a resonance effect, whereby the lone pair could be shared with the substituent group. Thus, the substituent, NO_2, for example, enhanced the ionization constant of phenol much more than that of benzoic acid. This complication led to attempts to separate inductive or field effects from resonance effects by Swain and Lupton [65] and by Taft and Lewis [66]. The latter expressed Hammet's σ according to

$$\sigma = F + R \tag{15}$$

where F represents the inductive effect and R the resonance effect. It was argued that F could be determined by measuring a modified σ, denoted by σ', obtained according to Eq. (13) but using 4-X-derivatives of bicyclo [2.2.2] octane-1-carboxylic acids where resonance effects were totally absent, instead of benzoic acids.

Although there have been many further modifications and extensions of the general ideas of Hammet, the above remarks seem adequate for the present purpose. Insofar as retention on graphite can be regarded as a reaction or equilibrium involving a "reactive centre which is influenced by a substituent," then the Hammett parameters should be relevant. Such a contention would be in line with Kaliszan's assertion that "a localised polar segment of the molecule was responsible for retention." However, we show below that this cannot account for PREG in general. It is likely that the whole electronic distribution of an analyte molecule and its polarizability is relevant and will have to be taken into account.

A. Single-Parameter Approach

In her very extensive study of retention of benzene derivatives by PGC, ODS silica gel, and PRP (polystyrene resin), Cocquart [45] measured log k' values over a range of methanol/water mixtures. In general, plots of log k' versus the volume fraction of organic component in methanol/water or acetonitrile/water mixtures were linear in accord with Eq. (2), so that value of log k' in pure water (i.e., values of log k'_w) could be determined either directly or by extrapolation. Coquart also carried out quantum mechanical calculations using a program called MOPAC [67] which evaluated the charge distributions on her test solutes, and from these she evaluated C_n for each of her test solutes. Hennion [68] has most kindly supplied us with the detailed calculations of these charge distributions. These have enabled us to calculate other charge related parameters, in particular S, U, and Kaliszan's Δ. Table 7 presents Coquart's original data and our own calculations of other parameters. The log P values are also included from Ref. 45. When Coquart [45] compared log k'_w with log P [69], excellent correlations were obtained when the stationary phase was either C_8 or ODS silica. With PRP-1, correlations were not as good, but were nevertheless adequate, as shown by Fig. 8.

Such correlations are well known and form the basis for the alternative determination of log P values by HPLC using reversed-phase hydrocarbon-bonded silica gels [70]. From these, it is deduced that the main interactions of the analytes with these supports were dispersive in accord with the hydrophobic retention theory of Horváth et al. [58]. With PGC/Hypercarb as the stationary phase, virtually no correlation was found, as shown by Fig. 9.

The data for the homologous alkyl benzenes are highlighted. It is noticeable that virtually all points lie above the line for the alkyl benzenes, confirming the additional retentive effect PREG.

Table 7 Data from Coquart [45] and Hennion [62] and Parameters Calculated Therefrom

Compound	C_n^1	S^2	U kJ/mol[3]	U″ kJ/mol[4]	Δ^5	Log P	Log k′ PRP-1	Log k′ PGC	PREG1[6]	PREG1′ kJ/mol[7]	PREG2[8]	PREG2′ kJ/mol[9]
Phenol	0.58	0.15	0.3	30.1	0.45	1.48	2.40	1.80	1.16	6.7	1.40	8.0
1,2-diOH-benzene Pyrocatechol	0.74	0.25	0.5	48.8	0.45	1.00	1.60	2.00	1.99	11.4	2.40	13.8
1,3-diOH-benzene Resorcinol	0.94	0.33	0.2	64.4	0.46	0.80	1.25	2.10	2.35	13.5	2.85	16.4
1,4-diOH-benzene Hydroquinone	0.79	0.25	0.5	49.2	0.44	0.59	0.83	2.05	2.58	14.8	3.22	18.5
1,2,3-triOH-benzene Pyrogallol	0.98	0.36	0.8	71.0	0.47		0.78	2.05			3.27	18.8
1,3,5-triOH-benzene Phloroglucinol	1.32	0.52	0.6	103.2	0.46		0.50	2.50			4.00	23.0
Nitrobenzene	1.03	0.61		120.9	0.89	1.86	3.70	2.30	1.16	6.7	0.60	3.4
1,2-diNitro-benzene						1.58	3.95	2.70	1.93	11.1	0.75	4.3
1,3-diNitro-benzene						1.49	4.00	3.60	2.95	16.9	1.60	9.2
1,4-diNitro-benzene						1.46	4.05	3.10	2.49	14.3	1.05	6.0
Benzaldehyde	0.66	0.25		48.6	0.60	1.50	2.92	1.75	1.09	6.2	0.83	4.8
Benzoic acid	0.96	0.47	0.6	93.3	0.77	1.87	3.30	2.60	1.45	8.3	1.30	7.5
1,2-diCOOH-benzene Phthalic acid	1.47	0.84	2.2	167.2	0.77	0.79	2.20	2.75	3.02	17.3	2.55	14.6
1,3-diCOOH-benzene Isophthalic acid	1.67	0.90	0.7	177.5	0.75	1.67	3.10	3.50	2.61	15.0	2.40	13.8
1,3,5-triCOOH-benzene Trimesic acid	2.41	1.37	0.8	270.7	0.76	1.15	2.60	4.50	4.29	24.7	3.90	22.4

Table 7 Continued

Compound	C_n^1	S^2	U kJ/mol[3]	U″ kJ/mol[4]	Δ^5	Log P	Log k′ PRP-1	Log k′ PGC	PREG1[6]	PREG1′ kJ/mol[7]	PREG2[8]	PREG2′ kJ/mol[9]
1,2,4-triCOOH-benzene Trimellitic acid	2.14	1.25		247.4	0.75		1.95	3.95			4.00	23.0
1,2,4,5-tetraCOOH-benzene Pyromellitic acid	2.60	1.54		305.5	0.71		1.15	3.85			4.70	27.0
1-COOH,2-OH-benzene	1.38	0.71	2.0	140.2	0.82	2.20	3.30	2.80	1.22	7.0	1.50	8.6
1-COOH,3-OH-benzene	1.17	0.55	0.8	109.2	0.75	1.60	2.40	2.60	1.80	10.4	2.20	12.6
1-COOH,4-OH-benzene	1.31	0.61	0.7	121.4	0.77	1.58	2.30	2.65	1.88	10.8	2.35	13.5
1-COOH,2,3-OH-benzene	1.40	0.68	1.5	135.5	0.78		2.40	2.90			2.50	14.4
1-COOH,2,6-OH-benzene	1.70	0.87	2.0	171.4	0.82		2.85	2.65			1.80	10.3
1-COOH,3,4-OH-benzene	1.40	0.68	0.6	135.6	0.76		1.50	2.95			3.45	19.8
1-COOH,3,5-OH-benzene	1.48	0.69	0.7	137.6	0.75		1.35	3.00			3.65	21.0
1-COOH,2-amino-benzene	1.27	0.60		118.5	0.75	1.21	2.85	2.90	2.61	15.0	2.05	11.8
1-COOH,3-amino-benzene	1.16	0.56		110.0	0.76		1.57	2.80			3.23	18.6
1-COOH,4-amino-benzene	1.20	0.57		112.4	0.76		2.02	2.85			2.83	16.3
Aniline	0.57	0.13		26.1	0.37	0.90	2.50	1.30	1.42	8.2	0.80	4.6
1,2-diAmino-benzene	0.78	0.23		45.6	0.39	0.15	1.70	1.40	2.50	14.4	1.70	9.8
1,3-diAmino-benzene	0.74	0.23		45.2	0.38		1.34	1.60			2.26	13.0
1,4-diAmino-benzene	0.74	0.23		44.8	0.38		1.05	1.50			2.45	14.1
2-Amino-phenol						0.52	1.70	2.05	2.67	15.3	2.35	13.5
3-Amino-phenol						0.17	1.30	1.70	2.78	16.0	2.40	13.8
4-Amino-phenol						0.04	1.10	1.58	2.83	16.2	2.48	14.2
4-Amino-2-Chloro-phenol							2.85	2.45			1.60	9.2

Compound										
Phenoxy-acetic acid				1.42	2.80	2.15	1.59	9.1	1.35	7.8
Benzene	0.35	0.04	8.3	0.12		1.52	0.03	0.2	0.12	0.7
Toluene	0.38	0.05	9.8	0.19		2.22	-0.00	-0.0	0.08	0.5
Ethyl benzene	0.36	0.04	8.2	0.15		2.85	0.02	0.1	0.05	0.3
Propyl benzene	0.39	0.04	8.6	0.15		3.50	-0.02	-0.1	0.00	0.0
Isopropyl benzene										
Cumene	0.39	0.04	8.2	0.14		3.40	0.05	0.3	0.10	0.6
				3.55	5.30		1.30	7.5		
Benzyl alcohol	0.69	0.24	46.8	0.56		2.00	1.86	10.7	1.70	9.8
1-Phenyl-ethanol	0.71	0.22	43.9	0.31	2.30	1.95			1.40	8.0
2-Phenyl-ethanol	0.71	0.21	42.5	0.42	2.55	2.15	1.46	8.4	1.50	8.6
1-Phenyl-propanol	0.74	0.23	44.9	0.52	2.65	2.70			1.30	7.5
2-Phenyl-propanol	0.70	0.20	40.1	0.52	1.10	3.30	2.80		1.50	8.6
3-Phenyl-propanol	0.73	0.22	43.1	0.52	1.52	2.90	1.61	9.2	1.48	8.5
					3.40					
					1.98	3.42				
Cyclohexane	0.09	0.00	0.4	0.02	3.44	1.90	-1.31	-7.5	-0.50	-2.9
Cyclohexanol	0.41	0.15	30.0	0.50	1.24	0.72	0.40	2.3	1.17	6.7
1,3,5-triOH-cyclohexane	1.06	0.45	89.6	0.51	-0.50	-0.50			2.00	11.5

[1] C_n in units are electronic charge; equation (8)
[2] S in units of (electronic charge) squared; equation (9)
[3] U by equation (11) and Table 4
[4] $U'' = S/(4\pi\varepsilon_o r)$ in kJ/mol.
[5] Δ = sum of largest positive and largest negative charges in electronic charge units
[6] PREG1 by equation (17)
[7] PREG1' = 2.303 RT PREG1
[8] PREG2 by equation (19)
[9] PREG2' = 2.303 RT PREG2

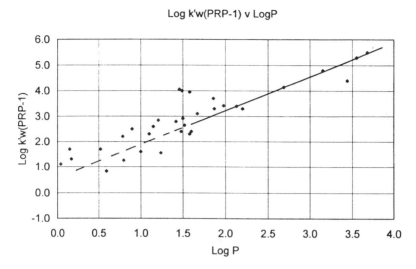

Fig. 8 Plot of log k'_w for PRP-1 versus log P. Data are from Table 7 [45]. Line drawn through points for alkyl benzenes.

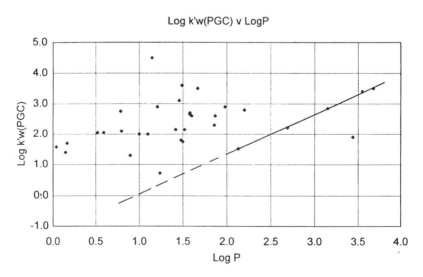

Fig. 9 Plot of log k'_w for PGC versus log P. Data are from Table 7 [45]. Line drawn through points for alkyl benzenes.

Although for PGC there was no correlation of log k'_w with log P, Coquart demonstrated a significant correlation between log k'_w and the total excess charge denoted by C_n [defined in Eq. (8)]. This is shown in Fig. 10. Because S shows a strong correlation with C_n, we obtain a very similar picture if log k'_w is plotted against S as shown in Fig. 11. What is particularly noticeable in both plots is that the points for the alkyl benzenes (highlighted) and phenyl alkanols show no correlation with C_n or S: Their C_n or S values are nearly identical, but (as expected) their log k'_w values are very different. If these points and those for cyclohexanols are omitted, the remainder do, nevertheless, fall within a band which is less than one $\log_{10}$ unit wide. Thus, a partial correlation appears to exist for benzene substituted by simple polar groups. The correlation fails for the alkyl benzenes (points above the band) and for the cyclohexanols (points below the band). The charge distribution with an analyte molecule is clearly not the whole story. When polar groups are associated with a benzene ring, PREG is large, but when they are associated with a saturated hydrocarbon, PREG still occurs but is much less (see earlier discussion of Refs. 42 and 43).

We have, therefore, attempted to find a more direct measure of PREG. The simplest way of allowing for hydrophobicity would be to subtract a term from log k' which allowed for hydrophobicity, such as log P. We therefore propose an equation of the form of Eq. (16).

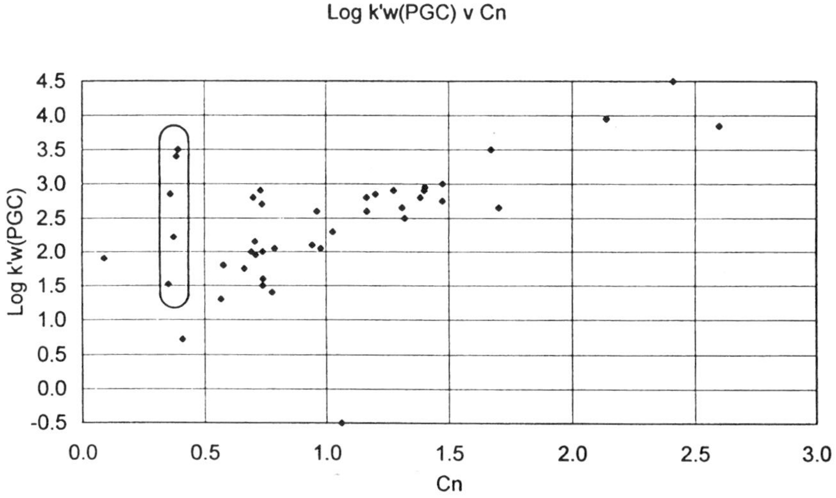

Log k'w(PGC) v Cn

Fig. 10 Plot of log k'_w for PGC versus C_n. Data are from Table 7 [45]. Points for alkyl benzenes are highlighted. For definition of C_n see Eq. (8).

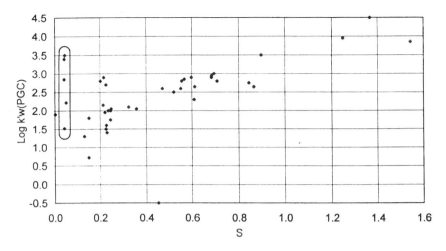

Fig. 11 Plot of log k'_w for PGC versus S. Data are from Table 7 [45]. Points for alkyl benzenes are highlighted. For definition of S see Eq. (9).

$$\text{PREG1} = \log k' - A \log P + B \tag{16}$$

or

$$\text{PREG1}' = 2.303RT\,(\text{PREG1}) \tag{16a}$$

where R (=8.314 J/K/mol) is the universal gas constant and T the absolute temperature. The constants A would be chosen on the basis of the known value of $d(\log k')/d(\log P)$ for nonpolar solutes, whereas B would be chosen so that PREG1 is zero for a homologous series, such as the alkyl benzenes. Equation (16a) expresses PREG in energy terms, enabling it to be compared with energies calculated in other ways.

Thus, from Table 5 it is seen that $d(\log k'_w)/d(\log P) = 1.31$ for PGC. Accordingly, for water as the eluant, $A = 1.31$. From the data of Cocquart (Table 7) it may readily be calculated that $B = 1.3$. Thus,

$$\text{PREG1} = (\log k'_w)(\text{PGC}) - 1.31 \log P + 1.3 \tag{17}$$

Alternatively, we can attempt to allow for the hydrophobic interactions by using another support, X, whose interactions are predominantly hydrophobic. Thus,

$$\text{PREG2} = \log k'(\text{PGC}) - A \log k'(X) + B \tag{18}$$

or, if we wish to express PREG in terms of free energy,

$$\text{PREG2}' = 2.303RT(\text{PREG2}) \tag{18a}$$

A is again chosen so as to eliminate any dependence of PREG2 on hydrophobic effects, and B is again chosen so that PREG2 = 0 for a homologous series such as the alkyl benzenes.

As noted in Table 5, the gradients $d(\log k')/d(\log P)$ are the same for PRP-1 and PGC, when pure water is the eluant, although they differ for methanol/water mixtures. It was also noted (Fig. 5) that b values, extrapolated to pure water, are essentially the same for the alkyl benzenes with these two stationary phases. With pure water, it would therefore appear that the hydrophobic retention by PRP-1 and PGC are very similar in terms of the energies of adsorption. Again, using Cocquart's data to determine B, we obtain

$$\text{PREG2} = \log k'_w (\text{PGC}) - \log k'_w (\text{PRP-1}) + 2.0 \qquad (19)$$

Figures 12 and 13 show plots of PREG1 and PREG2 versus S, respectively. Unfortunately, both correlation are poor. By definition, both PREG1 and PREG2 place the points for the alkyl benzenes and phenyl alkanols very close together. PREG1 correlates a little better with S than PREG2. However, this may be more due more to the paucity of values of P than any real advantage. The correlation of PREG2 shows an extreme range of three log units in PREG2 at an S value of about 0.6.

We are forced to conclude either (1) that equations (17) and (19) fail to give a proper measure of PREG or (2) that S and C_n are not appropriate descriptors of

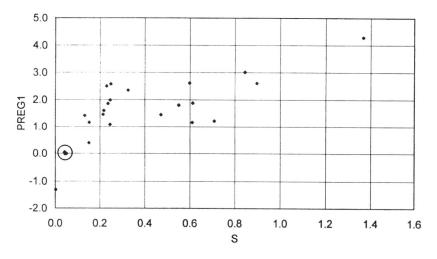

Fig. 12 Plot of PREG1 versus S. Points for alkyl benzenes are highlighted. For definitions, see Eqs. (17) and (9).

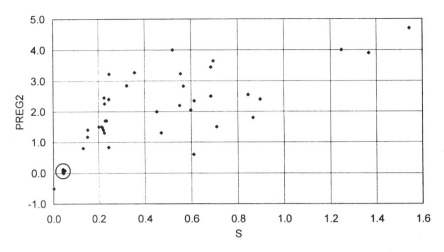

Fig. 13 Plot of PREG2 versus S. Points for alkyl benzenes are highlighted. For definitions, see Eqs. (19) and (9).

PREG. As already noted, there seems to be no good theoretical reason for regarding either S and C_n as having a fundamental significance, although both are clearly related to the charge separation in a molecule.

In considering other possible descriptors, it is worth noting that Kaliszan's parameter Δ [35], listed in Table 7, also gives no useful correlation. Typically, for any series of polysubstituted benzenes, for example the polysubstituted phenols or the polysubstituted acids or the hydroxy acids, all members have roughly the same value of Δ. This is because the largest charge differences within the molecules of any such series are determined by the the most polar group, not the number of such groups.

We have also calculated U (see Eq. (11)) for a number of Cocquart's compounds, and the results are shown in Table 7. Again, the results are disappointing, and U shows little correlation with any measure of PREG. First, the magnitudes of U, when compared with values of PREG', are generally much too small to account for the additional retentive effects on graphite over those expected from hydrophobic interaction. We require additional contributions to the interaction energy of up to 18 kJ/mol, where the calculated values of U are generally below 2 kJ/mol. The values of U'' (the energetic measure of S), on the other hand, are much too large.

A second problem, not obvious from the values given for U, is that U is strongly dependent on the configuration of the substituent group. Taking phenol as an example, the –OH group can, in principle, be orientated at any angle to the plane of the graphite surface. If we take Coquart's [45] and Hennion's [68] charges for phenol and assume that the ring is parallel to the plane of the carbon atoms and 3.5 Å away from it, the extreme values of U are 145 J/mol with the H atom pointing down into the graphite surface and 535 J/mol with the H atom up. For benzoic acid, similar calculations show that the energy has a minimum of 620 J/mol when the H atom is just below the plane of the ring and pointing away from the second oxygen atom, and a maximum of 2030 J/mol when the H atom is pointing toward the second O atom, again slightly below the plane of the ring. Such ranges of values, covering factors of about 3, make nonsense of any attempt to find close correlations between values of U and chromatographic parameters. To make progress in this direction it would be necessary to optimize the electronic configuration of each molecule with respect to the surface.

Simple models and theories evidently fail to explain the unusual retentive properties of graphite. More far-reaching and fundamental calculations will be required to explain PREG. These will have to take proper account of the polarization of both the graphite and the analyte and accommodate the possibility of electron transfer between the two before real progress can be made. These types of interaction energies are not normally allowed for in molecular modeling packages. It will also be necessary to take the solvent into account as Yarovsky et al. have done for alkyl silicas [62].

B. Factor Analysis Approach

Unless a fairly firm theoretical link exists between a physical or computed property and the retention parameter, it is possible that a weighted combination of physical properties will provide a better correlation than any single parameter. The combination is likely to depend on the class of analyte considered, but an examination of the optimum combinations for different classes of analyte may lead to a more fundamental understanding of the factors leading to retention. There are basically two methods of attack: factor analysis and principal component analysis. Both have been extensively used to interpret and predict retention data in chromatography, and the subject is dealt with at an elementary level by Kaliszan [35]. Both methods start from a listing of the experimental parameter(s) for a range of analytes, J. These will normally be log k' in chromatography or parameters derived from them, such as a and b of Eq. (1) or log k'_w and m of Eq. (2). A parallel listings of various physical parameters, I, which might correlate with the retention parameter(s) is then compiled.

In factor analysis, one attempts to find linear equations of the form

$$\log k'_j = a_1 F_{1,j} + a_2 F_{2,j} + a_3 F_{3,j} + \cdots + = \sum_i (a_i F_{i,j}) \qquad (20)$$

where the $F_{i,j}$ are the values of the physical parameters, I, appropriate to each solute J, and the a_i's are constants for the range of analytes. The values of the a_i are optimized to give the lowest standard deviation between the calculated values of log k'_j and their experimental values. By suitable statistical analysis, the standard deviations of each a_i can also be determined, so that factors of low significance can be excluded from the final equation. The weakness of the factor analysis method is that the different factors may themselves be strongly correlated. However, when more or less independent parameters dominate, factor analysis can give a good indication of which of the available physical parameters are important, and where one might look for alternative factors which might be of greater relevance.

Principle component analysis allows for factor correlation. By means of a matrix analysis of the known values of the factors, $F_{i,j}$ for the analytes, it is possible to derive so-called principal components, which are independent of one another (i.e., mutually orthogonal). Each principle component, P, is made up of a linear combination of the various experimental parameters. For an analyte, J, the Nth principal component, $P_{N,j}$ will be given by a linear sum of the factors, $F_{i,j}$:

$$P_{N,j} = v_{N,1}F_{1,j} + v_{N,2}F_{2,j} + v_{N,3}F_{3,j} + \cdots + = \sum (v_{N,i} F_{i,j}) \tag{21}$$

Finally, the values for log k' are calculated as a linear combination of the $P_{N,j}$

$$\log k'_j = b_1 P_{1,j} + b_2 P_{2,j} + b_3 P_{3,j} + \cdots \tag{22}$$

It may or may not be possible to associate a particular P_N with a particular molecular characteristic, because each P_N contains contributions from a range of physical parameters. Each principal component is said to be loaded with the various experimental parameters, the loading factors being the values of the $v_{N,i}$. It is possible to calculate what percentage of the total variance of the data can be accounted for by each of the principal components, and so to determine their importance.

Kaliszan [38] applied factor analysis to the log k' values of benzene derivatives, having a range of simple substituents, $-CH_3$, $-COCH_3$, $-Cl$, $-COOCH_3$, $-NO_2$, $-NH_2$, $-OCH_3$, $-CH_2OH$, and $-OH$, plus a range of alkyl-phenols, with alkyl groups of up to four carbon atoms, and pyridine. The chromatography was carried out with hexane as the eluant. Such an eluant would be expected to show little discrimination according to alkyl chain length, as indeed was found. The best correlation was with the submolecular parameter Δ, which suggested that "a localised segment of the molecule was responsible for retention rather than the π-orbitals of the aromatic ring." They also found that retention by graphite correlated well with retention on palladium metal spheres, suggesting that graphite behaved like a metal. This is perhaps not unexpected because the layers of carbon atoms in graphite are indeed two-dimensional conductors. Unfortunately, as noted above in considering Coquart's data, the correlation of retentivity with Δ does not hold for derivatives containing more than one functional group.

In a second article, Bassler, et al. [39] analyzed data from a similar group of compounds using the principal components method. Again, the eluant was hexane. If one simply looks at the tabulated values of log k', it is immediately clear that molecules of high polarity are much more retained than those of low polarity and that increase in the size or number of the alkyl substituents has less effect on retention than changes in polarity. This is not unexpected with a hydrocarbon as the eluant. It is also clear that the position of a substituent in say an alkyl phenol can have very significant effect on its retention, with para-substituents providing the highest retention. Surprisingly, the principal component which provided the best correlation with the experimental data was heavily loaded with factors which related to molecular size, and to a lesser extent those which related to molecular polarity (for example, dipole moment, the electron acceptor strength, and Δ). The second principal component was loaded with the heat of formation and "local dipole index." The evidence is, therefore, that retention from hexane is strongly influenced by both molecular size and by the distribution of molecular charge in the analyte.

Further extensive correlations have been made by Forgács, Cserháti, and coworkers [48–57] under reversed-phase conditions with methanol/water and acetonitrile/water as eluents. Their primary experimental measurements demonstrated a linear dependence of log k' on the percentage of organic component (methanol or acetonitrile) in the otherwise aqueous mobile phase [Eq. (2)].

Using multicomponent factor analysis they found that combinations of electronic parameters gave the best correlations with both log k'_w and m. They concentrated primarily on correlations with empirical parameters related to the Hammet σ. Using stepwise regression analysis, they examined the correlation of retention data for 45 barbiturates on PGC/Hypercarb using methanol/water eluants with a number of physico-chemical parameters [49]. The physico-chemical parameters selected were as follows: the Hansch–Fujita constant characteristic of hydrophobicity; proton acceptor and proton donor properties; molar refractivity; the Hammett constant σ, characterizing the electron withdrawing power of substitutents; the Taft constant, characterizing steric effects of substituents; and the two Swain–Lupton electronic parameters, F and R, which characterize the inductive and resonance effects of substituents, respectively. Forgács et al. measured log k' for a range of compositions of eluant to obtain values of log k'_w and m [Eq. (2)]. These values were themselves well correlated, indicating that the derivatives could be regarded as a "homogeneous series of compounds." The substituents (see original paper) varied widely, with R_1 and R_2 being hydrocarbons (alkyl, cyclohexenyl, phenyl, for example) and R_3 and R_4 contained alkyl, phenyl, benzoyl, and benzoly substituted with Cl, NO_2, and NH_2. The order of elution in many cases deviated from that expected from a reversed-phase packing material, with more polar derivatives eluting after less polar ones, particularly in the case of the nitro-derivatives. The physical parameter which correlated best with retention was the Swain–Lupton electronic parameter F, which characterizes the inductive effect, which accounts for 66% of the total variance.

Similar studies of the retention characteristics of phenols [50,53] again showed an elution order differing from those expected from simple reversed-phase behavior, with more polar analytes being more retained than otherwise expected. The phenols examined were were mono- or di-substituted with $-CH_3$, $-C_2H_5$;$C(CH_3)_2$, $-CH_2CH=CH_2$, $-N(CH_3)_2$, $-OH$, $-OCH_3$, $-CH_2OCH_3$, $-CH_2CN$, $-Cl$, $-Br$, and $-NO_2$. In general, the retention increased with the number of substituents, and for single substitution, the order of increasing retention was $F \cong CH_3 < Cl < CN < Br < NO_2$. Studies on amines [51,52] gave roughly a similar order of substituents, $CH_3 < Cl < Br < I < NO_2$. It was concluded that retention was mainly governed by steric parameters, electron-withdrawing power, and hydrogen donor capacity of substituents but not by lipophilicity ($\cong$ hydrophobicity).

IX. CONCLUSIONS

There is unquestionable experimental evidence that graphite shows unusual retentive behavior for polar analytes (PREG). As shown in Part B of this review [25], exploitation of PREG has led to a number of very unusual separations which can be successfully performed with PGC as stationary phase but are difficult or impossible with more conventional reversed-phase materials, such as ODS silica gels or polymeric materials. At this stage, the best that can be said in explanation of PREG is that it seems to be related in some way to the charge separation within molecules and probably involves the ready polarization of the electronic cloud of graphite. PREG is particularly pronounced when the polar groups are attached to a benzene ring, and presumably other larger aromatic systems. The effects cannot be explained by simple electrostatic models which regard graphite as a conductor. There has to be some type of orbital overlap between the conductivity electrons in graphite and the lone pair and/or π electrons in analytes.

Graphite is thus a totally unique chromatographic material which will no doubt be the subject of much further research.

REFERENCES

1. J. H. Knox and B. Kaur, in *High Performance Liquid Chromatography,* P. Brown and R. Hartwick, Eds., John Wiley & Sons, London, 1989, pp. 189–222.
2. J. H. Knox, K. K. Unger, and H. Müller, *J. Liq. Chromatogr., 6:* 1 (1983).
3. E. Lederer, *Chromatography,* Elsevier Publishing Company, Amsterdam, 1953.
4. L. R. Snyder and J. J. Kirkland, *Introduction to Modern Liquid Chromatography,* 2nd ed., John Wiley & Sons, New York, 1979.

5. N. Tanaka and M. Araki, in *Advances in Chromatography, Vol. 30,* Marcel Dekker, Inc., New York, 1989, pp. 81–122.
6. D. Berek and I Novák, *Chromatographia, 30:* 582 (1990).
7. A. V. Kiselev and Y. I. Yashin, *Gas Adsorption Chromatography,* J. E. S. Bradley, Trans. Plenum Press, London, 1969.
8. A. DiCorcia and A. Liberti, in *Advances in Chromatography, Vol. 14,* Marcel Dekker, Inc., New York, 1976, p. 305.
9. P. Ciccioli, R. Tappa, A. DiCorcia, and A. Liberti, *J. Chromatogr., 206*: 35 (1981).
10. H. Colin, C. Eon, and G. Guiochon, *J. Chromatogr., 119:* 41 (1976).
11. H. Colin, C. Eon, and G. Guiochon, *J. Chromatogr., 122:* 223 (1976).
12. H. Colin, N. Ward, and G. Guiochon, *J. Chromatogr., 149:* 169 (1978).
13. Z. Plzák, F. P. Dousek, and J. Jansta, *J. Chromatogr., 147:* 137 (1978).
14. V. Pazelova, J. Jansta, and F. P. Dousek, *J. Chromatogr., 148:* 53 (1978).
15. E. Smolkova, J. Zima, F. P. Dousek, J. Jansta, and Z. Plzák, *J. Chromatogr., 191:* 61 (1980).
16. K. Unger, P. Roumeliotis, H. Mueller, and H. Goetz, *J. Chromatogr., 202:* 3 (1980.)
17. J. H. Knox and M. T. Gilbert, UK Patent 2 035 282, 1978.
18. M. T. Gilbert, J. H. Knox, and B. Kaur, *Chromatographia, 16:* 138 (1982).
19. J. H. Knox, B. Kaur, and G. R. Millward, *J. Chromatogr., 352:* 3 (1986).
20. Shandon, *Hypercarb Guide,* Life Sciences International, Runcorn, UK, 1993.
21. I. Novák and D. Berek, Czech Patent 221197, 1982
22. E. Skutchanová, L. Feltl, E. Smolkova-Keulemansová, and J. Skutchan, *J. Chromatogr., 292:* 233 (1974).
23. Tonen Corporation, European Patent 0 458 548 A, 1990.
24. Nippon Carbon Co. Ltd., and Tosoh Corporation, European Patent 0 484 176 A, 1990.
25. P. Ross and J. H. Knox, in *Advances in Chromatography, Vol. 37,* Marcel Dekker, Inc., New York, 1996, pp. 121–162.
26. B. E. Warren, *Phys. Rev., 59:* 693 (1941).
27. J. D. Bernal, *Proc. Roy. Soc. (Lond.), A100:* 749 (1924).
28. A. A. Isirikyan and A. V Kiselev, *Russ. J. Phys. Chem., 36:* 618 (1962).
29. G. M. Jenkins and K. Kawamura, *Polymeric Carbons,* Cambridge University Press, Cambridge, 1976.
30. H. Dias, Ph.D. thesis, University of Edinburgh, 1990.
31. R. Atkins, *Physical Chemistry,* 3rd ed., Oxford University Press, Oxford, 1986.
32. D. M. Ruthven, in *Principles of Adsorption and Adsorption Processes,* John Wiley & Sons, New York, 1984, pp. 29–61.

33. E. A. Guggenheim, *Thermodynamics,* North-Holland, Amsterdam, 1949.
34. S. Ross and J. P. Oliver, in *On Physical Adsorption,* John Wiley & Sons, New York, 1964, p. 267.
35. R. Kaliszan, *Quantitative Structure–Chromatographic Retention Relationships,* John Wiley & Sons, New York, 1987.
36. B. Kaur, *LC–GC Int., 3:* 41 (1989).
37. L. R. Snyder, *Principles of Adsorption Chromatography,* Marcel Dekker, Inc., New York, 1968.
38. R. Kaliszan, K. Osmialowski, B. J. Bassler, and R. A. Hartwick, *J. Chromatogr.,* 499: 333 (1990).
39. B. J. Bassler, R. Kaliszan, and R. A. Hartwick, *J. Chromatogr.,* 461: 139 (1989).
40. M. F. Emery and C. K. Lim, *J. Chromatogr.,* 479: 212 (1989).
41. C. K. Lim, *Biomed. Chromatogr.,* 3: 92 (1989).
42. H. J. Möckel, A. Braedikow, H. Melzer, and G. Aced, *J. Liq. Chromatogr.,* 14: 2477 (1991).
43. N. Tanaka, T. Tanigawa, K. Kimata, K. Hosoya, and T. Araki, *J. Chromatogr.,* 549: 29 (1991).
44. J. Kříž, E. Adamcová, J. H. Knox, and J. Hora, *J. Chromatogr.,* 663: 151 (1995).
45. V. F. Coquart, Ph.D. thesis, University of Paris, 1993.
46. V. Coquart and M.-C. Hennion, *J. Chromatogr.,* 600: 195 (1992).
47. M.-C. Hennion and V. Cocquart, *J. Chromatogr.,* 642: 211 (1993).
48. E. Forgács, K. Valko, and T. Cserháti, *J. Liq. Chromatogr.,* 14: 3457 (1991).
49. E. Forgács and T. Cserháti, *J. Pharm. Biomed. Anal.,* 10: 861 (1992).
50. E. Forgács, T. Cserháti, and K. Valkó, *J. Chromatogr.,* 592: 75–83 (1992).
51. E. Forgács and T. Cserháti, *J. Chromatogr.,* 600: 43 (1992).
52. E. Forgács and T. Cserháti, *Chromatographia,* 33: 356 (1992).
53. E. Forgács, T. Cserháti, and B. Bordás, *Chromatographia,* 36: 19 (1993).
54. E. Forgács, T. Cserháti, and B. Bordás, *Anal. Chim. Acta,* 297: 115 (1993).
55. E. Forgács, K. Valk, T. Cserháti, and K. Magyar, *J. Chromatogr.,* 631: 207 (1993).
56. E. Forgács, *Chromatographia,* 39: 740 (1994).
57. E. Forgács and T. Cserháti, *Trends Anal. Chem.,* 14: 23 (1995).
58. C. Horváth, W. Melander, and I. Molnár, *J. Chromatogr.,* 125: 129 (1976).
59. D. E. Martire and R. E. Boehm, *J. Phys. Chem.,* 87: 1045 (1983).
60. A. Zangwill, in *Physical Chemistry of Surfaces,* Cambridge University Press, Cambridge, 1988, p. 185.
61. Biosystems Technology, *Discover Users Guide Version 2.9.0,* Biosystems Technology, San Diego, CA, 1993.

62. I. Yarovsky, A. Agular, and M. T. W. Hearn, Anal. Chem., 67: 2145 (1995); *J. Chromatogr.*, 660: 75 (1994).
63. L. P. Hammett, *Chem. Rev.*, 17: 125 (1935).
64. C. Hansch and A. Leo, *Substituent Constants for Correlation Analysis in Chemistry and Biology*, John Wiley & Sons, New York, 1979.
65. G. G. Swain and E. C. Lupton Jr., *J. Am. Chem. Soc.*, 90: 4328 (1968).
66. R. W. Taft and I. C. Lewis, *J. Am. Chem. Soc.*, 80: 2441 (1958).
67. Tripos Associates Inc., *Sybyl 5.41*, Tripos Associates Inc., St. Louis, MO, 1988.
68. M-.C. Hennion, private communication, 1995.
69. R. F. Rekker and R. Mannhold, *Calculation of Drug Lipophilicity*, VCH, Weinheim, 1992.
70. T. L. Hafkenscheud and E. Tomlinson, in *Advances in Chromatography, Vol. 25*, Marcel Dekker, Inc., New York, 1986, pp. 1–62.

3B
Carbon-Based Packing Materials for Liquid Chromatography:
Applications

Paul Ross *Hypersil, Runcorn, Cheshire, United Kingdom*
John H. Knox *University of Edinburgh, Edinburgh, United Kingdom*

I.	INTRODUCTION	122
II.	SEPARATIONS OF GEOMETRIC ISOMERS AND CLOSELY RELATED COMPOUNDS	124
	A. Simple Aromatic Molecules	125
	B. Pharmaceuticals and Natural Products	129
	C. Pesticides	132
III.	ENANTIOMER SEPARATIONS	132
	A. Separations of Diastereoisomers	133
	B. Separations Based on Nonbonding Interactions	134
	C. Separations Based on Inclusion Complexation	136
	D. Separations Based on Ion-Pair Formation	136
	E. Separations Based on Metal Complexation	138
IV.	SEPARATIONS OF SUGARS, CARBOHYDRATES, AND GLUCURONIDES	138
	A. Sugars and Carbohydrates	138
	B. Glucuronides	141
V.	RESIDUE ANALYSIS	142
	A. Polychlorinated Biphenyls	142
	B. Tissue Residues	143
	C. Solid-Phase Extraction and Water Analysis	143

VI.	SEPARATIONS OF IONIZED AND OTHER HIGHLY POLAR	
	COMPOUNDS	147
	A. Ionized Solutes	147
	B. Water Soluble Un-ionized Solutes	147
	C. Ion-Exchange Separations	151
Appendix 1:	Separations of Geometric Isomers and Closely Related Compounds	152
Appendix 2:	Enantiomer Separations	154
Appendix 3:	Analysis of Sugars, Carbohydrates, and Glucuronides	156
Appendix 4:	Residue Analysis	157
Appendix 5:	Separations of Ionized and Highly Polar Compounds	159
	REFERENCES	160

I. INTRODUCTION

When porous graphitic carbon (PGC) was first introduced as an adsorbent for high-performance liquid chromatography (HPLC) [1,2], it was thought that its main use would be as a substitute for reversed-phase silica gels in areas where these materials were unsatisfactory (e.g., at extreme pH and for compounds which interacted adversely with the unreacted silanol groups present in most bonded silica gels). PGC was regarded primarily as a strong reversed-phase packing material. Since that time, PGC has proved to have a number of unsuspected properties which have substantially enlarged its area of applications and opened up entirely new possibilities. Some of these features have been discussed in Part A of this review [3] (Chapter 3A of this volume), which covers the structure, physical properties, and general retentive behavior of PGC. The features which characterize PGC and distinguish it from other HPLC supports are the following:

1. Porous graphitic carbon, as originally envisaged, behaves primarily as a strong reversed-phase stationary phase, stronger than the alkyl-bonded silica gels. Increasing the hydrophobicity of a structure by the addition of $-CH_2-$ or other nonpolar groups increases retention. Nevertheless, by virtue of its great adsorptive strength, PGC can also be used with nonpolar eluants for the retention of polar analytes.

2. Porous graphitic carbon exhibits unexpected and largely unexplained retentive properties for molecules containing polar groups. Thus, with, say, an ODS silica gel, the addition of a polar group to a molecule will normally reduce retention in the reversed-phase mode. With PGC, retention is reduced to a much smaller extent or may even be increased. This has

been called the "polar retention effect on graphite" or PREG. The operation of PREG makes PGC particularly useful for the separation of highly polar compounds which would be difficult to retain on ODS phases. PGC is particularly suited to the separation of ionizable compounds, ionized solutes, carbohydrates, compounds containing numerous polar groups such as –OH, –COOH, –NH_2, >NH, and so forth.

3. The internal surface of PGC (around 120 m^2/g) comprises flat sheets of hexagonally arranged carbon atoms (as in a very large polynuclear aromatic molecule). This surface is crystalline and highly reproducible [4]. Being flat and highly adsorptive (heats of adsorption from the gas phase are substantially greater than heats of condensation [3]), it shows unusual discrimination of closely related stereoisomers. When an optically active additive is present, chiral separations are readily achieved. The flatness of the surface also reduces the retention of highly structured and rigid molecules which can contact the surface with only a small part of their surface. Molecular size will only confer strong retention when the molecule is very flexible and can adapt to the flat surface.

4. It is not possible to bond functionality onto PGC, as the graphitic surface, unlike that of silica, contains no reactive functional groups. However, modifiers of moderate to high molecular weight can be so strongly adsorbed that they are effectively bound irreversibly and become equivalent to chemically bonded groups. Such modifiers can partially or completely mask the properties of the underlying graphite; they enable it to be used much like a silica gel with similar chemically bonded groups. For example, ion-exchange properties can be conferred on graphite by adsorption of polyethylene imine; chiral selectivity can be conferred by adsorption of an optically active compound covalently bound to a large nonpolar "foot."

5. Chemically, PGC is extremely unreactive; it is stable under extreme conditions of pH (10 M acid to 10 M alkali), salt concentration, and temperature.

In this, chapter, we gather all published applications of PGC or Hypercarb® up to mid-1995 (Hypercarb® is the trade name for PGC marketed by Hypersil—formerly Shandon HPLC Ltd). The applications can be categorized in many different ways. We have chosen headings which seems to us to relate best to the applications so far available. These headings are listed in Table 1. Most pharmaceutical applications will be found under headings 1 and 2.

The applications are summarized in Appendices 1 to 5 at the end of this chapter; the majority of them are described in more detail below. Throughout this chapter, eluant compositions are given as volume percentages unless otherwise stated.

Table 1 PGC Application Areas

Appendix	Major area	Subsection
1	Geometric isomers and closely related compounds	a) Simple aromatics b) Pharmaceuticals and natural products c) Pesticides
2	Enantiomers	a) Separation of diastereoisomers b) Separation based on nonbonding interactions c) Separation based on inclusion complexation d) Separation based on ion-pair formation e) Separation based on metal complexation
3	Sugars, carbohydrates, and glucuronides	a) Sugars and carbohydrates b) Glucuronides
4	Residue analysis	a) Polychlorinated biphenyls (PCBs) b) Tissue and food residues c) Solid-phase extraction (SPE) and water analysis
5	Ionized and other highly polar solutes	a) Ionized solutes b) Water-soluble un-ionized solutes c) Ion-exchange separations

II. SEPARATIONS OF GEOMETRIC ISOMERS AND CLOSELY RELATED COMPOUNDS

The separation of positional isomers can often be difficult on conventional silica-bonded phases. Thus, for example, octadecyl (ODS) silicas cannot distinguish between m- and p-isomers of xylene or cresol, whereas PGC can [2]. Originally, this selectivity of graphite was explained on the basis of the simple stereochemical model first proposed by Kiselev and Yashin [5]. They argued that molecules of o- and p-xylenes, for example, can contact the flat surface of graphite with four of their C atoms, whereas the m-isomer can contact the surface with only three of its C atoms. The m-isomer therefore elutes first. Regrettably, this simple model does not stand up to detailed analysis, although general stereochemical principles still apply: Highly structured and rigid molecules will generally be less retained on PGC than flexible molecules of the same molecular weight. Flat polynuclear aromatic molecules are particularly strongly retained, as their rings match those in the graphite surface almost perfectly. Incorporation of a large aromatic "foot" is a good way to achieve strong adsorption of a modifier.

A. Simple Aromatic Molecules

Kříž et al. [6] carried out an extensive comparison of the retention of a group of 52 aromatic hydrocarbons on a number of stationary phases, including PGC. They noted that PGC was particularly selective with respect to positional isomers such as o-, m-, and p-derivatives, when compared to bonded silicas and the bare oxide adsorbents. Among other parameters, they determined the selectivity of the various packings with respect to the addition of CH_2 groups in alkyl benzenes and the substitution of benzene by successive CH_3 groups. The values of $b = \log k'_{n+1}/k'_n$ [see Eq (1) of Chapter 3A] for ODS silica gel were 0.17 for both addition and substitution, but with graphite $b(CH_2) = 0.22$ and $b(CH_3) = 0.46$, showing the much greater selectivity with respect to substitution than to addition. In this respect, PGC was similar to the bare adsorbents, alumina and silica with hexane as the eluant, but for them, $b(CH_2)$ was either zero or negative.

Wan et al. [7] determined the dependence of k' on pH for a range of di-substituted benzenes (amino benzoic acids, cresols, anisic acids, toluic acids, anisidines, and phenetidines). They found that these compounds were retained in the un-ionized form with (65/35) buffer/acetonitrile mixtures (the buffer being 0.01 M phosphoric acid plus triethylamine to adjust the pH) but were unretained in the ionized forms. The transition from high to zero retention followed the curve expected from the change in the degree of ionization. The effective pK_a values derived from the data were close to those taken from the literature, being about 0.5–1.0 units higher for acids and about 1 unit less for bases. Although all o-, m-, and p-trios could be separated, especially if the pH was suitably adjusted, the order of elution of the neutral forms followed no simple pattern: the best correlation was with the acid or base strength, with the strongest acid or base eluting first. Figure 1 shows two typical separations of aminobenzoic acids and of anisidines. Table 2 lists the elution orders of the three isomers for a range of disubstituted benzenes. Evidently, there is no universal order, and Kiselev's simple stereoselective model fails.

Similar results were obtained by Forgács and co-workers. Their results on phenols [8] substituted with a variety of hydrocarbon and polar groups agree with those of Wan et al. [7] where their data overlap. However, their range of 24 compounds is much greater, and their main objective was to relate retention to physico-chemical parameters. They concentrated on the dependence of retention on eluant composition (methanol/water) and found, in general, that k' depended on composition according to the simple volume-fraction relationship:

$$\log k' = \log k'_w + mC \tag{1}$$

where k'_w is the (extrapolated) value of k' in pure water, m is a constant, and C is the volume percentage methanol in the eluant.

Although $\log k'_w$ and m were well correlated, they showed little correlation

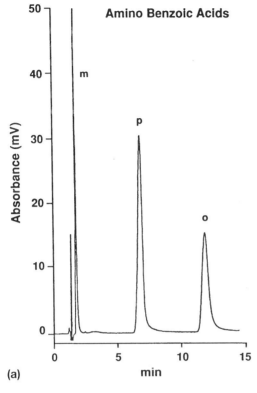

Fig. 1 Separations on PGC of (a) amino benzoic acids at pH 2.0 and (b) anisidines at pH 3.3. Eluant: acetonitrile/10 mM phosphate buffer (35/65). (Reproduced from Ref. 7 with permission of the authors and publisher.)

with analyte hydrophobicity. For example, the more hydrophilic nitrophenol elutes after the less hydrophylic halogenated phenols. This result, as explained in Chapter 3A [3] is in line with the general observations regarding PREG. Similar results were obtained with amines [9,10], where both methanol and acetonitrile were used as the organic component. They noted in particular that the peaks obtained with amines were highly symmetrical, in contrast to those normally obtained on bonded silica gels. Overall [11], they concluded that electronic parameters correlated better with retention than hydrophobicity (or lipophilicity). They also found that PGC often gave better separations than ODS silica gels.

A remarkable article by Deinhammer et al. [12] describe how the separation of a range of benzene- and naphthalene-sulphonic acids (SB and SN) on PGC and spherical glassy carbon can be influenced by the application of a small elec-

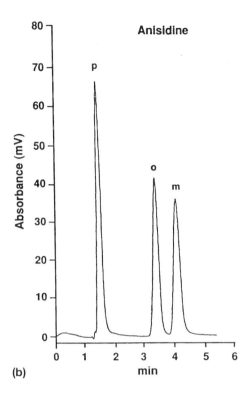
(b)

trical potential to the conducting carbon column packing. The compounds studied, denoting the substituents by $S = SO_3^-$, $H = OH$, $C = Cl$, $M = CH_3$, $E = C_2H_5$, and $V = C_2H_3$, were SB, 13DiSB, 14DiSB, 14HSB, 14MSB, 14ESB, 14CSB, 1S25DiHB, 14VSB, 1S345TriMB, 137TriSN, 136TriSN, 15DiSN, and 26DiSN. All but 1S345TriMB are resolved by PGC using as an eluant water/acetonitrile (96/4) 0.1 M in $LiClO_4$, as shown in Fig. 2. When an electrical potential is applied to the carbon packing, log k' decreases as the potential changes from about +0.6 V to –1.0 V, but the rate of decrease differs for different analytes, and selectivity can be adjusted by changing the applied potential. The basis for these remarkable results is discussed and is thought to be an alteration of the electron donor/acceptor properties of the graphite surface by the applied potential: a positive potential above the potential of zero charge, pzc (around –0.2 V relative to the Ag/AgCl/sat'd NaCl electrode), increases retention, whereas a potential below the pzc decreases retention. A simpler explanation would be that a Donnan potential is established at the surface of the particles, which encourages entry of anions when the potential is positive and rejects them when it is negative. This could be tested by using un-ionized test solutes. Whatever the explanation, it is

Table 2 Elution Orders for Di-substituted Benzenes

Compound	Elution order	Eluant	pH	Ref.
Xylenes	m,p,o	—	—	1, 6
Aminobenzoic acids	m,p,o	0.01M phosphate/CH$_3$CN 65/35 (v/v)	2.0	7
Cresols	m,p,o	0.01M phosphate/CH$_3$CN 65/35 (v/v)	2.0	7
Anisic acids	o,m,p	0.01M phosphate/CH$_3$CN 65/35 (v/v)	2.0	7
Toluic acids	o,m,p	0.01M phosphate/CH$_3$CN 65/35 (v/v)	2.0	7
Anisidine	p,o,m	0.01M phosphate/CH$_3$CN 65/35 (v/v)	3.3	7
	p,m=o	0.01M phosphate/CH$_3$CN 65/35 (v/v)	7.0	7
Phenetidine	p,o,m	0.01M phosphate/CH$_3$CN 65/35 (v/v)	3.3	7
	p,m,o	0.01M phosphate/CH$_3$CN 65/35 (v/v)	7.0	7
Nitroanilines	m,o,p	Methanol/water 97.5/2.5 (v/v)		10

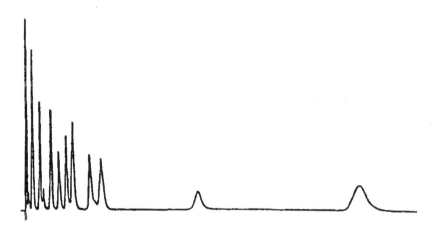

Fig. 2 Separation on PGC of benzene and naphthalene sulphonic acid derivatives. Elution order (for explanation of abbreviations see text): 13DiSB, 12DiSB, BS, 14HSB, 136TriSN, 137TriSN, 14MSB, 14ESB., 15DiSN, CSB, 1S25DiHB, 14VSB, and 26DiSn. Eluant: 0.1 M LiClO$_4$/acetonitrile (96/4). (Reproduced from Ref. 12 with permission of the authors and publisher.)

interesting that ionized sulphonic acids are readily retained and separated by PGC with an essentially aqueous eluant.

B. Pharmaceuticals and Natural Products

Forgács and Cserháti [13] studied a range of 54 barbiturates whose general formula is given in Fig. 3. Substituents R1 and R2 covered a wide range of hydrocarbon moeties, R3 covered H, various hydrocarbon moeities, benzoyl, nitro-, amino-, chloro-, and benzoyl-, and R4 was both O and S.

As with the phenols and amines, they found poor correlation with lipophilicity but reasonable correlation with the Swain-Lupton parameter, F, which measures inductive and resonance effects.

Overall, Forgács and co-workers concluded that electronic factors rather than lipophilicity governed retention on graphite. Their conclusions support the view (Chapter 3A) that a polar retention effect (PREG) exists with graphite which is absent from typical reversed-phase packings, such as ODS silica gels. The effect appears to arise from an interaction of the conductivity electrons of graphite with the π or lone-pair electrons of analytes.

A number of specific separations of pharmaceuticals have been achieved using PGC where standard silica-based materials have failed. Bell et al. [14] were able to separate the 3- and 4-hydroxy proline isomers of a new antifungal and antipneumocytis prodrug whose empirical formula was $C_{50}H_{80}N_8O_{20}PK$ (full formula given in the Ref. 14). The isomers could not be resolved on either oxide adsorbents or bonded silica phases. Good separation was quickly obtained on PGC using 20 mM pH 6.8 phosphate buffer/acetonitrile (55/45) as the eluant. This is a good example of the applicability of PGC to the separation of large molecular species which possess a multitude of polar groups and are likely to be fairly rigid.

Bassler and Hartwick [15] quote a remarkable separation achieved by Smith and Brennan of the isomers of a cephalosporin antibiotic developed by Glaxo R&D called Axetil E47. These compounds again have rigid and highly structured polar molecules. They could be well separated on PGC using a simple eluant consisting of acetonitrile/methanol/dioxan/water (35/35/10/20), as shown by Fig. 4. Equivalent separation on a bonded silica gel was difficult or impossible. Bassler and Hartwick also show separations of steroids of the cortisone group.

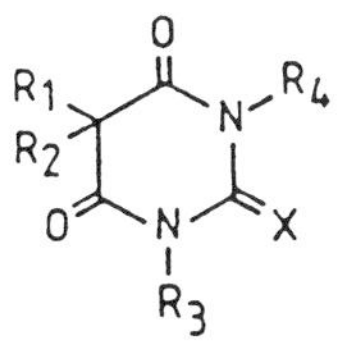

Fig. 3 Formula of barbiturates.

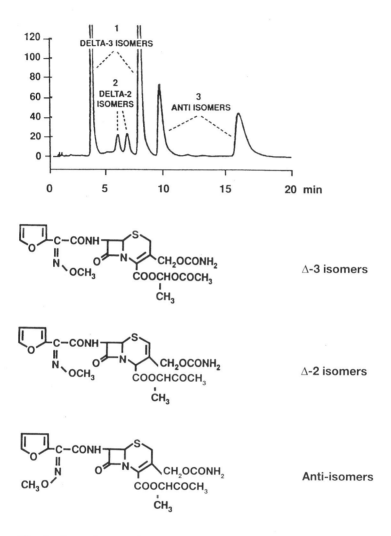

Fig. 4 Separation on PGC and formulas of isomers of Axetil, a cephalosporin antibiotic. Eluent: acetonitrile/water/methanol/dioxan (35/20/35/10). (Reproduced from Ref. 13 with permission of the authors and publisher.)

The cis- and trans-isomers of dithiopine hydrochloride were separated by Pawlak et al. [16, 17] using ethyl acetate/methanol/3% aq. ammonia (75/35/1) as the eluant. This separation was superior to that given by gas chromatography and was achieved in less than 8 min.

The separation of tiaconazole from its three closely related potential impurities [subject of a United States Pharmacopeia (USP) monograph] was studied by Berridge [18]. Existing methods are inadequate for the determination of small percentages of the impurities in the main component, although, as Berridge shows, individual peak maxima can be seen in the chromatograms using RP-18 silica gel and Hypersil phenyl. However, PGC, with an alkaline eluant consisting of tetrahydrofuran/water + 1% 0.88 ammonia (70/30), gives excellent separation with the baseline resolution of all components even when present at around 1%.

Mama et al. [19] showed good separations of amino acids and of diazopines using PGC. In the latter case, illustrated in Fig. 5, good peak shape and resolu-

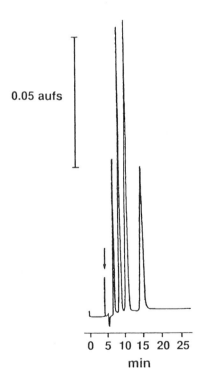

Fig. 5 Separation on PGC of benzodiazepines. Elution order: oxazepam, nitrazepam, diazepam, and medazepam. Eluant: acetonitrile/5 mM Na$_2$HPO$_4$, pH 10.6 (65/35). (Reproduced from Ref. 19 with permission of the authors and publisher.)

tion were obtained using a highly alkaline eluent, acetonitrile/5 mM phosphate pH 10.6 (65/35).

Fish [20] compared PGC and Vydac C_{18} for the separation of stereoisomers of a novel antiasthma drug. The drug was a derivative of butadiene, of the form R_1–CH=CH–CH=CH–R_2, and so had four cis-trans isomers (full formula in Ref. 20). Although they could be separated in 180 min on two coupled 25 cm Vydac C_{18} columns, separation on a 10 cm Hypercarb column took just 30 min.

Wutte et al. [21] showed that resolution of cis-trans isomers of proline containing dipeptides could be achieved on a number of stationary phases, including PGC and ODS silicas, as well as through ligand exchange chromatography on a phase loaded with the L-proline–copper complex.

Houghton and Woldemariam [22] compared a wide range of stationary phases for the HPLC separation of isomeric pairs of the alkaloids schumannificine and N-methylschumanificine. Only PGC was able to give resolution of the 7′α-OH– and 7′β-OH– isomers. The eluent used was tetrahydrofuran/20 mM NaH_2PO_4 + TFA pH 2.5 (50/50 w/v).

The examples discussed in this section indicate that PGC has a unique capability for the separation of geometric isomers and is often superior to bonded silica gels. This may be attributed to the fact that a flat surface of PGC will interact much more specifically with such isomers than a quasi-liquid structure such as a C_{18} chain which can readily accommodate itself to a variety of molecular shapes.

Early in the use of PGC it was feared that large molecules could be so strongly retained that they would be impossible to elute, even with neat organic solvents, and that this would limit the use of PGC to the separation of relatively small molecules. This concern has been shown to be unwarranted. PGC seems to be particularly suitable for the separation of large complex molecules having a high degree of functionality, especially if they cannot readily take up a flat configuration.

C. Pesticides

Forgács and Cserháti [23] have recently examined the retention of a range of 30 pesticides and herbicides by PGC. Using dioxan as an organic modifier, their data followed Eq. (1) over a concentration range 65/35 to 90/10 dioxan/water. The constants log k'_w and m are tabulated, and they correlated best with hydrophobicity of the analytes. These results contrast with their previous data on phenols, amines, and barbiturates, where polar effects dominated retention.

III. ENANTIOMER SEPARATIONS

Over the past decade, the requirements placed on the pharmaceutical industry to produce optically pure drugs has dramatically increased the pressure to devise separation methods for enantiomers. PGC has been shown to be effective in a number of such separations.

The chromatographic resolution of any enantiomeric pair requires the presence within the chromatographic system of a chiral selector which can interact differently with the two enantiomers. Such a selector must clearly be optically active itself. The selector may be present in the eluant or in the stationary phase (or, indeed, partitioned between the two). Both modes of selection have been extensively used with silica-based column packings, and the same is true for graphitic packings. As we have already noted, it is not possible to bond moeities chemically to graphite, but the strong adsorption, especially of large hydrophobic groups, makes it possible to achieve what is equivalent to chemical bonding by using a selector which is covalently bonded to a hydrocarbon "tail" or "foot."

Chiral recognition requires a very close interaction between the chiral center in the selector and the chiral center in the enantiomers to be separated. Basically, diastereoisomeric pairs must be formed which have different binding and/or partitioning properties in order to achieve separation. Following Wainer [24], the following basic types of interactions may be distinguished, on the basis of which chiral resolution can be achieved:

1. Covalent bonding to form diastereoisomers which are subsequently separated by chromatography.
2. Transient formation of diastereoisomeric complexes based on nonbonding interactions. These nonbonding interactions will normally be a combination of attactive interactions such as hydrogen bonding, π–π and dipole stacking, coupled with steric repulsion. Such interactions were elegantly exploited by Pirkle.
3. Formation of diastereomeric inclusion complexes or complexes with highly structured molecules, such as proteins. The most common example of the former is the use of cyclodextrins.
4. Formation of diastereoisomeric ion pairs.
5. Formation of covalent organometallic solute/selector complexes.

It may be expected that where the flat surface of the graphite takes part in the interaction process, PGC will provide better resolution than a corresponding bonded silica phase. However, if the specific interactions between the selector and the enantiomeric pairs are so dominant that the graphite plays a small role, the selectivity of the PGC and an equivalent silica-based phase are likely to be similar. Examples of both situations are found in the following examples.

A. Separations of Diastereoisomers

Josefsson et al. [25] separated a number of chiral cardioactive drugs by forming their (−)-menthyl chloroformate derivatives prior to chromatography. In general, good selectivity could be obtained with a range of normal-phase eluants containing dichloromethane (70% by volume) and a second component (ethanol, acetonitrile, ethyl acetate, acetic acid, or formic acid). Interestingly, the acids proved to be the strongest solvents but gave low selectivities. The best selectivities were given by acetonitrile and ethylacetate, or their mixtures with acetic or formic acids.

Chan et al. [26] derivatized standard amino acids with O-tetra-acetyl-β-D-glucopyranosyl isothiocyanate and showed that virtually all enantiomeric pairs could be resolved using Hypercarb as the column packing and an acetonitrile/water solvent gradient from 30/90 to 90/10, the eluent containing 0.1% TFA throughout.

B. Separations Based on Nonbonding Interactions

Dutton et al. developed a Pirkle-type phase [27] based on the adsorption of a large aromatic moeity or "foot" onto PGC. This foot, tetrabenzofluorene (TBF), was linked by a short alkyl chain to the chiral selector, R(N–3,5 dinitrobenzoyl)-α-phenylglycine. The formula is given in Fig. 6. Eluents were hexane solutions containing 5–10% of isopropanol or ethyl acetate. Good chiral selectivities were obtained for the aryl alchohols and Fmoc amino acids.

Overall, selectivities were slightly larger than those obtained with the corresponding silica-based phases. With this system, it appears that the underlying graphite plays little role in the interaction process and that the selector group is held well away from the graphite surface; the coupling between selector and analyte dominates the retention mechanism.

Although not strictly an example of the use of PGC in HPLC, Wilkins et al. [28] demonstrated superior enantiomer separations by supercritical fluid chromatography (SFC) using an adsorbed chiral selector. They used PGC coated with a selector derived from R,R tartaric acid which was linked by amide bonds to nitrobenzyl and an anthryl residue as "foot" (rather than TBF). This gave excellent enantiomeric separations of a range of tropic acid amides. The selectivities with the PGC phase were much greater (typically 1.4) than with a corresponding

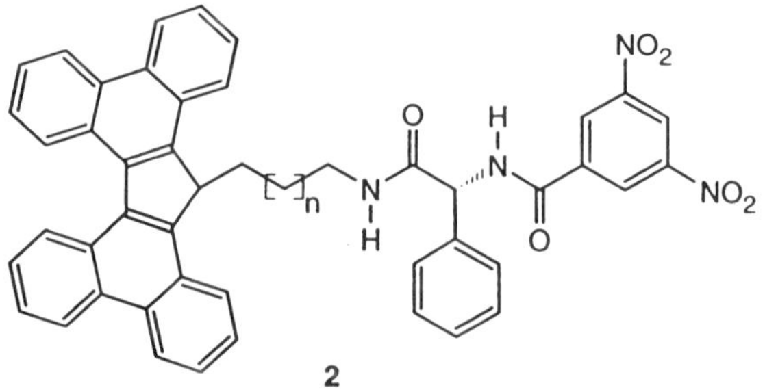

Fig. 6 Formulas of the chiral selector with a tetrabenzofluorene "foot."

bonded silica phase (typically 1.05). Here, one might infer that the graphite surface played a significant role.

Heldin et al. [29,30] developed remarkable separations of racemic compounds of the atropine group using (2S,3S)- and (2R,3R)-dicyclohexyl tartrate (DCHT) as the selector. With 0.25 mM DCHT in a phosphate buffer pH 2.8, 0.132 mmol/g (equivalent to about 1.4 µmol/m^2) of DCHT was adsorbed by the PGC [30]. Their separations were carried out in pure aqueous solution at a low pH, where the analyte amines would be predominantly in the ionized form. The separation achieved is shown in Fig. 7.

In this application, DCHT is strongly adsorbed and forms a monolayer coating on the PGC. So, does the graphite surface play a role in the separation or not? The separation is not based on ion-pair formation as DCHT is an ester. If PGC played no part in the retention, one would expect that the analytes being ionized would elute with the solvent front. Because this is not the case, we must conclude that there is indeed strong interaction between the ionized solutes and PGC and that this interaction allows secondary interactions of a nonbonding type between them

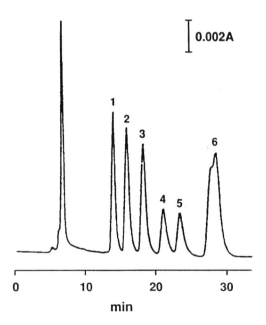

Fig. 7 Separation on PGC of atropine and analogs using (2S,3S)-dicyclohexyl-tartrate as the chiral selector: 1, (–)scopolamine; 2, (R)-homoatropine; 3, (S)-homoatropine; 4, (R)-atropine; 5, (S)-atropine; 6, racemic tropic acid. Eluant: 0.25 mM (2S,3S)-DCHT in a phosphate buffer pH 2.8. (Reproduced from Ref. 30 with permission of the authors and publisher.)

and DCHT. The strong retention of the analytes by PGC is in line with the "polar retention effect on graphite" discussed in Chapter 3A. It is interesting that both this separation and that of Wilkins et al. [28] use the tartaric acid moeity as part of the selector. Is this combination particularly effective in the presence of a graphite surface?

Grieb et al. [31] adapted the technology of coating silica with cellulose carbamate and applied it to PGC; 0.1g of cellulose *tris* (3,5-dimethylphenyl carbamate) (CDMPC) was dissolved in 20 mL of tetrahydrofuran (THF) and added to PGC. The THF was evaporated and the coating process repeated five times to provide 25% by weight of CDMPC on porous graphite. The material was then sieved and packed into an HPLC column. The material provided separations which were very similar to those obtained with a 15% by weight coated silica gel for a wide range of enantiomers. Selectivities with the coated PGC using hexane-based eluants containing small proportions of alcohols were generally somewhat greater than those obtained with the silica-based material. Peak shapes and symmetries were good.

Although this may not be considered to be a true application of porous graphite, it shows that graphite can replace silica gel as a support for a stationary phase deposited at a high weight percentage. Such procedures may be particularly relevant when it is desired to confer structural rigidity of an otherwise soft organic gel-type support, when this has to be combined with chemical inertness, and/or when extreme conditions deleterious to silica may have to be used. The use of a very high-percentage coating can effectively preserve the properties of the soft organic gel while masking the adsorptive properties of the underlying graphite.

C. Separations Based on Inclusion Complexation

Fell et al. [32] used 3 mM β-cyclodextrin in acetonitrile/5 mM phosphate buffer pH 10.6 to separate benzodiazepines on PGC. They compared their separations with those on crystalline cellulose triacetate and β-cyclodextrin bonded to silica. Although band dispersion left something to be desired, the resolution on PGC was much greater than on either of the other phases. Here, we are presumably seeing a specific role of graphite in the separation, with the flat surface providing strong steric selectivity for the differing diastereoisomeric complexes. Subsequently, Mama et al. [19] showed enantiomeric separations of metoprolol and of nomifensine maleate using acidic eluants of pH around 2. These two studies illustrate the use of PGC under pH conditions which could damage sensitive bonded-silica materials. The wide range of pH over which PGC can be used enables complete control of the ionization status of drugs of interest, allowing the degree of ionization to be optimized.

D. Separations Based on Ion-Pair Formation

Karlsson and Petersson [33,34] compared Lichrosorb DIOL with PGC as support for the separation of chiral amines such as propanolol and promethazine using N-benzoylcarbonyl-glycyl-L-proline (ZGP) as the chiral selector. The mobile phase was water-saturated dichloromethane containing 10 mM ZGP. The association between the selector and the amine was through ion pairing of the carboxy group in ZGP and the amino group in the drug. The comparison indicated that PGC was a preferred support for a number of reasons. First, it gave somewhat better peak shapes (lower HETPs); second, and more important, PGC columns could be much more rapidly equilibrated with eluant than the DIOL columns. This was thought to be due to the absence of –OH groups on PGC. It was also less necessary to add a protective background amine to achieve symmetrical peaks with PGC. In an addendum to the article, separations of R,S-promethazine and 1-phenyl-2-amino propane were demonstrated using 5 mM ZGP in an eluant of methanol/water (95/5 v/v), indicating that chiral resolutions could also be carried out under reversed-phase conditions.

This use of reversed-phase conditions was further exploited by Pettersson and Gioeli [35] who used the chiral acid (–)2,3:4,6-di-O-isopropylidene-2-ketogulonic acid (DIKGA) as the selector. A wide variety of eluants were used. Beta-blockers could be separated with 5 mM DIKGA in 50 mM acetate buffer (pH 4.7)/methanol (20/80), or with 20 mM DIKGA and 5 mM NaOH in isopropanol/acetonitrile (90/10), or with 5 mM DIKGA and 1–2.5 mM NaOH in water/methanol (20/80). This comprehensive study showed that increasing the ionization of the acid DIKGA increased retention and selectivity; increasing the concentration of DIKGA increased retention; changing the water content sometimes increased selectivity and sometimes decreased it. This study is important because it shows that hydration of the ionized sites does not rule out chiral separations based on ion-pair formation, contrary to previous belief.

This is further demonstrated by the recent work of Huynh et al. [36] in which ZGP and ZGGP (N-benzoylcarbonyl-glycyl-glycyl-L-proline) are shown to be effective under reversed-phase conditions, with PGC as the column packing. Their extensive study covered a wide range of chiral drug amines, under a wide range of experimental conditions. Typically, eluants contained 10–20 mM ZGP and 1–5 mM NaOH dissolved in methanol. k' for the test amines (alprenolol, sotalol, and terbutalin) increased linearly with concentrations of both ZGP and NaOH, but selectivities were only slightly changed, ranging from near unity to about 1.2. Replacement of NaOH by organic tertiary amines had relatively little effect on selectivity but did influence retention. Partial replacement of methanol by propanol had a significant effect on both retention and selectivity; for example, with trimipramine and promethazine, selectivity was substantially increased while retention decreased. Acetonitrile, on the other hand, had little effect on selectivity but decreased retention. Replacement of ZGP by ZGGP sometimes improved selectivity and sometimes impaired it. The authors discussed the relationship between molecular structure of the enantiomeric amines and selectivity but were not able to derive any clear predictive principles.

Reversed-phase systems are clearly much more convenient for biological samples, and their the use for chiral ion-pair separations should be further explored.

E. Separations Based on Metal Complexation

Knox and Wan [37] used copper complexation with amino acids to separate both simple amino acids and hydroxy acids. PGC was coated dynamically with N-(2-naphthalene-sulphonyl)-phenylalanine (NSPA) from methanolic solution, and samples were subsequently eluted with 2 mM aqueous copper acetate. The coating was sufficiently insoluble in the eluant that no change in retention occurred with some 8000 column volumes of eluant passed. The loading of NSPA was 0.14 μmol/m^2. This represents a complete monolayer coating of the surface and is very comparable to that achieved by Heldin et al. [30] using dicyclohexyl tartrate. The copper ion typically compexes with two amino acid anions; in this case, one is NSPA and the other an analyte. Resolution of the enantiomers depends on the differing complexation constants of the Cu-selector ion with the enantiomer ions of the analyte. The authors showed identical separations using the L- and D-isomers of the selector except that the orders of elution of enantiomers were reversed. Typical separations using 2 mM copper acetate pH 5.6 as the eluant are shown in Fig. 8. It is noteworthy that the separation system works with both amino acids and hydroxy acids.

IV. SEPARATIONS OF SUGARS, CARBOHYDRATES, AND GLUCURONIDES

A. Sugars and Carbohydrates

Carbohydrates are widely found in nature, and their separation poses a huge challenge to the chromatographer because of the great diversity of structure and composition of these materials. Often, carbohydrates are highly water soluble and therefore difficult to retain using bonded silica or polymeric phases. PGC promises to have the advantage that the polar retentive effect, PREG, can be exploited to retain and separate them using aqueous eluants. Because sugars can ionize at high pH there is also scope for the use of ion-pairing systems in strongly alkaline systems using PGC.

Davies et al. [38] demonstrated the separation of a range of alditols containing from two to five monosaccharides. Gradient elution was used from water to acetonitrile in the presence of 0.05% TFA. Generally, retention increased with the size of the oligosaccharide, and columns could be loaded with up to 2 mg, which allowed small-scale preparation of purified fractions. The order of elution of similar oligosaccharides depended on the nature of the linkages between the monosaccharide units, with 1,3-linked compounds eluting before 1,4-linked

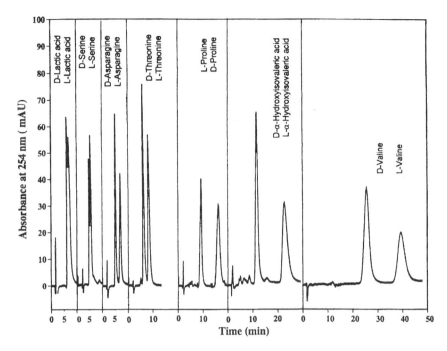

Fig. 8 Separation on PGC of enantiomers of amino and hydroxy acids using N-(2-naphthalene sulphonyl)-L-phenylalanine. Eluant: 2 mM aqueous copper acetate pH 5.6. Detection UV at 254 nm. (Reproduced from Ref. 37 with permission of the authors and publisher.)

compounds. The use of PGC enabled facile separation and identification of the chosen oligosaccharides and could be applied to identify glycopeptides from a tryptic digest. The authors conclude that PGC is a new and uniquely selective adsorbent for oligosaccharide analysis, whose mechanism of action involves interaction of the polar segments of the carbohydrate with the conductivity electrons of PGC.

Steffansson and Lu [39], in an unusual article, studied the separation of a range of mono- and di-saccharides, sugar acids, and sugar amines at pH up to 13. Generally speaking, sugars are un-ionized, except at very high pH (12–14) when they could be retained as ion pairs. Although silica-based materials are inapplicable, polymers such as PLRP-S and porous graphite are resistant. The authors showed that with PGC and aqueous eluants, the retention of typical sugars decreased as the NaOH concentration increased from 0.02 to 0.1 M. This was ascribed to the ionization of the sugars and their nonretention in the ionized state. Disaccharides were much more strongly retained than monosaccharides, sugar acids were somewhat more retained, and amino sugars similarly retained. With PGC, the addition of long-chain quaternary amines eliminated retention, although some retention was observed with tetrapropyl and tetrabutyl ammonium. It seems that the hydroxide ion pairs of the ammonium salts were preferentially adsorbed by Hypercarb and that OH$^-$ prevented ion pairing of the sugar anions with the tetra alkyl ammonium ions. With PLRP-S, the addition of long-chain alkyl ammonium ions did provide additional retention and useful separations.

Lu et al. [40], followed this article with a study of the reversed-phase separation of a range of di- and tri-saccharides on PGC using indirect UV detection with an added UV-absorbing component in the eluant, such as sorbic acid. This comprehensive study examined a wide range of variables, but, overall it showed that excellent separations could be achieved with good sensitivity resolution and peak shape. A chromatogram from their article is shown in Fig. 9.

Koizumi et al. [41] showed that monosaccharides could be directly eluted with water from PGC, with each peak being split into anomers. Disaccharides were eluted with water containing either 15% methanol or 4% acetonitrile, the peak for each sugar again being a doublet which coalesced on the addition of NaOH. Cyclomaltoses with six, seven, and eight units per ring were also separated using water/acetonitrile (85/15).

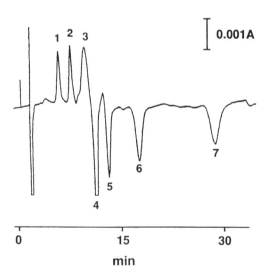

Fig. 9 Separation on PGC of di- and tri-saccharides using indirect detection with 0.025 mM sorbic acid added to the eluant. Solutes: 1, melibiose; 2, sucrose; 3, lactose; 4, systems peak of sorbic acid; 5, melizitose; 6, gentiobiose; 7, cellobiose. Eluant: 0.025 mM sorbic acid + 0.1 M NaOH/methanol (90/10). (Reproduced from Ref. 40 with permission of the authors and publisher.)

B. Glucuronides

The formation of glucuronides and sulphates are the main ways in which the body metabolises and excretes lipophilic drugs. Their water solubility makes their analysis by liquid chromatography difficult, especially when it is desirable to monitor the parent drug in the same analysis. Because of the polar retention effect exhibited by graphite, retention of these water-soluble metabolites can be enhanced so that their elution along with their parent drugs becomes possible.

Steffansson and Hoffmann [42] required a method for determining the four diastereometric glucuronides which arise from the metabolism of a new antiarrhythmic drug, Almokalant, denoted R1, R2, S1, S2 (formulas in Ref. 42). The glucuronides were first isolated from urine by adsorption onto a terbium-loaded cation exchanger (the precolumn) and then desorbed by an acidic mobile phase. The desorbed glucuronides could not be retained by an ODS phase but were well retained by Hypercarb. They could then be eluted with an acidic eluent such as 0.1% acetic acid in methanol/water (30/70), which provided an excellent separation of the R and S pairs (α typically 1.5). Although the precolumn had to be replaced daily, the Hypercarb column proved very robust with no change in retention parameters over hundreds of injections. When the eluant was changed to 0.01 M H_2SO_4, significant improvement in performance was obtained with plate counts of around 4000 for the 100 mm long column. All four glucuronides could then be separated after prior isolation using the original system in a small-scale preparative mode. The final separation is illustrated by Fig. 10.

Ayrton et al. [43] have very recently developed a rapid and simple method for the separation of the glucuronides of AZT, chloramphenical, β-estradiol, and fluparoxan. Initially, various organic eluants were examined, but none of them

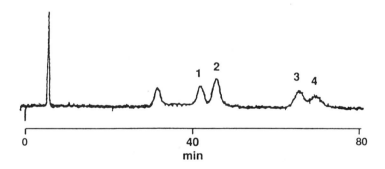

Fig. 10 Separation on PGC of four diastereoisomeric glucuronides of almokalant: 1, S1; 2, S2; 3, R1; 4, R2 from a pooled urine sample after extraction. Eluant: acetonitrile/10 mM sulphuric acid (25/75). (Reproduced from Ref. 42 with permission of the authors and publisher.)

eluted the glucuronides. Following Gu and Lim [60], they examined whether the addition of TFA would improve matters. This proved successful, and they achieved good separations with THF/water containing 0.5–1% TFA in the composition range 20/80 to 80/20. In most cases, the parent compound was eluted with the same eluant after the glucuronide, except for fluparoxan, where the drug eluted first.

These examples provide excellent illustrations of the strong retentive power of PGC for highly water-soluble carbohydrates and drug metabolites. The use of PGC in this area is far from fully explored, and further successful applications may confidently be expected.

V. RESIDUE ANALYSIS

A. Polychlorinated Biphenyls

Polychlorinated biphenyls (PCBs) were welcomed as a great boon for disinfection and were widely used following the Second World War. Regrettably they have created one of the worst man-made chemical disasters. The compounds are extremely resistant to degradation in the natural environment; they are highly toxic, and some are carcinogenic. Over the years, they have diffused into rivers, seas, and even into Antarctica. Their monitoring in the environment, in food, fish and animal tissues has now become a major activity for the environmental chemist.

Because of the relatively low concentrations in which the specifically toxic PCBs are found (the non-ortho planar and mono-ortho PCBs), multistep procedures are normally required to isolate them from environmental samples. The first stage is clean-up by liquid chromatography (LC) using various sorbents. Such procedures have proved difficult to reproduce and standardize. Activated carbons and Carbopack (a form of graphitized carbon black) have been widely used for the separation of polychlorinated dibenzofurans (PCDFs) and planar polychlorinated biphenyls. However, the separation of mono-ortho, and non-ortho PCBs on activated carbon requires a complex solvent program, and because of irreproducibility of the surface and the strongly nonlinear isothems exhibited by active carbons, the results are generally unsatisfactory. PGC being a much more homogeneous material has allowed some of these difficulties to be overcome.

Creaser and Al-Haddad [44] were the first to show that PCBs, PCDDs, and PCDFs could be separated by class on PGC using hexane as the eluant and that the PCB fraction could then be further fractionated by elution from PGC with acetonitrile/water (80/20) to give sharp peaks with little tailing. The elution order was determined by the chlorine content and by planarity, with the more planar PCBs elution later. Subsequent to separation by PGC, fractions could be subjected to gas chromatography/mass spectrometry (GC/MS) for further separation and identification.

Hong et al. [45] developed a method for isolation and separation of coplanar PCBs in human milk. As part of this procedure, a hexane extract was chromatographed on PGC in order to separate the planar from the nonplanar PCBs. The eluate containing the former was then concentrated and subjected to GC/MS. The chromatography on PGC showed clear baseline separation of the nonplanar from the planar PCBs.

Similar procedures are described by de Boer et al. [46] for the isolation, separation, and identification of non-ortho PCBs in fish and marine mammals, by Fuoco et al. [47] for the isolation of the ortho- and non-ortho-substituted PCBs in soils, and by Bohm et al. [48] for isolation of planar PCBs in food. Williams et al. [49] used similar methods for determining 2,3,7,8-tetrachlorodibenzo-*p*-dioxin in cormorant eggs.

For these applications, the PGC column is used primarily to separate the planar from the nonplanar PCBs before submitting fractions to GC/MS.

B. Tissue Residues

Tarbin and Shearer [50] compared the polystyrene phase PLRP-S with PGC for the determination of coccidiostat lasalocid (formula in Ref. 50) in chicken tissues and eggs. In order to exploit the natural fluorescence of lasalocid, it was necessary to use strongly basic eluants; therefore, silica-based packings were unsuitable. Both PGC and PLRP-S gave good separations following extraction and clean-up. However, PGC proved superior in that an interfering peak present in the extract from eggs could only be resolved from lasalocid using PGC. The eluants for PGC was acetonitrile containing 5% of 1,1,3,3-tetramethyl guanidine.

Liem et al. [51] have reviewed the whole area of the analysis of organic micropollutants in the lipid fraction of foodstuffs, including the uses of PGC.

C. Solid-Phase Extraction and Water Analysis

For the determination of pesticides in natural waters and effluents, the first task is to extract them from the water sample. Because the pesticides and residues are likely to be present at extremely low concentrations (0.1–10 µg/L) it is necessary to use fairly large volumes of water, say 10–1000 mL. The residues will normally be extracted by pumping the sample through a solid-phase extraction (SPE) cartridge containing some type of HPLC packing material. Subsequently, the collected residues are washed from the cartridge using a stronger eluant (e.g., water plus methanol or acetonitrile) and then analyzed by HPLC. A critical factor in the success of any extraction procedure is that none of the residue components should "break through" during the extraction procedure. The simplest way to determine the likely break-through volume is by chromatography using a column filled with the material proposed for the extraction cartridge. This procedure, which has been refined and used extensively by Hennion and co-workers

[52–57], involves determining k' for each compound of interest as a function of methanol or acetonitrile content. Then, from a plot of log k' against percentage of organic modifier, which normally follows Eq. (1), the extrapolated value, k'_w, for pure water may be determined. The break-through volume for the extraction cartridge can then be estimated and the most appropriate cartridge material selected.

Having extracted the residues from a water sample on an SPE cartridge, they then have to be desorbed and chromatographed to obtain a quantitative measure of each component. In the two-stage method, the residues are eluted from the cartridge with several cartridge volumes of a strong solvent; the solvent is evaporated, and the residue taken up in the eluant and injected onto the analytical column. In the on-line method, desorption and chromatography are carried out sequentially by washing the residues directly onto the chromatographic column, and then separating them using gradient elution. The on-line method will work satisfactorily only if one uses the same or a weaker adsorbent in the SPE cartridge than in the analytical column itself, as otherwise peak broadening and loss of resolution will occur. Recoveries and peak integrity are further improved if the desorption from the SPE cartridge is carried out by reversing the flow through the cartridge [55]. In this way, the more highly retained components are not forced to migrate through the entire precolumn but are desorbed directly from its front end.

It is found that ODS silicas and polystyrene resins are satisfactory as SPE materials for residues which are not too hydrophilic and have log P values above about 2 (P is the octanol/water partition ratio), but many compounds such as polyhydric phenols, atrazines and their degradation products, are very water soluble, and have log P values well below 2. With ODS and PRP materials, the breakthrough volumes are far too small. Hypercarb, with its remarkable retentive properties for polar compounds, has proved to be the only material so far developed which provides adequate retention.

As part of their work Hennion and co-workers have therefore carried out a large amount of HPLC using a variety of stationary phases including Hypercarb.

Cocquart and Hennion first reported the use of Hypercarb in 1992 [52]. They discovered the remarkable property of Hypercarb that it retained polyhydric phenols more than phenol itself, whereas ODS silica and the polystyrene PRP-1 showed lower retention, as expected, the more –OH groups there were in the molecule. This finding enabled them to extract these highly polar polyhydroxy phenols from water samples using a Hypercarb SPE cartridge and, subsequently, to elute and separated them on either an ODS, PRP, or Hypercarb HPLC column.

Major problems had been experienced with the extraction and chromatography of the triazine-based herbicides, as these are highly polar and have log P values ranging from 2.7 for atrazine itself down to negative values for cyanuric acid, ammelidine, and ammeline [53]. Because retention by ODS silicas and polystyrene phases more or less follows the log P values as seen from plots for a range of benzene derivatives given by Hennion and Cocquart [54], such compounds cannot be extracted by the conventional materials. The authors again determined log k' as a function of methanol content, but here the dependence was not linear and it was not possible to determine log k'_w. However, even at 10% methanol, log k' for the most polar solute ammeline was 1.5. As with other polar compounds, the order of retention differed from that of the log P values, as seen by the data of Table 3, but more importantly, the range of log k' values on Hypercarb (2.1 units) was much less than the range of log P values (3.9 units). Hypercarb was thus the choice for SPE. Subsequent chromatography was, however, carried out on an ODS silica column. Further work [55] confirmed that Hypercarb was highly effective for a wide range of polar pesticides having log P values in the range of –0.5 upward when used as a SPE material, especially if reverse flushing was used to desorb the materials. Indeed, this was essential if high recoveries were to be achieved. By using back-flushing, a minimal amount of methanol could be used, which could subsequently be evaporated and the residues taken up in the eluant for chromatography.

Table 3 log P and log k' Values for Triazine Derivatives

Compound	Symbol	Substituents			log P	log k' (30/70) MeOH/water
		2	4	6		
Simazine	S	NH–iPr	NHEt	Cl	2.3	2.7 (ext)
Atrazine	A	NHEt	NHEt	Cl	2.7	2.4 (ext)
Deisopropyatrazine	DIA	NH_2	NHEt	Cl	1.2	2.2 (ext)
Hydroxyatrazine	OHA	NHiPr	NHEt	OH	1.4	1.8
Deethylatrazine	DEA	NHiPr	NH_2	Cl	1.6	1.7
Hydroxydeethylatrazine	OHDIA	NHiPr	NH_2	OH	1.4	1.6
Deethyl-disoprophylatrazine	DAA	NH_2	NH_2	Cl	0.0	1.3
Hydroxydeisopripyl-atrazine	OHDEA	NH_2	NHEt	OH	0.2	1.2
Cyanuric acid	ACY	OH	OH	OH	–0.2	0.9
Ammelide	ADE	NH_2	OH	OH	–0.7	0.8
Ammeline	AME	NH_2	NH_2	OH	–1.2	0.6
Notes					1	2

Notes:
1. Values are estimated by authors of Ref. 54.
2. log k' values are those for 30% methanol; "(ext)" indicates values extrapolated from Figure 1 of Ref. 54.

Source: Ref. 54.

In a very recent article, Guenu and Hennion [56] have developed an on-line procedure for a range of polar pesticides in water using Hypercarb as both the SPE cartridge packing and the HPLC column packing. Impressive results for the most polar pesticides are demonstrated in Fig. 11 taken from Ref. 56. They have also developed an on-line procedure for other pollutants with very low log P values such as 2-chloro-4-aminophenol, chloroanilines, aminophenols, and

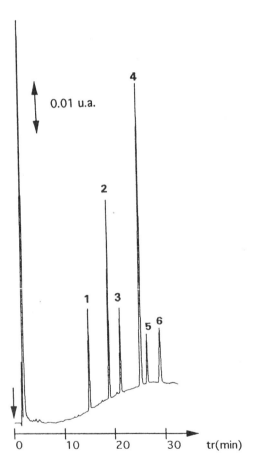

Fig. 11 Separation on PGC of polar pesticides after on-line preconcentration by PGC, 150 ng of each pesticide in 25 mL ultrapure water. Solutes (log P values in brackets): 1, oxamyl (-0.5); 2, methomyl (~1); 3, DIA (metabolite of atrazine) (1.1); 4, monocrotophos (n.v.); 5, fenuron (~0.8); and 6, metamitron (0.8). Eluant step gradient acetonitrile/5 mM phosphate butffer pH 7 (10/90), (15/85), (55/45). (Reproduced from Ref. 56 with permission of the authors and publisher.)

cyanuric acid [57], again using PGC for both the SPE cartridge and the HPLC column.

The comprehensive work of Hennion and co-workers has undoubtedly established the important role of PGC for the isolation by SPE and subsequent chromatography by HPLC of highly polar pesticides in water samples. Extension of work in this area may be confidently expected.

VI. SEPARATIONS OF IONIZED AND OTHER HIGHLY POLAR COMPOUNDS

A. Ionized Solutes

Lim and co-workers [58–60] were early pioneers in the use of PGC, and they achieved some remarkable separations of inorganic ions and complex ions using PGC. TcO_4^- and RhO_4^- ions (pertechnetate and perrhenate ions) are highly hydrophobic anions but would be expected to be unretained on most reversed-phase materials. Graphite, however, can apparently interact by an electron-transfer mechanism (the likely source of PREG) with molecules containing lone-pair electrons or aromatic-ring electrons. Lim [58] showed that these ions could be retained and separated by PGC using aqueous TFA as the eluant. In a second article, Emery and Lim [59] showed that various organometallic charged complexes of technetium and rhenium could also be separated using TFA solutions containing 2–10% acetonitrile. Work on further complexes of technetium showed that cationic and anionic complexes could be separated in the same chromatogram. The sign of ionic charge was not important for retention. Apparently, graphite can behave equally well as an electron donor or an electron acceptor. Lim coined the term "electronic interaction chromatography" for this type of separation.

B. Water-Soluble Un-ionized Solutes

Gu and Lim [60] then studied the retention of highly water-soluble organics using PGC. With the standard reversed-phase supports, oxalic acid could not be retained without ion pairing, but with PGC, it was totally retained unless a competing acid such as TFA was added to the eluant. Again, they argued that this was an example of "electronic interaction chromatography." Oxalic acid, creatine, and creatinine could readily be separated from human urine with PGC as stationary phase, using 3% acetonitrile in 0.1% TFA as the eluant. To determine creatine and creatinine in serum, protein was first precipitated with 10% TFA and the supernatant cleaned up on a C_{18}-bonded silica-gel cartridge. Examples of these separations are shown in Fig. 12. The authors note that these separations could not be obtained using bonded silica gels, which provided inadequate retention.

Dutton et al. [61] subsequently developed a rapid method for determining oxalate in liquid culture media of wood-rotting fungi. The separation was

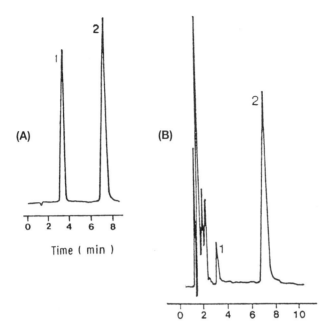

Fig. 12 Separations of oxalic acid, creatine, and creatinine. (A) Creatine (1) and creatinine (2); (B) creatine (1) and creatinine (2) in baby urine; (C) oxalic acid (1) and creatinine (2) in adult urine. Eluants: (A, B) 0.1% aqueous TFA/acetonitrile (97/3); (C) 0.08% aqueous TFA. (Reproduced from Refs. 33 and 60 with permission of the authors and publisher.)

achieved using aqueous 0.2 M orthophosphoric acid on a Hypercarb column. Oxalate was retained with a $k' = 2.5$, and well separated from acetate with $k' = 0.35$. The authors note that this separation would not have been possible on acid-labile silica-based columns.

Krause [62] applied PGC to the determination of ethylene thiourea (ETU), which is a metabolite of ethylenebisthiocarbamate fungicides. ETU is extremely polar and water soluble. Electrochemical detection was preferred, but its use required the presence of phosphoric acid. With bonded silica-gel columns, this has to be added following chromatography to avoid damaging the column, but with polymer and PGC columns, it could be incorporated in the eluant, which simplifies the procedure. Five different bonded silica-gel columns, polymer columns, and PGC were investigated. The polymer columns seemed to cause decomposition of the ETU, but PGC provided long-term stability and good quantitation. ETU was eluted as a sharp symmetrical peak with $k' = 1.7$, being eluted with acetonitrile/8 mM in phosphoric acid (5/95).

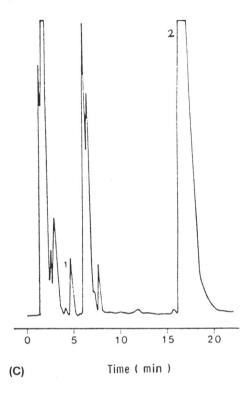

Ehrsson et al. [63] recently used PGC to separate *cis*-platin, *trans*-platin, and their hydrated complexes. *Trans*-platin is $(NH_3)_2PtCl_2$ with each pair of substituents diagonally opposed in the square planar complex, whereas the cis-isomer has each pair on the same side of the square. In vivo, the platins can be mono- or di-hydrated, replacing one or both –Cl groups by –OH. Accordingly there are six possible compounds. Furthermore, the hydrated forms can be singly or doubly protonated. Using 1 m*M* aqueous NaOH as the eluant, *cis*- and *trans*-platin are readily separated as are the mono- and di-hydrated complexes although the monohydrated complexes of *cis*- and *trans*-platin are not resolved, nor are the dihydrated complexes. Again, we see a remarkable separation of unusual but important clinical compounds under conditions which could not be achieved with silica-based materials. This separation is illustrated in Fig. 13.

The separations in this section are among the most remarkable that have been obtained using PGC, as the solutes are normally unretained on reversed-phase materials. They illustrate the remarkable ability of graphite to interact with molecules having lone-pair or aromatic-ring electrons, possibly by some kind of electron-transfer mechanism. The nature of this process requires fur-

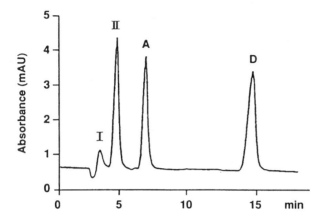

Fig. 13 Separation on PGC of platins. Solutes: (I) monohydrated platins; (II) dihydrated platins; (A) *cis*-platin; (D) *trans*-platin. Eluant: acetonitrile/0.02% aqueous acetic acid (50/50). (Reproduced from Ref. 63 with permission of the authors and publisher.)

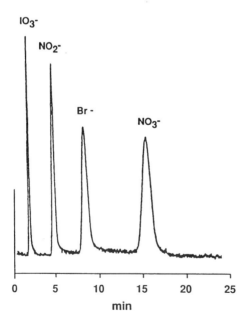

Fig. 14 Ion-exchange separation on PGC of anions on PGC coated with monolayer of polyethylene imine. Eluant: 20 mM sodium phosphate buffer pH 7. (Reproduced from Ref. 64 with permission of the authors and publisher.)

ther investigation in order that a proper physico-chemical explanation can be developed, which can then form the basis of quantitative prediction of retention phenomena.

C. Ion-Exchange Separations

Knox and Wan [64] coated PGC with polyethylene imine (PEI) by equilibrating with a 0.1% aqueous solution. The PEI was then rendered insoluble by flushing the column with phosphate buffer (typically 0.02 M pH 7). The column had an ion-exchange capacity of 1.5 µmol/m^2 (160 µmol/g) and gave good separations of simple anions (iodate, nitrite, bromide, and nitrate) as shown by Fig. 14. The dependence of retention on pH and phosphate concentration indicated that the mechanism was simple exchange of analyte ions with the two phosphate ions $H_2PO_4^-$ and HPO_4^{2-}. At high pH retention was reduced as the polyethylene amine became deionized. Quaternization of the PEI would eliminate this and allow ion-exchange properties to be preserved at high pH.

Appendices 1–5 follow

Appendix 1: Separations of Geometric Isomers and Closely Related Compounds

Ref.	Author	Analyte	Mobile Phase	Comment
	a. Simple Aromatic Molecules			
1	Knox, Kaur, Millward (1986)	Xylene isomers	(95/5) Methanol/water	Elution order m-, o-, p- isomers
6	Kříž, Adamcová, Knox, Hora (1994)	52 Aromatic hydrocarbons	Methanol	High selectivity esp w.r.t o-substitution by CH_3 in benzene
7	Wan, Shaw, Davies, Barrett (1995)	Various o-, m-, and p- derivatives of benzene	(35/65) Acetonitrile/phosphate with pH varied from 2 to 12	Good separations of all test o-, m-, p- mixtures. Variation of k' with pH follows pK_a
8	Forgács, Cserháti, Valkó (1992)	24 Phenol derivatives	Wide range of composition of water/methanol or acetonitrile	Good separations
9,10	Forgács, Cserháti (1992)	16 Ring-substituted aniline derivatives	Various eluents investigated	No buffers required for good peak shape and resolution
12	Deinhammer, Ting, Porter (1995)	14 Benzene and naphthalene sulphonic acids	(96/4) Water/acetonitrile 0.1M in $LiClO_4$	Excellent separation based mainly upon hydrophilicity. Both retention and elution order altered by applying emf to packing.
	b. Pharmaceuticals and Natural Products			
13	Forgács, Cserháti (1992)	54 Barbituric acid derivatives	(95/5) to (80/20) Methanol/ water eluent mixtures	Buffered eluents not required for good resolution
14	Bell, Tsai, Ip, Mathre (1994)	Isomers of a 3-hydroxyproline C50 prodrug	(55/45) 20 mM Phosphate buffer pH 6.8/acetonitrile mixture	Isomers not resolved on oxide adsorbents or bonded silica phases

#	Authors	Analyte	Mobile phase	Comments
15	Bassler, Hartwick (1989)	Steroids	Gradient: (10/90) to (80/20) acetonitrile/water over 30 min	Simple mobile phase—improved resolution
15	Smith, Brennan (quoted Bassler, Hartwick)	Cephalosporin antibiotic (Axetil)	(35/35/10/20) Acetonitrile/methanol/dioxan/water	Equivalent separation on bonded silica difficult or impossible
16,17	Pawlak, Kay, Clark (1989, 1990)	Dothiepin hydrochloride and geometric isomers	(75/35/1) Ethyl acetate/methanol/3% aq ammonia	Normal phase method—not possible with other phases. Superior to GC
18	Berridge (1988)	Tioconazole from closely related impurities	(70/30) THF/water containing 1% ammonia, 40°C	ODS and phenyl columns not able to resolve all the impurities and main degradation product.
19	Mama, Fell, Clark (1989)	Amino acids	(65/35) Acetonitrile/25 mM phosphate pH 1	Good separations at extreme pH
		Diazopines	(65/35) acetonitrile/ 3 mM phosphate pH 10.6	
20	Fish (1993)	Cis- and trans-isomers of anti-asthma compound	(68/32/7) Methanol/dichloromethane/TFA	Four cis-/trans-isomers better separated using PGC than Vydac C18
21	Wutte, Gubitz, Friebe, Kraus (1994)	Proline-containing di-peptides		Cis- and trans-isomers resolved
22	Houghton, Woldemariam (1995)	Chromone alkaloid isomers	(50/50 w/v) THF/20 mM phosphate buffer	Pairs of 7′ α-OH- and 7′β-OH-isomers separated by PGC, inseparable with other phases

c. Pesticides

#	Authors	Analyte	Mobile phase	Comments
23	Forgács, Cserháti (1995)	30 commercial pesticides	(65/35) to (10/90) Dioxan/water	Log k′ follows equation (1), retention mainly governed by hydrophobicity.

Appendix 2: Enantiomer Separations

Ref.	Author	Analyte	Mobile Phase	Comment
	a. Diastereoisomers			
25	Josefsson, Carlsson, Norlander (1993)	(−)-Methyl chloroformate derivatives	70/30 Mixtures of dichloromethane with acetonitrile, ethanol, ethyl acetate, acetic, or formic acids.	Acids were the strongest eluents but gave poorest selectivity.
26	Chan, Mickelwright, Barrett (1995)	Standard amino acids derivatized with O-tetra acetyl-β-D-glucopyranosyl isothiocyanate	Gradient (30/90) to (90/10) acetonitrile/water containing 0.1% TFA	Virtually all diastereoisomeric pairs of standard 26 amino acids resolved.
	b. Separations based upon non-bonding interactions			
26	Dutton, Knox, Radisson, Ritchie, Ramage (1995)	Aryl alcohols and Fmoc amino acids	(90/10) Hexane/isopropanol or ethyl acetate, on PGC coated with monolayer of Pirkle type phase linked to tetrabenzoyl fluorene (TBF) "foot"	Resolution of enantiomers similar to that on silica-based Pirkle phases.
28	Wilkins, Taylor, Smith (1995)	SFC of tropic acid amides	SC−CO_2 with added methanol or propanol on PGC coated with monolayer of chiral selector based upon tartaric acid having anthryl "foot"	Resolutions with α-values of about 1.4; equivalent silica phase gave about 1.05
29,30	Heldi, Huynh, Petterson (1991)	Atropine, scopolamine, homatropine, atropine and tropic acid	0.25 mM (2S,3S) Dicyclohexyltartrate in phosphate buffer at pH 2.8	Excellent resolution using an aqueous eluant.
31	Grieb, Matlin, Belenguer, Ritchie, Ross (1995)	Wide range of enantiomers	PGC coated with 25% by weight of cellulose carbamate	Performance similar to that of Dionex materials. Good peak shapes.

	c. Separations based on inclusion complexation			
32	Fell, Noctor, Mama, Clark (1988)	Metroprolol, benzodiazepines, nomifensine maleate,	β-Cyclodextrin (3 mM) in methanol/or acetonitrile/phosphate mixtures pH 2 to 12	Wide pH range allows suppression of ionisation of basic and acidic drugs giving improved peak efficiency and symmetry.
32	Fell, Noctor, Mama, Clark (1989)	Enantiomers of oxazepam, lorazepam	β-Cyclodextrin (3 mM) in acetonitrile/5 mM phosphate buffer (pH 12)	Improved resolution over silica bonded cyclodextrin columns and cellulose triacetate column.
	d. Separations based upon ion-pair formation			
33	Karlsson, Pettersson (1991)	Amine enantiomers, propanolol, etc.	10 mM N-benzoxy-carbonylglycyl-L-proline (ZGP) in dichloromethane (containing 80ppm of water)	Short equilibration times, triethylamine gives improved peak shape.
35	Petterson, Gioeli	Enantiomeric amines, bupivacaine, etc.	2,3,4,6 di-O-isopropyliene-2-keto-L-gulonic acid (−)-DIKGA dissolved in aqueous/organic eluents.	Aqueous mobile phases avoiding organics, allows direct injection of plasma. Ethanolamine improves symmetry
36	Huynh, Karlsson, Pettersson (1995)	Wide range of basic drugs	1–20 mM ZGP and 1–10 mM NaOH in methanol and other solvents	Good selectivity with α-values ranging from 1.0 to 2.0
	e. Separations based upon metal complexion			
37	Knox, Wan (1995)	D- and L-amino- and hydroxy-acids	2 mM Aqueous copper acetate pH 5.6 on PCG coated with monolayer of naphthalene-sulphonyl-D- or L-phenylalamine	Excellent resolution of enantiomers

Appendix 3: Analysis of Sugars, Carbohydrates, and Glucuronides

Ref.	Author	Analyte	Mobile Phase	Comment
a. Sugars and Carbohydrates				
38	Davie, Smith, Harbin, Hounsell (1992)	Neutral oligosaccharide alditols; Sialylated oligosaccharide alditols	Step gradients/ 0–15% and 0–40% acetonitrile in 0.05% aqueous TFA	Improved resolution of oligosaccharide alditol isomers differing by only a (1→3) or (1→4) linkage. Previously high pH phosphate buffer required.
39	Stefansson, Lu (1993)	Monosaccharides, disaccharides, sugar alcohols, sugar acids, amino sugars	$0.02M$–$0.1M$ NaOH	Retention very pH-dependent. Compared to monosaccharides, amino sugars similarly retained disaccharides and sugar acids more retained.
40	Lu, Steffansson and Westerlund (1995)	Di- and tri-saccharides	(90/10) 0.025 mM sorbic acid, 0.1 M NaOH/methanol (90/10)	Excellent separations in a strongly alkaline eluent using indirect UV-detection.
41	Koizumi, Okada, Fukuda (1991)	Mono- and oligo-saccharides	Water, and water containing up to 15% methanol and acetonitrile	Anomers resolved
b. Glucuronides				
42	Steffansson, Hoffman (1992)	Diastereomeric glucuronides of Almokalant	Coupled column separation: pre-column: terbium (III)-loaded polystyrene SCX; analytical column: PGC with (70/30) 0.1 M Acetic Acid pH 2.8/acetonitrile	Separation not possible with Nucleosil C18 or PLRP-S columns
43	Ayrton, Evans, Harris, Plumb (1995)	AZT, chloramphenicol, β-estradiol, fluparoxan and their glucuronides	(30/70/0.5) Tetrahydrofuran/ water/TFA	Good separations with glucuronide eluting before parent (except for fluparoxan)

Appendix 4: Residue Analysis

Ref.	Author	Analyte	Mobile Phase	Comment
	a. Polychlorinated biphenyls (PCBs)			
44	Creaser, Al-Haddad (1989)	PCB, PCDFS, PCDD (class separation) with further fractionation of the PCBs	Hexane (class separation) (80/20 v/v) Acetonitrile/water (PCB separation)	Increased reproducibility over activated carbon.
45	Hong, Bush, Xiao, Fitzgerald (1992)	PCBS from PCDDs and PCDFs in human milk (class separation)	Hexane	Increased reproducibility over activated carbon and simpler eluents.
46	De Boer, Stronk, van der Valk, Wester, Dault (1992)	PCBs in fish	Fractionation using PGC with sequential non-polar solvent mixtures of hexane, dichloromethane, toluene.	Greater selectivity for planar, unsaturated compounds with PGC than with silica-based phases.
47	Fuoco, Colombini, Samcova (1993)	Ortho and non-ortho PCBs from soil	Fractionation using PGC with hexane	Improved reproducibility and standardization
48	Bohm, Schulte, Thier (1991)	PCBs in food	Gradient: (0/100) to (40/60) toluene/cyclohexane	PGC superior to other packings
49	Williams, Giesy, Verbrugge, Jurzysta, Stromborg (1995)	PCBs in cormorant eggs	Fractionation using PGC	
	b. Tissue and Food Residues			
50	Tarbin, Shearer (1992)	Lasalocid in chicken tissue	(5:95) Tetramethyl-guanidine in acetonitrile	Strongly basic eluent giving improved resolution and detection.

Appendix 4: Continued

Ref.	Author	Analyte	Mobile Phase	Comment
51	Liem, Baumann, de Jong, van der Velde, van Zoonen (1992)	Micropollutants in foods	General review including use of PGC	
	c. Solid-phase extraction (SPE) and water analysis			
52	Coquart, Hennion (1992)	Polar phenolics–polyhydroxy benzenes	SPE followed by HPLC (both on PGC) gradient: (25/75) to (100/0) methanol/0.05 M perchloric acid–lithium perchlorate (pH 4) in 40min. electro chemical detection	Strong retention of di- and tri-hydroxy phenols, the opposite elution order to that on ODS and polymer phases.
53	Hennion, Pichon, Chen, Guenu (1994)	Polar pesticides–triazine group	SPE on PGC followed by HPLC on ODS silica	PGC ideal for SPE of very polar residues.
56	Guenu, Hennion (1995)	Polar pesticides–triazine group and others.	SPE followed with on-line HPLC, both with PGC	Excellent separations with good retention of highly polar pesticides
57	Pichon, Chen, Guenu, Hennion (1994)	Polar organics incl. chlorophenols, chloroanilines aminophenols and cyanuric acid	SPE followed with on-line HPLC, both with PGC. Various methanol/water and acetonitrile/water mixtures	Good gradient separations

Appendix 5: Separations of Ionized and Highly Polar Compounds

Ref.	Author	Analyte	Mobile Phase	Comment
	a. Ionized solutes			
58	Lim (1989)	Pertechnetate and Perrhenate ions	Aqueous TFA	Good retention on PGC, no retention on ODS-silica without ion pairing
59	Emery, Lim (1989)	Technetium and Rhenium organometallic complexes	2–10% Acetonitrile in aqueous TFA	Anionic and cationic complexes separated in same analysis
	b. Water-soluble unionized solutes			
60	Gu, Lim (1990)	Oxalic acid, Creatine and Creatinine in serum	3% Acetonitrile in 0.1% TFA	Oxalic acid completely retained without added TFA
61	Dutton, Rastall, Evans (1991)	Oxalate in liquid culture media	0.2 M Orthophosphoric acid	Oxalate well separated from acetate, impossible on silica-based columns.
62	Krause (1989)	Ethylene thiorea	(5/95) Acetonitrile/0.025N phosphoric acid	PGC provided excellent long term stability and quantitation of ETU.
63	Ehrsson, Wallin, Andersson (1995)	Cis-platin, trans-platin and hydrated platins	1 mM NaOH	Good separations in strongly alkaline eluent.
	c. Ion exchange separations			
64	Knox, Wan (1995)	Iodate, bromide, nitrite, nitrate	0.02 M Phosphate buffer on PGC coated with monolayer of polyethylene imine (PEI)	Good separation of inorganic anions, retention pH dependent due to de-ionization of PEI at high pH

REFERENCES

1. M. T. Gilbert, J. H. Knox, and B. Kaur, *Chromatographia, 16:* 138 (1982).
2. J. H. Knox, B. Kaur, and G. R. Millward, *J. Chromatogr., 352:* 3 (1986).
3. J. H. Knox and P. Ross, in *Advances in Chromatography*, Vol. 37, Marcel Dekker, Inc., New York, 1996, pp.
4. Shandon, *Hypercarb® Guide,* Life Sciences International. Runcorn, UK, 1993.
5. A. V. Kiselev, Y. I Yashin, and T. Bradley, Plenum Press, London, 1969.
6. J. Kříž, E. Adamcová, J. H. Knox, and J. Hora, *J Chromatogr. A, 663:* 151 (1994).
7. Q-H. Wan, P. N. Shaw, M. C. Davies, and D. A. Barrett, *J. Chromatogr. A, 697* (1995).
8. E. Forgács, T. Cserháti, and K. Valkó , *J. Chromatogr., 592:* 75 (1992).
9. E. Forgács and T. Cserháti, *Chromatographia, 33:* 356 (1992).
10. E. Forgács and T. Cserháti, *J. Chromatogr., 600:* 43 (1992).
11. E. Forgács and T. Cserháti, *Trends Anal. Chem., 14:* 23 (1995).
12. R. S. Deinhammer, E.-Y. Ting, and M. D. Porter, *Anal. Chem., 67:* 237 (1995).
13. E. Forgács and T. Cserháti, *J. Pharm. Biomed. Anal., 10:* 861 (1992).
14. C. Bell, E. W. Tsai, D. P. Ip, and D. J. Mathre, *J. Chromatogr. A, 675:* 248 (1994).
15. B. J. Bassler and R. A. Hartwick, *J. Chromatogr. Sci., 27:* 162 (1989).
16. Z. Pawlak and B. J. Clark, *J. Pharm. Biomed. Anal. 7:* 1903 (1989).
17. Z. Pawlak, D. Kay, and B. J. Clark, *Roy. Soc. Chem. Anal. Proc., 27:* 16 (1990).
18. J. C. Berridge, *J. Chromatogr., 449:* 317 (1988).
19. J. E. Mama, A. F. Fell, and B. J. Clark, *Roy. Soc. Chem. Anal. Proc., 26:* 71 (1989).
20. B. J. Fish, *J. Pharm. Biomed. Anal., 11:* 517 (1993).
21. A. Wutte, G. Gubitz, S. Friebe, and G.-J. Kraus, *J. Chromatogr., A, 670:* 186 (1994).
22. P. J. Houghton and T. Woldemariam, *Phytochem. Anal., 6:* 85 (1995).
23. E. Forgács and T. Cserháti, *Analyst, 120:* 1941 (1995).
24. I. W. Wainer, *Clin. Pharmacol., 18:* 139 (1993).
25. M. Josefsson B. Carlsson, and B. Norlander, *Chromatographia, 37:* 129 (1993).
26. W. C. Chan, R. Micklewright, and D. A. Barrett, *J. Chromatogr. A, 697:* 213 (1995).
27. J. K. Dutton, J. H. Knox, X. Radisson, H. J. Ritchie, and R. Ramage, *J. Chem. Soc. Perkin Trans. I,* 2581 (1995).
28. S. M. Wilkins, D. R. Taylor, and R. J. Smith, *J. Chromatogr. A, 697:* 587(1995).

29. E. Heldin, N. H. Huynh, and C. Pettersson, *J. Chromatogr., 585:* 35 (1991).
30. E. Heldin, N. H. Huynh, and C. Pettersson, *J. Chromatogr., 592:* 339 (1992).
31. S. J. Grieb, S. A. Matlin, A. M. Belenguer, H. J. Ritchie, and P. Ross, *J. High Resolut. Chromatogr.,* in press.
32. A. F. Fell, T. A. G Noctor, J. E. Mama, and B. Clark, *J. Chromatogr., 434:* 377 (1988).
33. A. Karlsson and C. Pettersson, *J. Chromatogr., Biomed. Appl., 494:* 157 (1989).
34. A. Karlsson and C. Pettersson, *J. Chromatogr., 543:* 287 (1991).
35. C. Pettersson and C. Gioli, *Chirality, 5:* 241 (1993).
36. N. H. Huynh, A. Karlsson, and C. Pettersson, *J. Chromatogr., 705:* 275 (1995).
37. J. H. Knox and Q.-H. Wan, *Chromatographia, 38:* 1 (1995).
38. M Davies, K. D. Smith, A.-M. Harbin, and E. F. Hounsell, *J. Chromatogr., 609:* 125 (1992).
39. M. Steffansson and B. Lu, *Chromatographia, 35:* 61 (1993).
40. B. Lu, M. Steffansson, and D. Westerlund, *J. Chromatogr., A, 697:* 317 (1995).
41. K. Koizumi, Y. Okada, and M. Fukuda, *J. Carbohydrate Res., 215:* 67 (1991).
42. M. Steffansson and K.-J. Hoffmann, *Chirality, 4:* 509 (1992).
43. J. Ayrton, M. B. Evans, A. J. Harris, and R. S. Plumb, *J. Chromatogr., B, 667:* 173 (1995).
44. C. S. Creaser and A. Al-Haddad, *Anal. Chem., 61:* 1300 (1989).
45. C.-S. Hong, B. Bush, J. Xiao, and E. F. Fitzgerald, *Chemosphere, 24:* 465 (1992).
46. J. de Boer, C. J. N. Stronck, F. van der Valk, P. G. Wester, and M. J. M. Dauldt, *Chemosphere, 25:* 1277 (1992).
47. R. Fouco, M. P. Colombini, and E. Samcova, *Chromatographia, 36:* 65 (1993).
48. V. Boehm, E. Schulte, and H. -P. Thier, *S. Lebensm. Unters. Forsch., 192:* 548 (1991).
49. L. L. Williams, J. P. Giesy, D. A. Verbrugge, S. Jurzysta, and K. Stromborg, *Arch. Environ. Contam. Toxicol, 29:* 327 (1995).
50. J. A. Tarbin and G. Shearer, *J. Chromatogr., 579:* 177 (1992).
51. A. K. D. Liem, R. A. Baumann, A. P. J. M. de Jong, E. G. Van der Velde, and P. van Zoonen, *J. Chromatogr., 624:* 317 (1992).
52. V. Coquart and M. -C. Hennion, *J. Chromatogr., 600:* 195 (1992).
53. M.-C. Hennion, V. Pichon, L. Chen, and S. Guenu, *Environ. Sci. Technol. 28:* 576A (1994).

54. M.-C. Hennion and V. Coquart, *J. Chromatogr., 642:* 211 (1993).
55. S. Guenu and M.-C. Hennion, *J. Chromatogr, A665*: 243 (1994).
56. S. Guenu and M.-C. Hennion, AMI J. in press.
57. V. Pichon, L. Chen, S. Guenu, and M.-C. Hennion, *J. Chromatogr. A711*: 257 (1995).
58. C. K. Lim, *Biomed. Chromatogr. 3:* 92 (1989).
59. M. F. Emery and C. K. Lim, *J. Chromatogr., 479:* 212 (1989).
60. G. Gu and C. K. Lim, *J. Chromatogr., 515:* 183 (1990).
61. M. V. Dutton, R. A. Rastall, and C. S. Evans, *J. Chromatogr., 587:* 297 (1991).
62. R. T. Krause, *J. Liq. Chromatogr., 12:* 1635 (1989).
63. H. C. Ehrsson, I. B. Wallin, and A. S. Andersson, *Anal. Chem., 67:* 3608 (1995).
64. J. H. Knox and Q.-H. Wan, *Chromatographia, 42*: 83 (1996).

4
Directly Coupled (On-Line) SFE–GC: Instrumentation and Applications

Mark D. Burford and Keith D. Bartle
University of Leeds, Leeds, United Kingdom
Steven B. Hawthorne
University of North Dakota, Grand Forks, North Dakota

I.	INTRODUCTION	164
II.	TECHNIQUES FOR COUPLING SFE–GC	166
III.	EXTERNAL TRAPPING OF ANALYTES	168
IV.	INTERNAL ACCUMULATION OF ANALYTES	171
V.	CONSTRUCTION OF SFE–GC INSTRUMENTATION	174
VI.	OPTIMIZATION OF SFE–GC CHROMATOGRAPHY	178
	A. Extraction Flow Rate	178
	B. Column Trapping Temperature	182
	C. Column Stationary-Phase Thickness	186
VII.	QUANTITATIVE CONSIDERATIONS FOR SFE–GC	187
VIII.	OPTIMIZATION OF EXTRACTION CONDITIONS FOR SFE–GC	192
IX.	SFE–GC APPLICATIONS	198
X.	CONCLUSIONS	198
	REFERENCES	200

I. INTRODUCTION

The saying "time is money" is becoming ever more common in the laboratory environment. There has always been a need to control analytical costs in terms of solvent usage and disposal, consumable expenditure, and administration overheads. However, there is currently an even greater emphasis on the time an analysis takes, and this has led to the development of automated commercial extraction methods using solvent, microwave-assisted, and supercritical fluid extraction (SFE) systems. To fully automate the analytical process, the extraction and analysis procedures need to be combined to produce a coupled or on-line technique that minimizes the sample handling and preparation steps which contribute to analyte loss, degradation, and/or contamination. This direct transfer of extracted analytes into the analyte separation and detection system produces a more sensitive analysis and increases the laboratory sample throughput. For example, the extraction and chromatographic analysis of real-world samples with low (ppm–ppb) analyte concentrations can be achieved within 1 h using an on-line approach [1–3].

Conventional extraction methods such as Soxhlet and sonication techniques do not readily lend themselves to convenient means of direct sample introduction for the most popular method of analysis, namely chromatographic separation. These liquid-solvent extraction methods generally lack selectivity, producing extracts that often contain species which interfere with the chromatographic analysis of the target analytes, and so require a clean-up step prior to the analysis. This sample preparation step makes the coupling of the extraction and analysis procedures complicated. Furthermore, the typical injection volumes of the chromatographic techniques such as capillary supercritical fluid chromatography (SFC, <0.1 µL), capillary gas chromatography (GC, ~1 µL), or high-performance liquid chromatography (HPLC, ~10 µL) column are incompatible with the large solvent volumes used in the extraction process. Even the relatively large, 100 µL injection volume which is now possible with capillary GC [4–6] is still only a small fraction of the original solvent extract. So an additional preconcentration step is required prior to the chromatographic analysis and this can result in the coevaporation of the more volatile compounds [7].

The volume of organic solvent is a particular problem for trace analysis where the concentration of the analytes in the extract has to be significantly enhanced, but the solvent volume cannot be accurately reduced to a volume of less than ~100 µL. Therefore, to obtain adequate sensitivity in the analysis of analytes in the low ppm to ppb concentration range, the sample size (and corresponding collection time for samples such as air particulates) has to be increased. Commercial on-line extraction/analysis systems such as headspace, purge-and-trap, thermal desorption, and pyrolysis capillary GC techniques are suitable for trace analysis because they quantitatively transfer the entire extract to the analytical system without introducing large volumes of liquid into the chromatographic column. Although these on-line techniques are valuable extraction methods capable of recovering volatile analytes from a wide range of samples and are relatively easy procedures to combine, because the extraction fluid is in the same phase as the chromatographic mobile phase (e.g., a gas), they are limited by the requirement of high analyte volatility. For example, headspace and purge-and-trap capillary GC techniques are routinely used to analyze petroleum-contaminated samples containing very volatile gasoline range organics (e.g., C_4–C_{12} boiling-point range). However, when the sample contains only moderately volatile kerosene and diesel fuel organics (e.g., C_9–C_{25} boiling-point range), the fuels are poorly detected [8,9]. Thus, for the moderate-to-low-volatility analytes, conventional liquid-solvent extraction methods are required, but these methods often yield low recoveries for the volatile analytes and, as stated above, are not easily combined with chromatographic systems.

Ideally, what is required is an extraction technique which combines the ability of a liquid-solvent extraction method (e.g., Soxhlet) to quantitatively recover a broad range of analytes, with a gas-phase extraction method (e.g., thermal desorption) which can quantitatively transfer the analytes into a chromatographic system. Such a technique, which can successfully bridge the gap between the extraction and analysis steps, is supercritical fluid extraction. Supercritical fluids have solvent strengths approaching that of a liquid so that a wide of analytes can be extracted, and yet at ambient conditions the fluid becomes a gas and is therefore amenable to a range of chromatographic techniques. Under supercritical fluid extraction conditions (typically 400 atm and 60°C), the supercritical fluid has an order-of-magnitude lower viscosity (10^{-4} versus 10^{-3} N s/m^2) and higher solute diffusivity (10^{-4} versus 10^{-5} cm^2/s) than liquid solvents, so that quantitative analyte recoveries can usually be achieved within 30 min rather than several hours or days. This rapid mass transfer of analytes into the supercritical fluid enables the extract to be efficiently transferred to the chromatographic column, thus minimizing the potential for band broadening and poor chromatographic separation.

The solvent strength of the supercritical fluid can also be varied with temperature, pressure, and/or polarity (e.g., by the addition of a modifier or more polar supercritical fluid) so that a cleaner, more selective extract can be obtained with fewer chromatographically interfering components. Using these various extraction conditions, supercritical fluids have been used to quantitatively extract a broad range of analytes, including polycyclic aromatic hydrocarbons [10–12], polychlorinated biphenyls [2], pesticides [13,14], hydrocarbons [15,16], ionic surfactants [17], metals [18,19], polymer additives [20], essential oils [21,22], fatty acids [23,24], vitamins [25], soft resins [26], and caffeine [27]. Very pure supercritical fluids are commercially available as SFE–SFC grade solvents (99.999% pure [28]) and this has enabled fluids such as carbon dioxide to be directly linked to several detection systems, such as flame-ionization (FID) [29], electron-capture (ECD) [30], ultraviolet absorption (UV) [31], mass spectrometric (MS) [32], Fourier transform infrared spectrometric (FTIR) [33], and atomic emission spectrometric (AES) [34] detectors. The majority of these detectors have a very low response to carbon dioxide and any interference is usually due to contamination problems in the pump or the associated plumbing of the system. The high-grade fluids can be directly depressurized into a chromatographic column without any detrimental effects. For example, the SFE–SFC-grade fluids do not strip off the column stationary phase, cause column contamination problems, or decrease the chromatographic resolution. The latter two problems are most often observed when low-grade fluids containing traces of hydrocarbons and/or water are used [30].

The supercritical fluid extraction process can be coupled to a wide range of chromatographic techniques, including GC [1–3], SFC [35–38], HPLC [39,40], thin-layer chromatography (TLC) [41], and gel-permeation chromatography (GPC) [42]. The choice of chromatographic system depends on the nature of the sample under investigation, but, generally, the analytes extracted using supercritical carbon dioxide are nonpolar to moderately polar in nature, with sufficient volatility to be analyzed by GC. Thus, the most popular supercritical fluid online technique is SFE-GC. This chapter will discuss the practical requirements needed to assemble and operate an SFE–GC system, with particular attention on the modifications needed to obtain both qualitative and quantitative analysis. It will be demonstrated that the coupling of the extraction and analysis procedures requires a detailed understanding of both the extraction and chromatographic parameters.

II. TECHNIQUES FOR COUPLING SFE–GC

To successfully couple the supercritical fluid extraction technique to gas chromatography, a series of procedures or steps must be undertaken to enable the analytes of interest to be quantitatively removed from the sample matrix and efficiently transferred and focused in the chromatographic column. Figure 1 details the steps involved. In the initial step, the extraction fluid is pumped into a heated extraction cell containing the sample of interest. Conditions inside the extraction cell are above the critical temperature and pressure of the extraction fluid (i.e., for CO_2 above 31°C and 73 atm), so that the fluid is supercritical. As the supercritical fluid comes into contact with the sample matrix and penetrates the material, the soluble analytes are partitioned into the bulk of the fluid, which is then swept out of the extraction cell and through a flow restrictor. The restrictor is used to maintain the pressure in the extraction system, so once the supercritical fluid has passed through the SFE system it is depressurized at the head of the chromatographic column, where the analytes are deposited in the stationary phase and

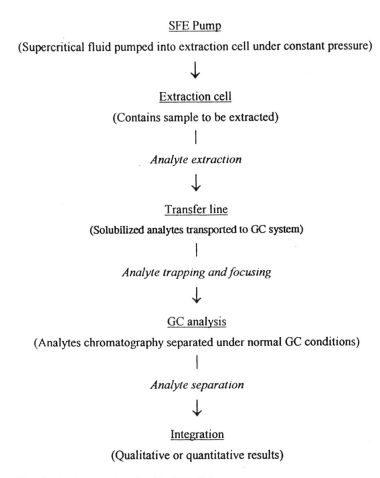

Fig. 1 Major steps involved in SFE–GC.

the gaseous extraction fluid is vented through the injection port. At the end of the extraction step, the flow of supercritical fluid through the sample is halted and the gaseous extraction fluid is allowed to dissipate through the column to allow a conventional GC separation to be performed in the normal manner.

To obtain quantitative SFE recoveries, the extraction conditions are usually optimized in terms of the extraction temperature, pressure, and supercritical fluid (e.g., nonpolar CO_2, polar Freons, or organically modified CO_2/methanol) [43–48]. However, to achieve optimum recoveries, the extraction and subsequent focusing of the analytes on the column requires ~30–60 min compared to the several seconds needed to introduce a liquid-solvent extract into the column using a conventional manual injection technique. Furthermore, the depressurization of the supercritical fluid generates a high gas flow rate at the head of the GC column which can result in poor focusing of the analytes and severe chromatographic band broadening. For example, a typical supercritical fluid extraction volume of 1 mL/min of liquid CO_2 expands to ~550 mL/min CO_2 gas which then has to be separated and removed from the analytes at the SFE–GC interface. Thus, the fundamental problem with coupling SFE to GC is how to efficiently collect and focus the extracted analytes over a long period of time and under the high gas flow rates generated from the depressurization of the supercritical fluid.

III. EXTERNAL TRAPPING OF ANALYTES

Several methods have been employed to overcome the problems encountered in retaining the extracted analyte for chromatographic analysis. These various approaches can be loosely divided into two main groups, as indicated in Table 1 and schematically shown in Fig. 2. The extracted analytes obtained from the SFE system can either be (1) collected on an analyte accumulation device external to the GC using an "analyte trap" or (2) collected internally in the GC, using the GC column itself. With an external accumulation device, the supercritical fluid extract is depressurized inside a cold trap (e.g., cryogenically cooled stainless-

Table 1 Analyte Collection Method in SFE–GC

Location of analyte collection relative to GC column	Collection method	Path taken by supercritical fluid (SCF)	Heated depressurization zone	Ref.
External	Cooled sample loop on switching valve	SCF vented out of sample loop	Yes	50
External	Cold trap	SCF vented out of purge vent on trap	Yes	49
External	Sorbent trap (Tenax–GC)	SCF vented out of purge vent on trap	Yes	51
Internal	Direct injection on head of GC column	SCF vented through GC column	Yes No	56, 57, 58
Internal	Cooled on column injection port	SCF vented through the GC column and injection port	No	2, 3
Internal	Split/splitless injection port	SCF vented through GC column and split vent	Yes	59–61
Internal/external	PVT injection port	SCF vented through GC column and split vent	No	62

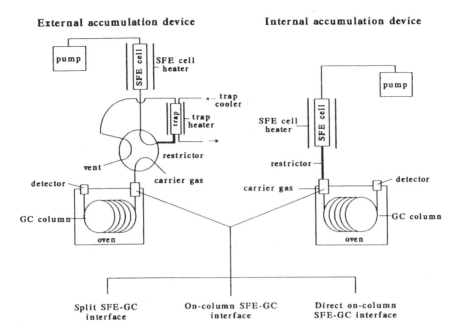

Fig. 2 Schematic of external and internal accumulation devices used in SFE–GC.

steel or silica tubing) [49,50] or a sorbent trap (e.g., Tenax) [51], with analytes selectively retained in the trap while the gaseous extraction fluid is vented out of the purge line at the base of the device (Fig. 2). In order to prevent plugging of the restrictor during the extraction (as depressurization of the supercritical fluid within the restrictor typically causes the analytes to precipitate and accumulate), the restrictor is either heated with the tip situated just inside the cold trap [49] or the restrictor and the trap are heated as is possible when using a sorbent resin [51]. For example, a Tenax–GC trap has been used at 50°C to simultaneously maintain the extraction flow rate and analyte collection during the SFE–GC analysis of pesticides [51]. On completion of the extraction, the analytes are thermally desorbed from the trap and transferred to the GC column by closing the trap purge vent and using a switching valve to divert the carrier gas through the rapidly heated trap and into the column. Alternatively, the analytes can be recovered by extracting the trap with a supercritical fluid so that, in effect, an SFE–SFE–GC analysis is undertaken [51].

The advantage of collecting the analyte in an external trap is that during the extraction step the depressurized gas is not vented through the GC column, so that potentially larger supercritical fluid extraction volumes can be used than are possible when using a direct SFE–GC interface (although this will partially depend on the volatility of the analytes involved). For example, the sorbent Tenax–GC can successfully retain extracted analytes over a relatively long period (e.g., 30 min) during the SFE step, after which the analytes can be quickly transferred to the GC (e.g., in , <10 min) by extracting the resin with a supercritical fluid [51]. Thus, the extraction fluid volume and the transfer time required to transport the analytes into the GC column can been significantly reduced and this enables the analytes to be more efficiently focused in the column. Alternatively, if the analytes are recovered from the trap using the GC carrier gas, so that none of the supercritical fluid is introduced into the column during SFE–GC, the potential chromatographic and detection problems from the supercritical fluid can be avoided. This is particularly important when an organically modified supercritical fluid is used, because if the extraction fluid is introduced into the column, the modifier will generate a "solvent" peak in the resulting chromatogram which could coelute with the target analytes. The use of an external trap such as Tenax–GC enables modified supercritical fluids like CO_2/methanol to be used without significant detection of the modifier [51], as the sorbent can selectively retain the semivolatile hydrocarbons during the extraction (as they have very large retention volumes [52]) but not retain polar compounds such as methanol, which are purged from the trap and diverted away from the GC column.

The disadvantage of external analyte traps is that they can have a limited collection efficiency compared to the wide range of analytes that are retained when depressurizing the extract directly into the chromatographic column. For example, volatile and semivolatile analytes which can be retained in an external fused silica tube cold trap operated at conventional 10–60 mL/min gas flow rates [53] are poorly retained when operated under SFE–GC conditions where gas flow rates of 300–600 mL/min are routinely used [49]. The cold trap also has a restricted SFE–GC working temperature range (typically –10°C to –50°C), because if too high a temperature is chosen, volatile analytes are lost, and if too low a temperature is used, carbon dioxide (below –78°C) or extracted water (below 0°C) will freeze and plug the trap.

These problems are not as severe for sorbent resins, which can retain a wider range of semivolatile analytes at ambient conditions [52]. Water has little impact on the affinity of the resin for organic compounds [54], but the sorbent will not retain the analytes indefinitely during an extraction and this may lead to problems when extracting difficult samples which require long extraction times. The sorbent resins are also unable to efficiently retain hazardous organic solvents such as alcohols, ketones, Freons, and chlorinated solvents which can be successfully analyzed when depressurized directly into the GC column [55]. External traps are therefore easier to optimize in terms of the SFE extraction conditions, such as extraction fluid volume and use of a modifier, but are somewhat limited by the range of possible analytes.

IV. INTERNAL ACCUMULATION OF ANALYTES

A simpler, more popular method of coupling the SFE and GC systems which avoids switching valves or external traps involves collecting the extract within the chromatographic system. By using the chromatographic column as the accumulation device, more quantitative and reproducible recoveries are obtained with little or no modifications required to the GC system. The same GC injection port can be used for conventional liquid-solvent injection or SFE–GC without conversion. A schematic of the process is shown in Fig. 2. This "internal accumulation method" which directly couples the SFE effluent to the GC column can be undertaken using either (1) "on-column" SFE–GC, where the SFE effluent is depressurized directly into the chromatographic column, so that all the extracted analytes are deposited either in the column stationary phase [2,51,63], or on a retention gap at the head of the column [56,57] or (2) "split" SFE–GC, where the SFE effluent is depressurized in a conventional split/splitless injection port operated in the split mode so that a fraction of the extract is deposited in the column stationary phase (or retention gap) and the majority of the extract is vented through the injection port split vent [55,59,61,64]. Table 1 summarises these two approaches and Fig. 3 illustrates the experimental designs.

Both the on-column and spilt SFE–GC approaches usually require the GC column to be cooled during the extraction so that the analytes (particularly volatile analytes such as n-pentane) can be as efficiently retained and focused in

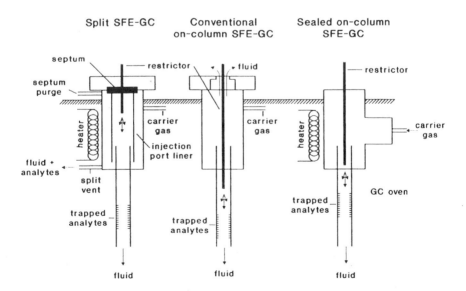

Fig. 3 Common SFE–GC interfaces.

the column stationary phase [55]. Alternatively, the injection port can be cooled using a programmed temperature vaporizer (PTV) injection system so that the analytes are retained on the cold injection port liner during the extraction [62]. The chromatographic peak shapes and peak area reproducibilities obtained from these SFE–GC techniques are comparable to those generated from the conventional on-column or split injection of the liquid-solvent extract. Relative standard deviations of replicate SFE–GC analyses have been reported to be in the region of 2–10% [3,55,61]. The appropriate use of the SFE–GC techniques to transfer part (split) or all (on-column) of the supercritical extract to the GC system will depend on the nature of the sample (e.g., water content), analyte concentration (e.g., ppb or ppm), and extraction conditions (e.g., flow rate, use of modifier, etc.). When only a small quantity of sample is available, such as a forensic or air particulate sample, the method of choice is on-column SFE–GC as all of the extracted analytes are transferred to the GC column, so that maximum sensitivity and absolute quantitation of the extract can be obtained, with low ppb (ng/g) detection limits for milligram samples [2]. However, by depositing all the extract in the column stationary phase, the non-GC-able components are also transferred to the GC system and this can lead to poor chromatographic peak shapes and reduce the working life of the column. The use of a retention gap at the head of the column may reduce the column contamination problem [56,57], but when a sample contains a high percentage of extractable components not amenable to GC, the most practical approach is to use the alternative split SFE–GC technique.

In split SFE–GC only a fraction of the extract is transferred into the GC column and the nonvolatile matrix components are retained in the injection port liner rather than the column stationary phase, so a "cleaner" chromatogram is obtained. A potential disadvantage of split SFE–GC is that the extract is divided in the injection port; a fraction of the extract enters the column while the remainder is flushed out the split vent, so the technique has a factor of 10–100 (depending on the split ratio) lower detection limit than on-column SFE–GC, although, in practice, this is rarely a drawback.

The composition of the sample matrix may also influence the choice of coupling technique. A cool on-column SFE–GC system is ideal for samples which contain thermally labile components, as the restrictor and injection port are not heated [2,3]. However, samples which contain a high percentage of extractable components, (e.g., elemental sulphur [65] or lipids and waxes [66]) are not suitable for this technique, because these analytes have a tendency to precipitate and plug in the unheated restrictor. In these instances, split SFE–GC [59–61] or heated on-column SFE–GC [56,57] are the more appropriate methods of analysis as the injection port and/or restrictor are heated to ~250–300°C and, at these temperatures, restrictor plugging is completely eliminated.

The presence of water in the sample may also create problems in both split and on-column SFE–GC. Most environmental samples contain at least trace quantities of water which, when extracted, can freeze and plug in the cooled GC column, because during the SFE step the column is usually kept below 0°C to efficiently trap the analytes. Heating the restrictor will ensure that a constant extraction flow rate is maintained with a wet sample matrix. However, this will not stop the extracted water freezing and plugging the top of the column. If the column becomes plugged during the extraction, the further transfer of analytes from the extraction cell to the GC column is prevented and only qualitative, rather than quantitative, results are obtained. To ensure that the extracted analytes are continuously transferred to the cooled GC column throughout the entire extraction, the sample can be placed on a bed of drying agent, so that during the SFE step, the water is selectively retained in the extraction cell and only the analytes are deposited in the column [55]. Alternatively, the column can be maintained above 0°C [61], but here the chromatographic trapping and focusing of the volatile analytes (e.g., C_4 to C_8 n-alkanes) is greatly reduced [55]. Problems have also been encountered when analyzing animal tissue such as crabmeat containing a high percentage of extractable fat which, if deposited on the column, can cause gross contamination of the GC system. By using a basic alumina sorbent, the fat can be selectively retained during the SFE step so that only the analytes of interest, namely PCBs, are deposited in the column stationary phase [66].

The maximum extraction flow rate that can be accommodated by the GC system will depend on whether all (on-column) or part (split) of the extract is transferred into the chromatographic column. If all the gaseous extraction fluid is vented through the column (e.g., in sealed on-column SFE–GC which involves a gas-tight seal between the restrictor and column, see Fig. 3), then the maximum practical flow rate will depend on the inner diameter of the GC column. For example, when a narrow-bore (0.25 mm i.d.) column is used, low gaseous extraction fluid flow rates of ~5–10 mL/min are employed, as the small chromatographic void volume of the system limits the amount of fluid which can be vented through the column [57]. If higher flow rates are attempted with the narrow-bore column, a back pressure builds up in the GC system and this will lead to poor chromatographic peak shapes and possible failure of the GC fittings, which are not designed to withstand high pressures. By increasing the inside diameter of the column to 0.32 mm, higher gaseous flow rates of 30–80 mL/min can be used to efficiently focus the analytes [56]. With a conventional on-column injection port (Fig. 3), gaseous extraction fluid flow rates of up to ~350 mL/min can be used because there is no gas-tight seal between the restrictor and the injection port, and the effluent can vent back through the injection port and, to a lesser extent, through the column [3]. However, the highest flow rates are achieved with a split SFE–GC interface as the majority of the depressurised extraction fluid is vented through the split vent (split vent typically 100 : 1) so gaseous effluent flows of ~550 mL/min can be routinely used [55,61].

The flow-rate limitations of the various SFE–GC techniques also limit the maximum possible sample size, because the extraction flow rate can determine the time required to obtain quantitative recoveries. The optimum extraction flow rate will depend on the kinetics and mechanism of the extraction, the void volume of the SFE system, and the time allocated to the SFE step (typically 30 min). For many environmental samples the flow rate has a minimal effect on the analyte extraction rate as the rate-limiting step in SFE is the slow kinetics of the partitioning process between the sample matrix and the extraction fluid, and the diffusion of the analytes to the sample matrix surface [67,68]. However, the flow rate does affect the time taken to expel the void volume from the extraction system and this can have a significant effect on the efficiency of the SFE–GC system. For example, typical void volumes in a packed extraction cell are ~20–80%, so several cell volumes of supercritical fluid would be required for a quantitative extraction. If the extraction is to be completed in 30 mins, then a sealed on-column SFE flow rate of ~60 µL/min (~35 mL/min gaseous extraction fluid) involves only a total supercritical extraction fluid volume of ~1.8 mL, for extraction and transfer of the analytes into the GC system. This small extraction volume limits the sample size of sealed on-column SFE–GC to ~20–50 mg and requires the use of microextraction cells with small internal volumes (typically < 0.5 mL). Extraction cells with a larger volume can be used by periodically sampling the cell with a rotary valve so that a "heart cut" of the extract is transferred into the column [69]. This approach can also be used to obtain class-selective extracts such as the selective determination of alkane, alkene, and aromatic components in gasoline [70], although quantitation is difficult.

Conventional on-column injection ports enable higher extraction flow rates of ~0.6 mL/min (~350 mL/min gaseous extraction fluid) to be used so that larger sample sizes of up to 1 g can be investigated. However, the most versatile system is the split SFE–GC technique which can accommodate SFE flow rates of up to ~2 mL/min (~1100 mL/min gaseous effluent) [2,61] and analyze sample sizes up to 15 g [61]. Though the analyte concentration in the sample matrix must also be taken into account, for if the sample is highly contaminated, then only a small sample size can be used. For example, only 50 mg of a gasoline-contaminated charcoal filter containing 17 wt% extractable hydrocarbons could be analyzed, to avoid gross overloading of the column stationary phase, even though a high (100 : 1) split ratio was used during split SFE–GC analysis [71]. All of the coupled SFE–GC techniques can therefore use much smaller samples sizes (typically 0.1–1 g) than conventional solvent extraction methods (typically 25–100 g), but for the on-line methods, sample homogeneity is of even greater importance.

V. CONSTRUCTION OF SFE–GC INSTRUMENTATION

When constructing a SFE–GC system it is best to keep the apparatus as simple as possible so that contamination and void volume problems can be minimized. The main components of the system are illustrated in Fig. 4 and are described below: a reciprocating or syringe pump capable of pressures from 100 to 400 bars and delivering a constant flow rate from μL/min to mL/min; packless on–off valves to avoid contamination by lubricants and extractable species from the components such as the O-rings; high-purity supercritical fluids (e.g., SFE–SFC-grade CO_2), as previous studies have shown that impurities in the extraction fluid can cause artifact peaks in the SFE–GC-generated chromatograms [49,58,72]; transfer tubing constructed from narrow-bore 316 stainless steel (1/16 in. o.d. × 0.01 in. i.d.) fitted with zero dead volume metal-to-metal connections to reduce the void volume in the system (Teflon tape should be avoided due to potential contamination problems); a small 316 stainless-steel extraction cell (e.g., 0.1–1.0 mL) designed to withstand working pressures of typically 400 atm; fine frits (e.g., 2 μm mesh size) placed at either end of the extraction cell to retain the sample inside the vessel during the extraction; silanized glass beads (100 μm o.d.) added to either end of the sample to reduce cell dead volume and prevent outlet frit blockage (beads can also be mixed with the sample matrix to avoid sample compaction and the formation of an impervious plug [73]); finger-tight fittings to

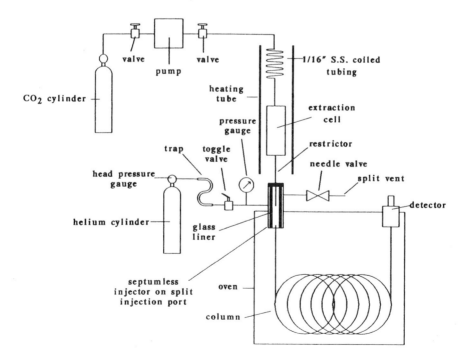

Fig. 4 Schematic of General SFE–GC Equipment.

connect the cell to the extraction system, as these fittings are much less prone to leaking despite extensive usage; a thermostatically controlled oven or heater (constructed from a steel pipe, a thermocouple, heating tape, and a temperature control unit) to heat the extraction cell; and a narrow-bore fused silica restrictor (i.e., 15–30 µm i.d.) [2,3] or crimped stainless tube (i.e., 0.01 in. i.d.) [62,70] to control the flow rate and, indirectly, the pressure inside the extraction system.

The internal diameter of the fused silica restrictor controls the rate at which the supercritical fluid can flow through the extraction cell: the larger the diameter, the higher the flow. For example, typical extraction conditions of 400 atm CO_2 at 40°C and a 10 cm, 20 µm i.d. fused silica restrictor result in a flow rate of ~0.4 mL/min (measured as liquid CO_2 at the pump), whereas a restrictor with a 30 µm i.d. provides a flow rate of ~1 mL/min of liquid CO_2. In general, larger-internal-diameter restrictors yield an initially faster extraction as the void volume of the extraction system is more rapidly transferred to the GC system, although the amount of analyte extracted decreases with time, as other rate-limiting steps such as the kinetic of the extraction become more significant. Consequently, there is a trade-off between optimizing the extraction flow rate to efficiently extract and transfer the analytes to the column, and the flow rate which can be accommodated in the GC system to enable well-resolved chromatograms to be obtained.

Once assembled, the SFE system can quickly be connected to the GC apparatus. The easiest and most reliable approach is to use split SFE–GC, as only a few minor modifications (depending on the gas chromatograph) need to be made to the GC instrumentation. The principal steps involve the insertion of the extraction cell restrictor into the split injection port so that the end of the restrictor is about 3 cm above the GC column inlet. In contrast, if a commercial SFE system is used, a heated transfer line is usually required to achieve the coupling due to the greater separation of the SFE and GC systems. However, if a homemade system is used, the extraction cell and heater can be mounted directly above the injection port (as shown in Fig. 4) so that over 90% of the restrictor can be inserted into the heated injection port liner, and no additional transfer line is required. This SFE–GC coupling can be laboriously achieved by loosening the injection port septum nut and carefully pushing the restrictor through the septum, and then retightening the septum nut. Unfortunately, to undertake this operation, the injection port must be cooled and reheated each time a restrictor is changed, so a delay of up to 60 min can be involved. A more convenient alternative is to use a septumless injector (e.g., Scientific Glass Engineering model SLI-M) which can be installed on a conventional injection port and can be used for both GC and SFE–GC analysis. With this system, the restrictor is passed through a cylindrical seal which can be opened and closed by simply pressing a lever, so that the coupling of the SFE and GC systems can be achieved within seconds.

With the restrictor installed inside the injection port liner, the split ratio is adjusted using the needle valve on the split vent (see Fig. 4). It is important to note that the restrictor must form a gas-tight seal with the injection port in order to maintain a constant split ratio throughout the extraction. Problems can arise if a mass flow controller (controlling the carrier gas flow rate) and a back-pressure regulator (controlling the column head pressure) are part of the GC instrumentation [55]. With this type of GC system, the split ratio is controlled by the amount of carrier gas supplied by the mass flow controller, and the back-pressure regulator allows the split flow to exit while still maintaining a constant column head pressure. Under SFE conditions, the high gaseous extraction fluid flow rates in the injection port can cause the head pressure regulator to open or close during the SFE step (in an attempt to maintain its set-point pressure), thus varying the split ratio. Although this does not affect the qualitative peak shapes, a change in split ratio during the SFE step will yield poor quantitative results [55]. In an attempt to maintain the split ratio, the mass flow controller and back pressure regulator have been isolated and replaced with a simple needle valve installed on the split vent and a manual head pressure regulator situated on the carrier gas inlet line (see Fig. 4) [55]. This simplified system used by many GC manufactures is easy to maintain and clean and results in good peak area reproducibility.

One additional modification is required regardless of the GC system used: a shut-off or toggle valve located prior to the injection port (Fig. 4). This valve is closed during the SFE step to prevent the SFE effluent "back flushing" into the carrier gas line and contaminating the carrier gas. This can be a problem in SFE–GC, as the internal volume of the vaporizing chamber (i.e., the glass liner) inside the injection port is relatively small (~0.5 mL) compared to the gaseous flow of the supercritical fluid (~250 mL/min of gaseous CO_2 at a typical CO_2 liquid flow rate of 0.5 mL/min). The pressure inside the injection port during the extraction can therefore be higher than the carrier gas head pressure. A glass liner with a larger void volume can alleviate the problem, but a shut-off valve is still recommended.

With the apparatus assembled, the following SFE-GC procedure is recommended:

1. Cool the GC oven to the appropriate cryogenic trapping temperature (e.g., 25°C to –50°C) using liquid nitrogen, liquid carbon dioxide, or ambient air.
2. Place the extraction cell into the heater above the injection port. Shut off the GC carrier gas and insert the restrictor into the injection port. The position of the restrictor should correspond to the syringe needle length appropriate for that injection port.
3. Pressurize the heated extraction cell and extract the sample for 10–30 min. A fraction of the gaseous extraction entering the injection port goes into the GC column for focusing in the stationary phase; the remainder of the extract is flushed out the split vent.
4. At the end of the extraction, the restrictor is withdrawn from the injection port and the extraction fluid allowed to dissipate from the GC system (this usually takes ~1 min).

5. Switch on the carrier gas and commence the GC analysis, rapidly heating the column to the initial starting temperature, after which the GC analysis is conducted in the usual manner.

VI. OPTIMIZATION OF SFE–GC CHROMATOGRAPHY

A. Extraction Flow Rate

The ability of SFE–GC to yield good peak shapes depends on the capability of the coupling method to efficiently collect and focus the analytes during the extraction step. This is dependent on several factors: the extraction flow rate; the column trapping temperature; and the type of column used. Previous studies have shown that successful SFE–GC is dependent on the extraction flow rate and extraction time [2,3,61]. High extraction flow rates may be desirable, as the potential sample size can be increased (e.g., to several grams) and the extraction time decreased [2,61], although additional factors such as the kinetics and mechanism of the extraction may also affect the extraction rate [47]. The impact of the extraction flow rate on the chromatography of the analysis can be investigated by comparing conventional split injections of a text mix, containing analytes with a broad range of volatilities [e.g., C_4 to C_{20} n-alkanes and benzene, toluene, ethylbenzene, and xylene (BTEX)], with the split SFE–GC analysis of the same test mix spiked onto an SFE compatible matrix (e.g., a sorbent resin such as Tenax–TA). Figure 5 shows the effect of the SFE flow rate on the peak shapes obtained by split SFE–GC using a thick film (30 m × 0.32 mm i.d., 5 µm film thickness) column, cryogenically cooled to –50°C [55]. At extraction flow rates of 0.6 mL/min (measured as liquid CO_2 at the pump, which corresponds to a gaseous flow rate of ~330 mL/min), good chromatographic peak shapes are obtained which are indistinguishable from those in a conventional split GC injection, indicating that SFE–GC does not introduce any significant splitter discrimination. However, differences in the absolute peak intensities do occur, with peak areas increasing with increasing extraction flow rate (Fig. 5). This is related to changes in the split ratio discussed later.

The symmetry or shape of the peaks obtained by SFE–GC at moderate extraction flow rates (0.2–0.6 mL/min) compare favorably to those obtained by split

---

Fig. 5 Effect of SFE extraction flow rate on the peak shapes of BTEX and C_4 to C_{20} n-alkanes obtained by split SFE–GC. The hydrocarbon mixture was either injected onto the capillary column (top chromatogram) or was spiked onto Tenax–TA and extracted on-line for 10 min with 400 atm, 60°C, CO_2 at various extraction fluid flow rates. The corresponding split ratios are given in Table 3. The 30 m × 0.32 mm i.d. (5 µm film) DB-1 capillary column was kept at –50°C during the injection or extraction step. After each extraction or 10 min after the injection, the GC oven was heated at ~50°C/min to 40 °C and then at 8°C/min to 300°C. (From Ref. 55.)

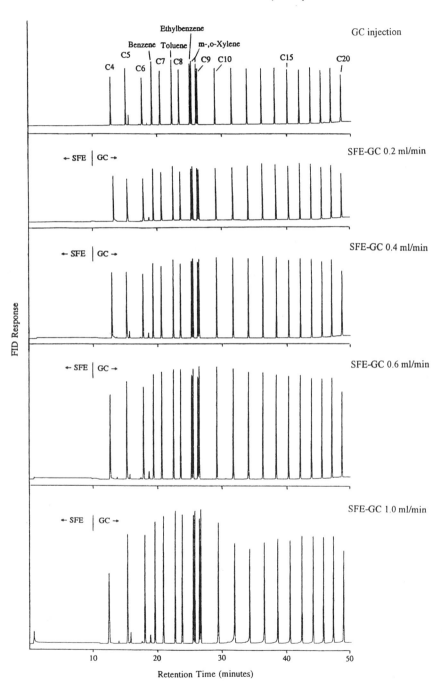

injection. As shown in Table 2, the peak width at half-height for both the very volatile (*n*-butane) and semivolatile (*n*-eicosane) analytes analyzed by SFE–GC are essentially identical to those obtained in a conventional solvent injection of the test mix. The slight band broadening of the early eluting peaks is a common occurrence and is to be expected with the coupled technique, as these are from the most volatile analytes, which are the hardest to chromatographically focus. The reproducibility of the SFE–GC and GC techniques are comparable, both methods having low relative standard deviations (RSD) of less than 5% (Table 2). This is a particularly striking value when it is considered that the RSDs of the SFE–GC method include all the possible errors associated with spiking procedure, the extraction of the analytes, the chromatographic focusing of the extract, the GC separation, and the chromatographic peak integration. Conversely, conventional solvent injection involves only the latter two variables and variations in the injection volume.

At high extraction flow rates (e.g., 1.0 mL/min liquid $CO_2 \equiv 550$ mL/min gaseous CO_2), SFE–GC can produce poor chromatographic peak shapes, with broadening of early eluting peaks and fronting of later eluting peaks (Fig. 5). One might initially conclude that the poor focusing of the analytes is due to the high gaseous extraction fluid flow entering the injection port during the SFE step, but this is not the case, as higher injection port flow rates can be tolerated under normal GC conditions (Table 3). It is more likely that poor peak shapes are associated with the high volumetric flow rate through the chromatographic column. For example, peak fronting occurs when the gaseous CO_2 flow rate through a 30 m × 0.32 mm i.d. column exceeds ~9 mL/min during the extraction (Table 3, Fig. 5). By increasing the split ratio so that more of the CO_2 can be vented

Table 2 Comparison of the Half-Height Peak Widths for a Volatile Analyte Test Mix Obtained by Split SFE–GC and Conventional GC Solvent Injection

Analyte	Peak width at half-height (s)[a]	
	Split injection	SFE–GC
n-Butane (C_4)	0.96(3)	1.18(3)
n-Hexane (C_6)	1.03(2)	1.09(4)
Benzene	0.97(2)	1.06(3)
Toluene	0.97(2)	1.04(3)
Ethylbenzene	0.97(2)	1.01(2)
m-Xylene	0.97(2)	0.99(1)
n-Decane (C_{10})	0.98(1)	0.97(3)
n-Pentadecane (C_{15})	0.95(3)	0.95(4)
n-Eicosane (C_{20})	1.09(2)	1.12(5)

[a]Values in parentheses are the percent relative standard deviations of triplicate 10 min extractions or triplicate manual solvent injections [55].

Table 3 Split Ratio Measured Under GC and Split SFE–GC Conditions

Analysis	Restrictor i.d. (μm)	SFE flow rate[a] (mL/min)	Column head pressure (psi)	Column volumetric flow[b] (mL/min)	Split vent volumetric flow[b] (mL/min)	Split ratio (column : split)
GC	—	—	15	5.1	789	1 : 155
Split SFE–GC[c]	15	0.18	1	0.8	86	1 : 107
Split SFE–GC[c]	22	0.38	4	2.8	267	1 : 95
Split SFE–GC[c]	25	0.58	9	5.0	400	1 : 80
Split SFE–GC[c]	30	0.96	16	9.2	577	1 : 63

[a]Flow rate measures as liquid CO_2 at the pump.
[b]Flow rate measured as volume of gas using a bubble flow meter.
[c]SFE–GC conditions: 400 atm, 60°C CO_2 depressurized into a thick phase (30 m × 0.32 mm i.d., 5 μm) column at −50°C.
Source: from Ref. 55.

through the split vent and less through the column, peak broadening and peak fronting can be eliminated even though the SFE flow rate and hence the total gas flow through the injection port remains the same [55]. The best approach for SFE–GC is to use the highest SFE flow rate which will result in good extraction efficiencies in a reasonable time and yet be compatible with the GC system to maintain good peak shapes. For split SFE–GC, the maximum practical SFE flow rate depends on the split ratio; that is, higher split ratios (and thus lower column flows) allow higher SFE flow rates. However, a point is reached where the split ratio required to give the appropriate gaseous extraction fluid column flow rate will be so large that splitter discrimination may become a problem.

Although the use of the appropriate split ratio will ensure that the analytes are adequately focused at the top of the column, it can be seen that as the extraction flow rate is increased, the chromatogram starts to "spread out" and the retention time of the volatile analytes decrease and the analyte peak widths increase (Fig. 5). This suggests that at the high extraction flow rates, the analytes are less focused in the analytical column. A short (<1 m) retention gap of deactivated fused silica can be used as a method of obtaining good peak shapes, as it enables the analytes to be refocused in the chromatographic column at the start of the GC analysis [40].

For high extraction flow rates (>1 mL/min), several minor modifications need to be made to the GC apparatus. For example, even though most of the SFE effluent (typically 95–99%) is vented through the split vent, the gas flow through the column may still be sufficient to extinguish a FID flame. Therefore, the hydrogen flow may need to be increased to sustain the detector signal. The mass spectrometer (MS) is also sensitive to the extraction flow rate, as high flow rates in the column can cause excessively high back pressures (e.g., 10^{-4} torr) in the MS ion source. However, this can be avoided in both on-column and split SFE–GC by either using a longer and narrow GC column (e.g., a 60 m × 0.25 mm i.d. column instead of a 30 m × 0.32 mm i.d. column), or by placing a 1 m section of a smaller-diameter (e.g., 0.15 mm i.d.) transfer line between the column and the ion source [2]. For split SFE–GC, the use of a narrower column or transfer line results in an increased split ratio and, thus, poorer detection limits. It is therefore more convenient to adjust the split ratio until an acceptable MS ion source pressure is obtained.

B. Column Trapping Temperature

The cryogenic trapping temperature employed during the extraction step will also affect the ability of SFE–GC to efficiently focus the analytes, particularly the volatile components. Figure 6 shows the effect of the cryogenic trapping

---

Fig. 6 Effect of the cryogenic trapping temperature on the retention of BTEX and C_4 to C_{20} n-alkanes on a thick film (5 μm) 30 m ×0.32 mm i.d. DB-1 capillary column during split SFE–GC analysis. The hydrocarbon mixture was extracted from Tenax–TA using 400 atm, 60°C, CO_2 at 0.6 mL/min (fluid flow) for 30 min. After each extraction, the GC oven was heated at ~50°C/min to 40°C and then at 8°C/min to 300°C. (From Ref. 55.)

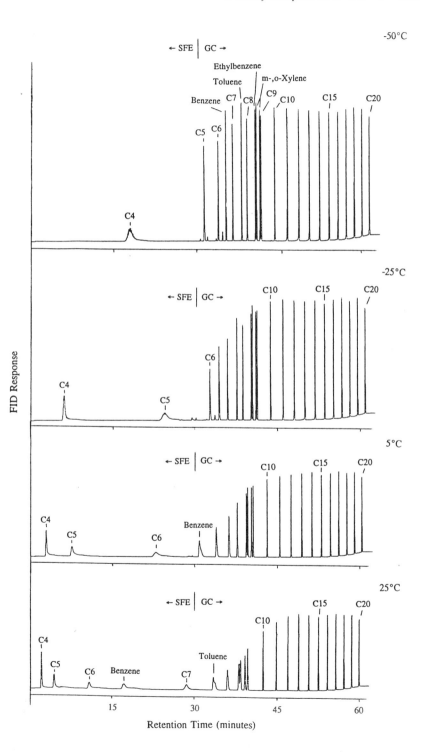

temperature on the chromatographic peak shape of a test mix (representing the major components in gasoline and diesel fuel) analyzed by split SFE–GC using identical extraction conditions (e.g., 0.6 mL/min liquid CO_2) and a 30 m × 0.32 mm i.d., 5 μm film-thickness column [55]. To efficiently retain the volatile analytes in the C_4 to C_9 n-alkane boiling-point range, the column was cryogenically cooled to ~ –50°C. However, even at this very low trapping temperature, the most volatile analyte, n-butane, is not indefinitely retained in the column stationary phase, as the analyte will pass through the column when the extraction time is increased from 10 (Fig. 5) to 30 (Fig. 6) mins. The rest of the analytes (e.g., C_5 to C_{20} n-alkanes) are efficiently focused over at least a 60 min extraction time (Fig. 7). It is not possible to increase the trapping efficiency of the column by using a lower trapping temperature (e.g., <–50°C), as the column stationary phase is a solid below –60°C [74,75], so the column will become more characteristic of a retention gap because analytes can no longer partition into the stationary phase. Furthermore, at ~ –78°C, the CO_2 freezes and plugs the column.

When the cryogenic trapping temperature is raised from –25°C to 25°C, the trapping efficiency of the column stationary phase correspondingly decreases, and an increasing number of analytes pass through the column during the extraction step (Fig. 6). The choice of cryogenic trapping temperature will depend on the analyte composition. For example, a sample with gasoline range organics (defined as compounds in the C_6 to C_{10} boiling-point range [8]) will require a column trapping temperature of –50°C to –25°C, whereas a kerosene-contaminated sample (C_9 to C_{16} boiling-point range) requires a less rigorous procedure with a trapping temperature in the region of 5°C. It is interesting to note that with no cryogenic cooling during the extraction and the column maintained at room temperature (25°C), analytes as volatile as n-decane can still be efficiently retained. Peak shapes of the retained analytes are generally broader for more volatile species (e.g., toluene to n-nonane), but less volatile analytes (e.g., C_{10} to C_{20} boiling-point range) have very good chromatographic peak shapes, so that the 25°C trapping temperature is suitable for the retention of analytes in the diesel range (C_{10} to C_{25}).

Using column trapping temperatures above 0°C may be advantageous, as wet samples can be analyzed without the problem of extracted water freezing and plugging the column. However, if the wet sample contains volatile analytes, then quantitative analysis of the whole extract is no longer possible, as the

---

Fig. 7 Effect of capillary column stationary-phase thickness on the chromatographic peak shape of BTEX and n-alkanes during split SFE–GC. The hydrocarbon mixture was extracted from Tenax–TA using 400 atm, 60°C, CO_2 at 0.6 mL/min (fluid flow) for 60 min. The extracted analytes were cryogenically focused onto a 5 mm film DB-1 capillary column (30 m × 0.32 mm i.d.), 1 mm film DB-5 capillary column (30 m × 0.32 mm i.d.), or a 0.25 mm film DB-5 capillary column (20 m × 250 mm i.d.). After each extraction, the GC oven was heated at ~50°C/min to 40°C and then at 8°C/min to 300°C. (From Ref. 55.)

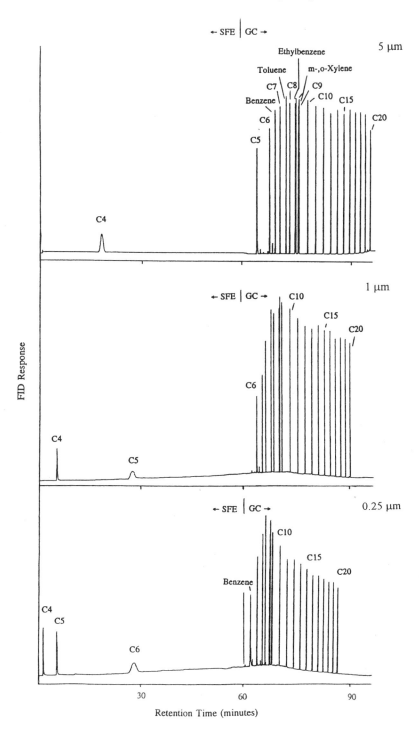

volatile fraction will have eluted through the column before the GC analysis has commenced. An alternative procedure, as mentioned earlier, is to place the sample on a bed of drying agent inside the extraction cell, so that the water is selectively retained by the reagent inside the cell while the analytes of interest are transferred to the GC column. The most popular drying agents in SFE with a high water capacity but low affinity for the analytes of interest are hydromatrix (diatomaceous earth) [76,77], anhydrous magnesium sulphate [15,16], and molecular sieves 3A and 5A [73]. The use of molecular sieve 3A has enabled very wet (25% w/w water) gasoline-contaminated samples to be successfully analyzed by split SFE–GC at a low cryogenic trapping temperature of –50°C, the problems of frozen water in the GC system having been eliminated [71]. To ensure that the drying agent is efficiently retaining the water, it is suggested that the column gas flow rate is monitored during the entire extraction. This is easily achieved by switching off the detector and placing a gas flow meter at the end of the column, and if FID is used, this simply means inserting an appropriate adapter in the cooled detector and measuring the flow directly.

Monitoring the column flow during the SFE step is a very important and yet often underestimated procedure. Poor SFE–GC quantitation and reproducibility are often blamed on inefficient column analyte trapping or irreproducible split ratios, when, in fact, the problem is with frozen water or extracted analytes slowly building up at the head of the column, so that during the extraction, the column flow rate decreases. For example, at very low cryogenic trapping temperatures (e.g., –50°C), a diesel and/or motor oil extract has a "waxy" consistency which causes the column to become plugged on commencing the extraction. However, a continuous column flow is achieved if a less severe cryogenic trapping temperature of –25°C is used [71]. Similar plugging problems have been encountered when analyzing plant material containing cuticle waxes and these problems have been overcome by increasing the column trapping temperature from –50°C to –30°C [63]. It is often preferable to start a SFE–GC investigation using a –25°C cryogenic trapping temperature, as this experimental condition allows a wide range of analytes to be correctly focused and retained while still ensuring a continuous gaseous extraction flow rate through the column.

C. Column Stationary-Phase Thickness

Another important experimental parameter is the capillary column stationary-phase film thickness, for as the film thickness increases, the ability of the column to retain and focus the analytes during the extraction also increases. For example, Figure 7 shows the split SFE–GC analysis of a volatile analyte test mix using thick (5 µm), medium (1 µm), and thin (0.25 µm) film columns. The results demonstrate that the column with the thickest stationary phase is the most efficient in retaining the analytes, because the 5 µm column is able to give good peak shapes for analytes as volatile as n-pentane, even after 60 min of SFE [55].

When chromatographic columns are used with thinner stationary-phase films, the trapping efficiency of the SFE–GC decreases and more analytes pass through the column during the SFE step (Fig. 7). The advantage of using a thinner stationary-phase film is that a wider boiling-point range of components can be resolved. For example, the thick-film 5 μm column can resolve analytes up to the C_{25} boiling-point range in a typical 30 min GC run, but the thin-film 0.25 μm column can resolve analytes up to ~C_{40} in the same analysis time. It follows that the choice of stationary phase will depend on the requirements of efficient trapping of the volatile analytes during the extraction step and the ability to elute the semivolatile analytes during the GC analysis.

The importance of selecting the appropriate column stationary-phase film thickness can be demonstrated with the analysis of fuel-contaminated samples. A thick-film (5 μm) column was used in split SFE–GC to investigate both a kerosene- and motor-oil-contaminated sediment (Fig. 8). For both samples, quantitative fuel recoveries were possible, as the hydrocarbon recoveries from a 15 min SFE–GC extraction were comparable to those obtained by sonicating the samples in methylene chloride [71]. By rerunning the GC analysis without an extraction or injection (i.e., a blank run) it can be seen that the kerosene extract had completely eluted from the thick-film column. However, this was not the case for the motor oil extract, as the less volatile analytes accumulated on the column to produce a rising baseline during subsequent GC analysis (Fig. 8). If several motor oil samples were investigated using the thick-film column, then poor quantitative results would have quickly resulted. The degradation of the column performance can be rectified by trimming ~15 cm off the injection end of the column. A more practical alternative is to use a thin-film (0.25 μm) column which can efficiently elute all the components in the sample.

VII. QUANTITATIVE CONSIDERATIONS FOR SFE–GC

After optimization of the chromatographic conditions, a quantitative assessment of the SFE–GC technique can be undertaken. The ability of the on-line method to enable a quantitative and reproducible transfer of the analytes from the SFE system to the GC column has been investigated using both spike recoveries [1,2,51,55,64] and comparisons of SFE–GC recoveries with conventional liquid-solvent extraction [1,3,10,11,61]. Ideally, a calibration method is required in which the calibration standard is introduced into the GC system in the same way as the real sample (i.e., through the SFE restrictor). This would enable the SFE–GC collection parameters to be investigated independent of the SFE extraction efficiencies. The calibration standard should consist of a matrix spiked with the analytes of interest and should be easy to prepare, convenient to handle, and stable. Previous off-line SFE collection efficiency studies have used an inert matrix such as silanized glass beads [68,78], but this unretentive matrix is unsuit-

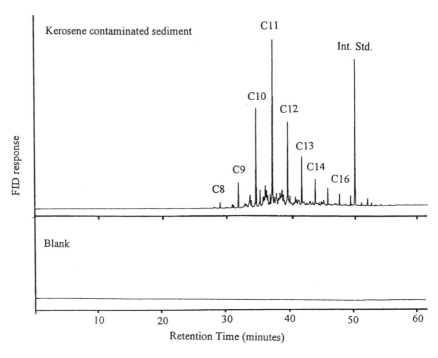

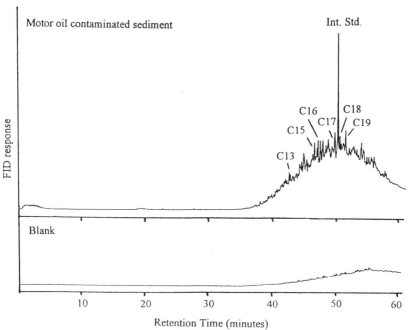

able for use with the volatile analytes due to analyte losses occurring during the spiking process through volatilization [71]. Conversely, sorbent resins such as Tenax–TA have proved to be ideal matrices for a calibration standard, as volatile and semivolatile analytes have large gas retention volumes on the resin at ambient conditions [52], yet analytes can be quantitatively recovered by SFE [10,51,55]. As shown in Table 4, quantitative SFE–GC recoveries for a wide range of analytes (e.g., C_4 to C_{20} n-alkanes) can be achieved by a 10 min supercritical fluid extraction of spiked Tenax–TA. In addition, the sorbent also has a good "shelf life" as the analytes are retained by the Tenax for at least 24 h at room temperature and pressure (Table 4). This enables the preparation of several calibration standards which can be stored for use throughout the working day.

Depending on the method of sample introduction, either an internal or external calibration method can be used to quantify the analyte recoveries achieved by SFE–GC. For sealed on-column SFE–GC, both calibration methods are possible, as all the analytes are transferred into the GC column, because a gas-tight seal exists between the column and the syringe needle or SFE restrictor. On-column SFE–GC peak areas can therefore be compared to those obtained with either a conventional GC injection of the external calibration standard or can be compared to the internal standard which is added to the sample matrix and coextracted with the sample analytes for GC analysis. However, the use of an external calibration method is often not a suitable method for split SFE–GC, as the on-line split ratio can significantly differ from that of conventional GC. For example, as shown in Table 3, a split ratio of ~155 : 1 during GC analysis decreases to ~65 : 1 under split SFE–GC conditions, even though the needle valve (which controls the split) has not been adjusted and the pressure inside the injection port is similar [55]. Therefore, to quantify the split SFE–GC analysis, an internal calibration method is recommended, as any variation in the split ratio between the techniques will not affect the analyte response factors (e.g., area ratio of analyte to internal standard) [71] (see Table 5). However, an external calibration method can be used if the analytes are not introduced into the GC using a conventional GC injection but are introduced under the same conditions as the extracted analytes (i.e., through the SFE restrictor) and this is achieved by using a Tenax sorbent resin

---

Fig. 8 Memory effect from low- and high-molecular-weight petroleum hydrocarbons during split SFE–GC analysis. The samples (1 g) and a piece of filter paper spiked with the internal standard (octahydroanthracene) were placed on a bed of drying agent (molecular sieve 3A) inside an extraction cell and extracted for 15 min with 400 atm, 60°C, CO_2 at 0.6 mL/min fluid flow. The extracted analytes were trapped onto a thick-film (5 μm) 30 m × 320 μm i.d. DB-1 capillary column at –25°C. After the extraction, the GC oven was heated to ~50°C/min to 40°C and then at 8°C/min to 300°C. To assess column contamination, the GC temperature program was repeated without undertaking a blank extraction. (Reproduced with permission from Ref. 71.)

Table 4 SFE–GC Recovery of a Volatile Analyte Test Mix 5 min (fresh) and 24 h (aged) After Spiking onto Several Matrices

	Percent recovery (RSD)[a]			
	Glass beads		Tenax–TA	
Analyte	Fresh	Aged	Fresh	Aged
n-Butane (C_4)	61 (20)	ND[b]	109 (6)	94 (5)
n-Pentane (C_5)	69 (18)	ND	105 (4)	109 (4)
n-Hexane (C_6)	91 (9)	ND	103 (3)	107 (3)
Benzene	93 (8)	ND	97 (3)	100 (4)
Toluene	97 (8)	ND	97 (3)	96 (6)
Ethylbenzene	101 (4)	5 (59)	96 (1)	95 (5)
m-Xylene	101 (3)	7 (65)	96 (1)	95 (4)
o-Xylene	101 (3)	8 (69)	98 (3)	95 (4)
n-Decane (C_{10})	100 (0)	52 (53)	100 (0)	100 (0)
n-Pentadecane (C_{15})	100 (5)	94 (7)	98 (2)	102 (9)
n-Eicosane (C_{20})	100 (2)	97 (7)	99 (5)	100 (11)

[a]Value in parentheses is the percent relative standard deviation of triplicate 10 min extractions using 400 atm, 60°C, CO_2.
[b]ND = Not detected.
Source: Data from Ref. 55.

spiked with the calibration standard and analyzed by SFE–GC, as discussed earlier. It should be stressed that the split ratio during the extraction or GC injection is constant, as good quantitative reproducible raw peak areas are achievable by both methods. Furthermore, no splitter discrimination between the high- and low-molecular-weight analytes is observed, as a similar chromatographic profile is obtained by both split SFE–GC and conventional GC analysis (Fig. 5).

It is thought that this difference between the GC and SFE–GC split ratios is due to significant differences in the gaseous flow profiles generated in the injection port during the analysis [55,64]. In split GC analysis, a relatively small volume of solvent is introduced into the injection port in a short period of time. Conversely, in split SFE–GC, a relatively large volume of solvent (CO_2) is continually introduced into the injector port at near-supersonic velocities [79]. During the SFE step, the highest linear velocities in the injection port will be in the center of the restrictor exit and flows will be slower toward the wall of the injection port liner (Fig. 9). As the column inlet is directly "inline" with the restrictor outlet, it seems likely that a greater proportion of the CO_2 is vented into the column than if a more laminar flow had been generated. Thus, as the SFE flow rate is increased, a greater proportion of the CO_2 enters the column, resulting in a decrease in the split ratio (see Table 3). Furthermore, if a "packed" injection port liner is used (e.g., containing glass wool or a porous glass frit) to disperse the ex-

Table 5 Comparison of Peak Area Reproducibility for Split SFE–GC Versus a Conventional Injection of a Volatile Analyte Test Mix

Analyte	Average GC area counts[a]	Average SFE–GC area counts[b]	Average GC int. stand. value[c]	Average SFE–GC int. stand. value[c]	Recovery based on raw area (%)	Recovery based on int. stand (%)
n-Butane	988088(4)[d]	1643680(5)[d]	0.88 (4)[d]	0.94(5)[d]	166	107
n-Hexane	1066732(4)	1706508(3)	0.95(1)	0.97 (3)	160	102
Benzene	1383139(4)	2060167(1)	1.23(1)	1.18 (2)	149	96
Toluene	1330885(5)	1992884(2)	1.18(1)	1.14 (2)	150	97
Ethylbenzene	1239959(4)	1856394(3)	1.10(1)	1.06 (1)	150	96
m-Xylene	1269779(4)	1902309(3)	1.13(1)	1.08 (1)	150	96
n-Decane (I.S.)[e]	1124422(4)	1753399(3)	0.00 (0)	0.00(0)	—	—
n-Pentadecane	970689(1)	1430895(5)	0.86 (3)	0.82(2)	147	95
n-Eicosane	1036660(2)	1587179(1)	0.92 (3)	0.91(4)	153	99

[a] Split ratio during GC analysis ~155 : 1.
[b] Split ratio during split SFE–GC analysis ~80 : 1.
[c] Int. stand. value = internal standardization value (ratio of peak area to internal standard area).
[d] Values in parentheses are the percent relative standard deviations of triplicate manual solvent injection of a volatile analyte test mix, or triplicate 10 min extraction of Tenax–TA spiked with a volatile analyte test mix.
[e] I.S. = internal standard.
Source: Data from Ref. 55.

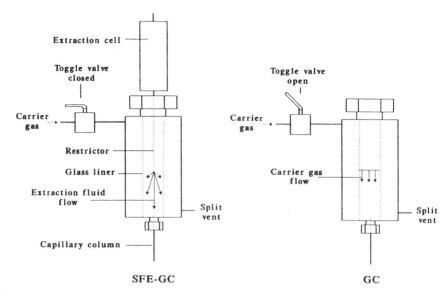

Fig. 9 Flow profile during split SFE–GC and GC analyses.

panding stream of CO_2 existing the restrictor, a split ratio comparable to that of conventional GC analysis can be achieved and raw peak areas can be obtained in split SFE–GC similar to those in GC [1,61].

VIII. OPTIMIZATION OF EXTRACTION CONDITIONS FOR SFE–GC

For quantitative recoveries in SFE–GC, the supercritical fluid must be able to solvate the analyte of interest and overcome the interactions between analyte and sample matrix, so that a favorable partitioning of the analyte into the supercritical fluid can occur. With these considerations in mind, the most popular supercritical fluid, namely CO_2, which is a nonpolar fluid, has proved to be particularly suited for the extraction of hydrocarbons, as these analytes have a fairly high solubility in CO_2. Both selective and quantitative hydrocarbon recoveries are possible by varying the temperature and pressure of the supercritical fluid [80]. The selectivity of CO_2 has been used in SFE–GC to fractionate polycyclic aromatic hydrocarbons (PAHs). As shown in Fig. 10, progressively higher-molecular-weight material was extracted as the density of the extraction fluid was increased. Low extraction pressures (e.g., CO_2 at 80 atm, 50°C) recovered two- and three-ring PAHs, whereas higher extraction pressures (e.g., CO_2 at 200 atm, 50°C) recovered predominately four-ring and larger aromatics [57]. Al-

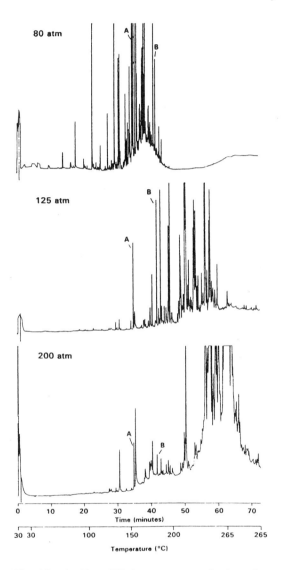

Fig. 10 Capillary GC chromatograms of polycyclic aromatic fractions obtained by on-column SFE–GC. Polycyclic aromatic compounds were extracted from glass beads at 50°C with a 0.005 mL/min fluid flow at several extraction pressures. The extracts were trapped at 30°C onto a 15 m × 0.25 mm i.d. (0.25 μm film) SE-54 capillary column fitted with a 30 cm × 0.53 mm i.d. deactivated fused silica retention gap. After the extraction, the GC oven was maintained at 30°C for 2 min then temperature programmed at 4°C/min to 265°C. Compounds A and B are marked in each fraction to facilitate comparison. (Reproduced with permission from Ref. 57.)

though some overlap of the components occurred in the different fractions, the results indicate that SFE–GC is a feasible method for undertaking a class-fractionation analysis.

Using more extreme extraction conditions (typically CO_2 at 400 atm and 50°C), quantitative SFE–GC of a wide range of analytes has been carried out for various sample matrices, including fuel-contaminated sediments [55,61], PCBs in biological samples [66], PAH from Tenax–GC [10] essential oils in plant material [63], *N*-heterocyclic compounds from cigarette smoke particulates [81], and phenolics from wood smoke particulates [3]. However, several of these "quantitative recoveries" were determined by using a second sequential SFE extraction performed under conditions identical to the first to demonstrate that no further analytes were present in the sample matrix. Although this approach may be valid for some samples, it is often more reliable to reextract the sample using more rigorous extraction conditions or to compare the recoveries to those from a conventional solvent extraction method [47]. SFE–GC analysis of certified reference material (quantitative recoveries based on an exhaustive Soxhlet extraction) has proved an ideal means of assessing the quantitative capability of the technique. Two certified reference material from the National Institute of Standards and Technology (NIST), namely a marine sediment and an urban air particulate sample, have been analyzed by both split and/or on-column SFE–GC. The results are given in Table 6. Using CO_2 at a high pressure (375 atm) and moderate temperature (50°C), quantitative SFE–GC recoveries of both low- and high-molecular-weight PAHs were achieved from a 50 mg urban air particulate sample. What is particularly interesting is that the SFE–GC analysis could be completed within the hour, compared to several days for the certified method. However, poor PAH recoveries have been obtained when a larger urban air particulate sample size is used (e.g., 320 mg [47]) and when the PAHs are present in a more retentive or "less accessible" sample matrix, such as a marine sediment [61]. Poor PAH recoveries were obtained from the sediment with CO_2 and quantitative recoveries were only achievable when a more polar fluid such as nitrous oxide (N_2O) was used (Table 6). The difference in PAH recoveries is related to the sample matrix and demonstrates the importance of the matrix–analyte interactions. The results also highlight the versatility of the coupled technique, in that other fluids besides CO_2 can be used. N_2O is ideally suited for the on-line method, as it is a gas at ambient conditions and does not give an appreciable response in the FID. However, caution must be exercised, as the fluid is an oxidant and should not be used with high organic content samples due to a possible explosion hazard. Other supercritical fluids such as sulphur hexafluoride (SF_6) have also been successfully used with split SFE–GC to obtain selective *n*-alkane extractions from a gasoline-contaminated soil [82].

Several analytes and sample matrices are not quantitatively extracted by pure supercritical CO_2 and it has often proved necessary to use a small amount

Table 6 SFE–GC Analysis of Certified Reference Material

	Concentration (μg/g)					
	Urban air particulates (NIST 1649)			Marine sediment (NIST 1941)		
PAH	Certified value[a]	CO_2[b]	N_2O[c]	Certified value[d]	CO_2[e]	N_2O[f]
Fluoranthene	7.1±0.5	7.2 ±0.9	7.3±1.0	1.22 ±0.24	1.37± 0.26	1.45±0.14
Benz[a]anthracene	2.6± 0.3	2.6±0.8	2.6± 0.8	0.55±0.08	0.38 ±0.01	0.60±0.08
Indeno[1,2,3-cd]pyrene	3.3± 0.5	3.4±0.6	3.0± 0.5	0.57±0.04	0.12 ± 0.03	0.56±0.08
Benzo[ghi]perylene	4.5± 1.1	3.9±1.0	3.6± 0.9	0.52±0.08	0.15 ±0.05	0.56±0.09

[a]NIST certified values based on 48 h Soxhlet extraction of 1 g samples using methanol or methylene chloride as the extraction solvent.
[b]Split SFE–GC analysis of a 50 mg sample using a 30 min, 375 atm, 50°C, CO_2 extraction [1].
[c]On-column SFE–GC analysis of a 2 mg sample using a 20 min, 350 atm, 45°C, N_2O extraction [2].
[d]NIST certified values based on 16-h Soxhlet extraction of 7–25g samples using both hexane and methylene chloride as the extraction solvents.
[e]Split SFE–GC analysis of a 30 mg sample using a 10 min, 400 atm, 50°C, CO_2 extraction [61].
[f]Split SFE–GC analysis of a 30 mg sample using a 10 min, 400 atm, 50°C, N_2O extraction [61].

(typically 10% v/v) of organic modifier to achieve quantitative recoveries. The addition of the modifier enhances the solvation strength of the CO_2, so that the analyte solubility and/or extraction fluid–matrix interaction are increased. The modifier is normally introduced into the SFE–GC system by using a dual-pump delivery system, a premixed cylinder, or by adding the modifier to the sample inside the extraction cell. Depending on the modifier concentration, a retention gap or thick-film capillary column is usually required to enable the solvent to be eluted from the column [82]. The use of a modifier may cause problems because organic solvents are generally not volatile under ambient conditions and appear as a large solvent peak in the chromatogram. Depending on the extract, the modifier may be eluted as a discrete peak away from the analytes of interest (see Fig. 11) and a range of solvents including methanol, hexane, carbon disulphide [83], ethanol [62], benzene, propylene carbonate [82], acetone, chloroform, methylene chloride, toluene [84], and formic acid [1] have all been successfully used in SFE–GC. To minimize the amount required, the modifier is usually added to the sample inside the extraction cell and then an initial static extraction is carried out, followed by a short dynamic extraction [82,83]. With this approach, a small (50 µL) volume of modifier was successfully used to enhance the extraction efficiency of aromatic analytes from a sludge/fly ash sample matrix [82] (Table 7). Modifiers such as carbon disulphide [83] and formic acid [1] are also of particular interest to on-line SFE–GC techniques, as these solvents have a very low response in FID and therefore have a minimum impact on the chromatography part of the method.

To increase the range of analytes which can be investigated by SFE–GC, on-line derivatization of "non-GC-able" components has been incorporated into the method of the coupled technique [85,86]. For example, a supercritical fluid reactor (SFR) consisting of a packed bed of alumina pretreated with methanol has been used in situ to transesterify extracted triglycerides from oilseeds [85]. The SFR is placed in line between the extraction cell and GC system so that during the extraction, the triglycerides are transesterified to their fatty acid methyl esters as they pass through the SFR under supercritical fluid conditions. This allows the

Table 7 Use of Modifiers in SFE–GC

Analyte	Percent recovery			
	CO_2	CO_2/10% propylene carbonate	CO_2/10% benzene	CO_2/10% methanol
Ethylbenzene	74	96	95	80
Cumene	72	96	96	79
2-Chloronaphthalene	93	93	92	82
1,2,4-Trimethylbenzene	71	96	96	84

Note: SFE step: 225 mg of sludge/fly ash, 375 atm, 50°C, CO_2, 10 min static, 7 min dynamic extraction, 50 µL modifier in a 500 µl extraction cell [82].

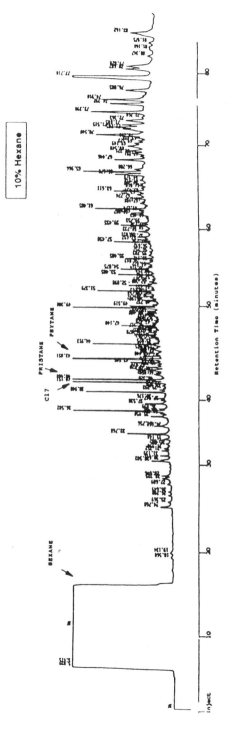

Fig. 11 Characterisation of shale rock using split SFE–GC with modifier. The sample was initially extracted statically for 30 min with 350 atm, 60°C, CO_2 and 50 μL of hexane in the 500 μL extraction cell. The sample was then dynamically extracted for 10 min using pure CO_2. The extract was trapped onto a 50 m × 0.20 mm i.d. PONA column with a temperature of 100°C (10 min) to 330°C at 5°C/min and flame ionization detection. (Reproduced with permission from Ref. 83.)

resulting methyl esters to be collected and analyzed by GC [85]. Alternatively, a derivatization reagent such as pentafluorobenzylbromide (PFBBr) with potassium carbonate can be added directly to the extraction cell, so that during the extraction polar compounds such as phenoxycarboxylic acids can be alkylated to produce esters which can be analyzed by GC [86]. In-line derivatization and reaction processes may prove to be a popular means of pretreating and analyzing a wide range of samples.

A recent development has been the use of high-temperature (e.g., 150°C) supercritical fluids such as CO_2 [87,88] or CO_2 modified with organic solvents [89] to enhance the recovery of environmental pollutants like PAHs from real-world samples. This approach, as yet, has not been extensively applied to SFE–GC, but it is envisaged that the high-temperature extraction fluid will be compatible with the current SFE–GC techniques. The increased recoveries and faster rate of extraction achieved by this approach make the use of high-temperature fluids in SFE–GC an attractive area of research for further investigation.

IX. SFE–GC APPLICATIONS

Supercritical fluid extraction–gas chromatography has been used to analyze a wide variety of analytes and sample matrices ranging from natural products such as flavor and fragrance compounds in plant material, to organic pollutants in soil, sediment, air particulate matter, animal tissue, and even water. These diverse application are shown in Table 8 and demonstrate the ability of SFE–GC to yield good chromatographic peak shapes and quantitative results. It is clear that SFE–GC is a very versatile technique which is still rapidly developing as new applications are continually being reported in the literature. The method is limited by the ability of the GC system to chromatographically resolve all the analytes in the SFE extract. However, as discussed earlier, the SFE procedure can be linked to a wide range of chromatographic and spectroscopic techniques to enable the majority of analytes to be investigated.

X. CONCLUSIONS

On-line SFE–GC is an attractive approach to coupling the extraction and analysis procedures of real-world samples and offers the analyst a rapid, sensitive, quantitative, and yet selective analytical method. A quantitative SFE–GC analysis, including the loading and assembly of the extraction cell, performing the extraction and GC analysis, can be routinely undertaken in less than 1 hr. Comparable recoveries obtained by conventional solvent extraction usually require several hours.

The optimization of the SFE–GC method can initially appear to be a complicated procedure and is inherently more involved than off-line SFE. However,

Table 8 Applications of SFE–GC

Sample matrix	Analytes	Extraction fluid	Ref.
Environmental Samples			
Fuel-contaminated sediment	Petroleum hydrocarbons	CO_2	55, 61, 82
		SF_6	82
		CO_2/5% modifier (MeOH, toluene, CH_2Cl_2, $CHCl_3$, acetone, hexane)	84
Shale rock	Petroleum hydrocarbons	SF_6	90
		CO_2/10% modifier (MeOH, CS_2, hexane)	83
Heavy hydrocarbon wax	Hydrocarbons	CO_2/10% SF_6	83
Solid waste	Hydrocarbons	CO_2/0.3% formic acid	1
Dry sludge	Hydrocarbons	CO_2/20% SF_6	90
Sludge/fly ash	Aromatics	CO_2/10% modifier (MeOH, benzene, propylene carbonate)	82
Biosludge	PAHs	CO_2	10
Lamp black	PAHs	N_2O	2
Marine sediment	PAHs	CO_2 and N_2O	61
Marine sediment	PCBs	CO_2	93
River sediment	PAHs, alkanes	CO_2	61
River sediment	PCBs	CO_2/5% modifier, (acetonitrile CS_2, propylene carbonate)	83
Sediment	PCBs	CO_2/2% MeOH	58
Sediment	Purgable halocarbons	CO_2	29
Urban air particulates	PAHs	CO_2	1
		N_2O	2
Wood smoke particulates	Guaiacol and syringol derivatives	CO_2	1
Cigarette smoke particulates	Phenols, nicotine, N-heterocycles	CO_2	56, 81
		N_2O	2
Polyurethane foam (PUF) diesel exhaust, coal gasification water	Alkanes, alkyl benzenes, PAHs, phenols, N-heterocycles	CO_2	81, 61
Spiked pesticides, herbicides, HCB, PCBs	Pesticides, herbicides HCB, PCBs	CO_2	49, 51
Herbicides on C_{18}	Herbicides	CO_2/derivatization	86
Organic solvents on Tenax–TA	Hazardous organic solvents	CO_2	71
Water	Water contaminants	CO_2	91

Table 8 Continued

Sample matrix	Analytes	Extraction fluid	Ref.
Plant and Animal Tissue			
Crab and cod tissue	PCBs	CO_2	66
Fir needles	Essential oils	CO_2	56
Thyme (*Thymus vulgaris* L.)	Essential oils	CO_2	50
Eucalyptus leaves	Essential oils	CO_2	1
Broccoli	Pesticides	CO_2	90
Oilseed	Triglycerides	CO_2/derivatization	85
Brewing hops	Essential oils	CO_2	61
Lemon peel	Essential oils	CO_2	3
Spices (e.g., rosemary, cloves thyme, cinnamon, oregano, basil, allspice)	Essential oils	CO_2	63, 81, 92

the utilization of the GC column for analyte collection and focusing has sufficient advantages over off-line techniques to warrant the extra time and effort in assembling the on-line system. By optimizing the most important experimental parameters (namely the SFE conditions, the cryogenic trapping temperature, and the column stationary-phase thickness), quantitative SFE–GC results can quickly be achieved. On-column SFE–GC is recommended when maximum sensitivity is required, whereas split SFE–GC is more suitable for larger sample sizes and high analyte concentrations. External analyte traps are advantageous when the sample requires the use of modified supercritical fluids.

Supercritical fluid extraction–gas chromatography unquestionably offers an excellent alternative to conventional sample preparation and has the potential to become a routine problem-solving tool for the analytical chemist. It is envisaged that the technique will continue to be developed and become a more reliable, simple, and automated analytical procedure which could realistically replace purge-and-trap analysis of volatiles and liquid-solvent extraction of semivolatiles.

REFERENCES

1. J. M. Levy, R. A. Cavalier, T. N. Bosch, A. M. Rynaski, and W. E. Huhak, *J. Chromatogr. Sci.*, 27: 341 (1989).
2. S. B. Hawthorne and D. J. Miller, *J. Chromatogr.*, 403: 63 (1987).
3. S. B. Hawthorne, D. J. Miller, and M. S. Krieger, *J. Chromatogr. Sci.*, 27: 347 (1989).

4. S. Ramalho, T. Hankemeier, M. de Jong, U. A. Th. Brinkman, and R. J. Vreuls, *J. Microcol. Sep., 7:* 383 (1995).
5. J. Staniewski and J. A. Rijks, *J. High Resolut. Chromatogr., 16:* 182 (1993).
6. F. D. Rinkema, A. J. H. Louter, and U. A. Th. Brinkman, *J. Chromatogr., 678:* 289 (1994).
7. K. Grob and E. M ller, *J. Chromatogr., 404:* 297 (1987).
8. J. L. Parr, G. Walters, and M. Hoffman, *Hydrocarbon Contaminated Soils and Groundwater,* Enseco Inc., Arvada, CO, 1991, Chap. 8.
9. Tekmar, *Technical Bulletin,* Tek/Dat B021062, Tekmar, Cincinnati, OH, 1982.
10. S. B. Hawthorne and D. J. Miller, *J. Chromatogr. Sci., 24:* 258 (1986).
11. S. B. Hawthorne and D. J. Miller, *Anal. Chem., 59:* 1705 (1987).
12. B. W. Wright, C. W. Wright, and J. S. Fruchter, *Energy Fuels, 3:* 474 (1989).
13. J. R. Wheeler and M. E. McNally, *J. Chromatogr. Sci., 27:* 534 (1989).
14. R. M. Campbell, D. M. Meunier, and H. J. Cortes, *J. Microcol. Sep., 1:* 302 (1989).
15. V. Lopez-Avila, J. Benicto, N. S. Dodhiwala, and R. Young, *J. Chromatogr. Sci., 30:* 335 (1992).
16. S. E. Eckert-Tilotta, S. B. Hawthorne, and D. J. Miller, *Fuel, 72:* 1015 (1993).
17. J. A. Field, D. J. Miller, T. M. Field, S. B. Hawthorne, and W. Giger, *Anal. Chem., 64:* 3161 (1992).
18. Y. Lin, R. D. Brauer, K. E. Laintz, and C. M. Wai, *Anal. Chem., 65:* 2549 (1993).
19. J. Wang and W. D. Marshall, *Anal. Chem., 66:* 1658 (1994).
20. Y. Hirata and Y. Oikamoto, *J. Microcol. Sep., 1:* 46 (1989).
21. K. Sugiyama and M. Saito, *J. Chromatogr., 442:* 121 (1988).
22. R. M. Smith and M. D. Burford, *J. High Resolut. Chromatogr., 32:* 265 (1994).
23. J. W. King, *J. Chromatogr. Sci., 27:* 355 (1989).
24. B. E. Berg, E. M. Hansen, S. Gjorven, and T. Greibrokk, *J. High Resolut. Chromatogr., 16:* 358 (1993).
25. M. A. Schneiderman, A. K. Sharma, and D. C. Locke, *J. Chromatogr. Sci., 26:* 458 (1988).
26. British Patent Specification 1 388 581, 1975.
27. K. Sugiyama, M. Saito, T. Hondo, and M. Senda, *J. Chromatogr., 332:* 107 (1985).
28. Air Products plc, Crewe United Kingdom.
29. J. M. Levy and A. C. Rosselli, *Chromatographia, 28:* 613 (1989).
30. M. W. F. Nielen, J. A. St b, H. Lingeman, and U. A. Th. Brinkman, *Chromatographia, 32:* 543 (1991).

31. K. Jinno, T. Hoshino, T. Hondo, M. Saito, and M. Senda, *Anal. Chem., 58:* 2696 (1986).
32. R. D. Smith, H. R. Udseth, and B. W. Wright, *J. Chromatogr. Sci., 24:* 238 (1986).
33. S. B. French and M. Novotny, *Anal. Chem., 58:* 164 (1986).
34. K. Jinno, *Hyphenated Techniques in Supercritical Fluid Chromatography and Extraction* K. Jinno, Ed., Journal of Chromatography Library Series, Elsevier Science Publishers, Amsterdam 1992, Vol. 53, Chap. 9.
35. L. Baner, T. B cherl, J. Ewender, and R. Franz, *J. Supercrit. Fluids, 5:* 213 (1992).
36. M. Ashraf-Khorassani, D. S. Boyer, and J. M. Levy, *J. Chromatogr. Sci., 29:* 517 (1991).
37. Q. L. Xie, K. E. Markides, and M. L. Lee, *J. Chromatogr. Sci., 27:* 365 (1989).
38. R. W. Vannoort, J.-P. Chervat, H. Lingeman, G. J. Dejong, and U. A. Th. Brinkman, *J. Chromatogr., 505:* 45 (1990).
39. M. H. Liu, S. Kapila, K. H. Nam, and A. A. Elseewi, *J. Chromatogr., 639:* 151 (1993).
40. I. L. Davies, M. W. Raynor, J. P. Kithinji, K. D. Bartle, P. T. Williams, and G. E. Andrews, *Anal. Chem., 60:* 683A (1988).
41. E. Stahl, W. Schilz, E. Sch tz, and E. Willing, *Angew. Chem. Int. Ed. Engl., 17:* 731 (1978).
42. D.L. Stalling, S. Saim, K.C. Kuo, and J.J. Stunkel, *J. Chromatogr. Sci., 30:* 486 (1992).
43. S.B. Hawthorne, *Anal. Chem., 60:* 633A (1990).
44. M.R. Anderson, J.W. King, and S.B. Hawthorne, *Analytical Supercritical Fluid Chromatography and Extraction*, M.L. Lee and K.E. Markides, Eds., 1990, Chap. 5.
45. J.R. Dean and M. Kane, *Applications of Supercritical Fluids in Industrial Analysis*, D.R. Dean, Ed., CRC Press, Boca Raton, FL, 19, Chap. 3.
46. M-L. Riekkola, P. Manninen, and K. Hartonene, *Hyphenated Techniques in Supercritical Fluid Chromatography and Extraction*, Journal of Chromatography Library Series, Elsevier Science Publishers, K. Jinno, Ed., Amsterdam, 1992 Vol. 53, Chap. 14.
47. M.D. Burford, S.B. Hawthorne, and D.J. Miller, *Anal. Chem., 65:* 1497 (1993).
48. V. Janda, K.D. Bartle, and A.A. Clifford, *J. Chromatogr., 642:* 283 (1993).
49. M.W.F. Nielen, J.T. Sanderson, R.W. Frei, and U.A. Th. Brinkman, *J. Chromatogr., 474:* 388 (1989).
50. K. Hartonen, M. Jussila, P. Manninen, and M.-L. Riekkola, *J. Microl. Sep., 4:* 3 (1992).
51. J.H. Raymer and G.R. Velez, *J. Chromatogr. Sci., 29:* 467 (1991).

52. J.F. Pankow, *Anal Chem.*, *60*: 950 (1988).
53. X. Liang and C.N. Hewitt, *J. Chromatogr.*, *627*: 219 (1992).
54. S.D. Cooper and E.D. Pellizzari, *J. Chromatogr.*, *498*: 41 (1990).
55. M.D. Burford, S.B. Hawthorne, and D.J. Miller, *J. Chromatogr.*, *685*: 79 (1994).
56. M. Lohleit and K. B chmann, *J. Chromatogr.*, *505*: 227 (1990).
57. B.W. Wright, S.R. Frye, D.G. McMinn, and R.D. Smith, *Anal Chem.*, *59*: 640 (1987).
58. F.I. Onuska and K.A. Terry, *J. High Resolut. Chromatogr.*, *12*: 527 (1989).
59. J. Levy and A.C. Rosselli, *Chromatographia*, *28*: 613 (1989).
60. S.A. Leibman, E.J. Levy, S. Lurcott, S. O'Niel, J. Guthrie, T. Ryan, and S. Yocklovich, *J. Chromatogr. Sci.*, *27*: 118 (1989).
61. S.B. Hawthorne, D.J. Miller, and J.J. Langenfeld, *J. Chromatogr. Sci.*, *28*: 2 (1990).
62. R.J. Houben, H.-G.M. Janssen, P.A. Leclercq, J.A. Rijks, and C.A. Cramers, *J. High Resolut. Chromatogr.*, *13*: 669 (1990).
63. S.B. Hawthorne, M.S. Krieger, and D.J. Miller, *Anal. Chem.*, *60*: 472 (1988).
64. X. Lou, H.-G. Janssen, and C.A. Cramers, *J. High Resolut. Chromatogr.*, *16*: 425 (1993).
65. S.M. Pyle and M.M. Setty, *Talanta*, *38*: 1125 (1991).
66. H.R. Johansen, G. Becher, and T. Greibrokk, *Fresenius Z. Anal. Chem.*, *344*: 486 (1992).
67. S.B. Hawthorne, D.J. Miller, M.D. Burford, J.J. Langenfeld, S. Eckert-Tilotta, and P.K. Louie, *J. Chromatogr.*, *642*: 301 (1993).
68. J.J. Langenfeld, M.D. Burford, S.B. Hawthorne, and D.J. Miller, *J. Chromatogr.*, *594*: 297 (1992).
69. M. Lohleit, R. Hillmann, and K. B chmann, *Z. Anal. Chem.*, *339*: 470 (1991).
70. J.M. Levy and J.P. Guzowski, *Fresenius Z. Anal. Chem.*, 330 (1988) 207.
71. M.D. Burford, S.B. Hawthorne, and D.J. Miller, *J. Chromatogr.*, *685*: 95 (1994).
72. J.C. Wallace, M.S. Krieger, and R.A. Hites, *Anal. Chem.*, *64*: 2655 (1992).
73. M.D. Burford, S.B. Hawthorne, and D.J. Miller, *J. Chromatogr.*, *657*: 413 (1993).
74. A. Hagman and S. Jacobsson, *J. Chromatogr.*, *448*: 117 (1988).
75. J.W. Graydon and K. Grob, *J. Chromatogr.*, *254*: 265 (1983).
76. M.L. Hopper and J.W. King, *J. Assoc. Off. Anal. Chem.*, *74*: 661 (1991).
77. N.L. Porter, A.F. Rynaski, E.R. Campbell, M. Saunders, B.E. Richter, J.T. Swanson, R.B. Nielsen, and B.J. Murphy, *J. Chromatogr. Sci.*, *30*: 367 (1992).
78. M.D. Burford, S.B. Hawthorne, and D.J. Miller, *J. Chromatogr.*, *609*: 321 (1992).

79. R.D. Smith, J.L. Fulton, R.C. Petersen, A.J. Kopriva, and B.W. Wright, *Anal. Chem., 58*: 2057 (1986).
80. K.D. Bartle, A.A. Clifford, S.A. Jafar, and G.F. Shilstone, *J. Phys. Chem. Ref. Data., 20*: 713 (1991).
81. S.B. Hawthorne, D.J. Miller, and M.S. Krieger, *J. High Resolut. Chromatogr., 12*: 714 (1989).
82. J.M. Levy and M. Ashraf-Khorassani, *Hyphenated Techniques in Supercritical Fluid Chromatography and Extraction*, K. Jinno, Ed., Elsevier Science Publishers, New York, 1992, Chap. 11.
83. J.M. Levy, E. Storozynsky, and M. Ashraf-Khorassani, *Recent Advances in Supercritical Fluid Technology*, M.E. McNally, Ed., American Chemical Society, Washington, DC, 1991.
84. J.M. Levy, L. Dolata, R.M. Ravey, E. Storozynsky, and K.A. Holowczak, *J. High Resolut. Chromatogr., 16*: 368 (1993).
85. J.W. King, J.E. France, and J.M. Snyder, *Fresenius Z. Anal. Chem., 344*: 474 (1992).
86. R. Hillmann and K. Bachmann, *J. High Resolut. Chromatogr., 17*: 350 (1994).
87. S.B. Hawthorne and D.J. Miller, *Anal. Chem., 66*: 4005 (1994).
88. J.J. Langenfeld, S.B. Hawthorne, D.J. Miller, and J. Pawliszyn, *Anal. Chem., 67*: 727 (1995).
89. Y. Yang, A. Gharaibeh, S.B. Hawthorne, and D.J. Miller, *Anal. Chem., 67*: 641 (1995).
90. J.M. Levy, E. Storozynsky, and R.M. Ravey, *J. High Resolut. Chromatogr., 14*: 661 (1991).
91. J.B. Pawlyszin and N. Alexandrou, *Water Pollut. Res. J. Can., 24*: 207 (1989).
92. C.K. Huston and H. Ji, *J. Agric. Food Chem., 39*: 1229 (1991).
93. J.M. Levy, A.C. Rosselli, E. Storozynsky, R. Ravey, L.A. Dolata, and M. Ashraf-Khorassani, *LC–GC, 10*: 386 (1992).

5
Sample Preparation for Gas Chromatography with Solid-Phase Extraction and Solid-Phase Microextraction

Zelda E. Penton *Varian Chromatography Systems, Walnut Creek, California*

I.	INTRODUCTION	205
II.	CONVENTIONAL SOLID-PHASE EXTRACTION	206
	A. Introduction	206
	B. Method Development	207
	C. Batch Processing	214
	D. SPE/GC Applications	215
III.	SOLID-PHASE MICROEXTRACTION	218
	A. Background	218
	B. Principles of SPME	220
	C. Optimizing SPME Sampling	226
	D. Applications	229
IV.	CONCLUSIONS	232
	REFERENCES	233

I. INTRODUCTION

Gas Chromatography (GC) has often been described as a "mature" technique, but since 1980, there have been continuing major innovations [1]. The most important was the replacement of packed columns with fused silica columns for most applications. The movement to fused silica columns spearheaded the design of sophisticated injectors, with temperature-programming capability and electronically controlled pneumatics. The development of detectors with improved selectivity and sensitivity, autosamplers in which virtually all injection parameters can be controlled, and impressive improvements in data handling systems should also be noted. At the present time, sample preparation is the area where a great deal of work needs to be done [2]. In many applications, analytes must be determined in difficult matrices such as soil, sludge, blood, fuel oils, and food. Liquid–liquid extraction (LLE), the traditional method of treating samples prior to GC injection, is no longer acceptable for many reasons. A typical LLE procedure requires several steps, making the cleanup process tedious and difficult to automate, the highly purified solvents that are required are expensive to purchase and to dispose of; many of these solvents are suspected of endangering the health of laboratory workers. Several alternative methods which reduce or eliminate the use of solvents are now being used to prepare samples for GC analysis. These include static and dynamic headspace for volatile compounds and supercritical fluid extraction and solid-phase extraction (SPE) for semivolatiles. Although SPE has been used for preparing samples for analysis by high-performance liquid chromatography (HPLC), there are many applications where the technique is used prior to GC analysis. Recently, a new variation of SPE has been developed—solid-phase microextraction (SPME). This new technique is normally used prior to GC analysis. Both of these techniques will be described here.

II. CONVENTIONAL SOLID-PHASE EXTRACTION

A. Introduction

In conventional solid-phase extraction (SPE), a liquid sample is passed over a solid or "sorbent" that is packed in a medical-grade polypropylene cartridge or embedded in a disk. As a result of strong attractive forces between the analytes and the sorbent, the analytes are retained on the sorbent. Later, the sorbent is washed with a small volume of a solvent that has the ability to disrupt the bonds between the analytes and the sorbent. The final result is that the analytes are concentrated in a relatively small volume of clean solvent and are therefore ready to be injected into a GC without any additional sample workup. In addition, extraneous compounds are usually removed.

The sorbent normally consists of a silica substrate bonded to an organosilane compound. The functional groups on the organosilane determine the selective properties of a particular sorbent. Recently, polymeric-based substrates such as methacrylate resins have been reported [3]. A typical SPE cartridge is shown in Fig. 1; the process is summarized in Fig. 2.

A variation of the extraction cartridge is the disk where the sorbent (on a

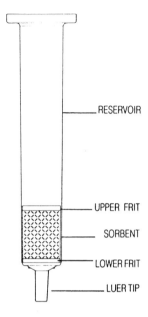

Fig. 1 Schematic of a solid-phase extraction cartridge. (From Ref. 4.)

polymer or silica substrate) is embedded in a "web" of Teflon™-based fibrils. The sorbent particles are smaller than those in cartridges (8 μm in diameter rather than 40 μm). The short sample path and small particle size allow efficient trapping of analytes with a relatively high flow rate through the sorbent as compared to the cartridges. The disks are primarily used to reduce analysis time when handling large volumes of aqueous environmental samples. Disks are available in several different diameters with the larger volumes allowing faster flow rates. The solid-phase extraction process is quantitative. It has been used to prepare samples for GC analysis of biological, environmental, and industrial applications. Government agencies such as the United States Environmental Protection Agency (USEPA) have approved SPE for sample cleanup.

B. Method Development

In general, when developing an analytical procedure, it will usually save a great deal of time to begin with a literature search and then adapt an existing procedure according to the requirements of an individual laboratory. The object of this section is to give some general guidelines for developing a SPE method to be followed by GC analysis. Specific methods from the literature will be discussed in Section II.D.

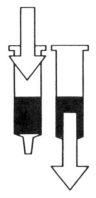

CONDITIONING
Conditioning the sorbent prior to sample application ensures reproducible retention of the compound of interest (the isolate).

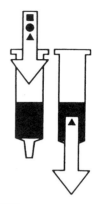

RETENTION
■ Adsorbed isolate
● Undesired matrix constituents
▲ Other undesired matrix components

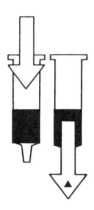

RINSE
▲ Rinse the columns to remove undesired matrix components

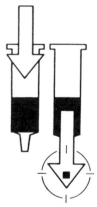

ELUTION
● Undesired components remain
■ Purified and concentrated isolate ready for analysis

Fig. 2 The four basic steps in the solid-phase extraction process. (From the *Varian Sample Preparations Products Catalog*, reprinted with permission.)

Selection of the SPE Cartridge and Elution Solvent

When developing a SPE method, a number of factors must be considered:

- A sorbent should be chosen with **selectivity** characteristics such that the analytes will be retained and most of the interferences will not be retained.
- The **sorbent capacity** should be sufficient to retain all of the analytes.
- The **elution solvent** should be capable of recovering virtually all of the analyte from the sorbent in as small a volume as possible while leaving any interfering compounds on the sorbent.

Sorbent Selectivity. Sorbent–analyte interactions fall into three categories—nonpolar, polar, and ionic. In order to optimize the retention of compounds of interest on the sorbent, one must consider the polarity of the analytes and the sample matrix. In a majority of applications, nonpolar and slightly polar analytes are dissolved in water, a highly polar solvent. For these applications, nonpolar sorbents are selected. Analytes containing very polar functional groups such as hydroxyl, carbonyls, amines, and sulfhydryls will be retained on polar sorbents (i.e., those sorbents exhibiting a strong tendency to form hydrogen bonds). Retention of polar compounds on polar sorbents is facilitated by nonpolar solvents. Analytes that are capable of forming cations or positively charged ions include amines; analytes with the potential to form anions or negatively charged ions include carboxylic and sulfonic acids and phosphates. For retention to occur with ionic interactions, an anionic sorbent should be selected to retain cations, and a cationic sorbent to retain anions. For maximum retention, the pH of the matrix should be 2 pH units below the pK_a of the cation and 2 pH units above the pK_a of the anion [4]. The diagrams in Fig. 3 summarize the various interactions between sorbents and analytes.

Sorbent Capacity. Solid-phase extraction cartridges are available in a wide range of sizes with volumes ranging from less than 1 mL to over 50 mL. When selecting the optimum cartridge size for a particular application, factors to be considered are the ability to retain all of the analytes in the sample, volume of original sample, and final volume of the purified sample after elution. In general, the mass of the analytes and interfering compounds retained by the sorbent should be less than 5% of the mass of the sorbent. A large-capacity cartridge would appear to be desirable for maximum analyte retention and to accommodate a large-sample volume. However, the ultimate goal of the SPE process is to concentrate the analyte into as small a volume as possible. This is particularly important for GC applications where injection volumes are normally 1–2 µL. In order to minimize the elution volume (and avoid the necessity of concentrating the sample by evaporating solvent), the cartridge volume should be as small as

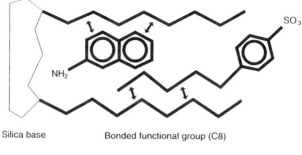

Fig. 3 Schematic of sorbent interactions. (From the *Varian Sample Preparations Products Catalog*, reprinted with permission.)

possible. A rule of thumb is that the elution volume should be two to five times the bed volume* of the cartridge [4].

Testing the SPE Procedure with Standards. After selecting one or more possible SPE cartridges for the analysis, prepare the samples for application to the cartridges and then prepare a blank matrix that closely resembles the sample matrix at this point. Next, spike the blank matrix with known quantities of the analytes. For example, if tissue containing drugs is to be homogenized in a buffer, the buffer should be spiked with drug standards; if soil is to be extracted with a mixture of organic solvents prior to pesticide determination, a blank mixture of the above solvents should be spiked with the pesticides expected in the samples. At this point, the procedure continues as outlined in the first three steps of Fig. 2.

- **Condition the sorbent:** Pass one or two column volumes† of an appropriate solvent through the sorbent to wet it and prepare its functional groups for eventual interaction with the analytes; then pass 10–20 bed volumes of the blank matrix through the column.
- **Retention:** Apply the spiked matrix to the column and collect the liquid that elutes.
- **Rinse:** Wash the column with several additional bed volumes of the blank matrix and collect the eluting liquid. The sorbent should not be allowed to dry out during these steps.

Test the eluants for the presence of the analytes. If present, consider a different sorbent phase or a higher-capacity cartridge.

Optimizing the Wash and Elution Steps. The elution solvent must have the ability to disrupt the interactions between the sorbent and the analytes. Often, mixtures of solvent are most effective. As mentioned in he section on sorbent capacity, two to five bed volumes of elution solvent should elute close to 100% of the analytes from the sorbent. To optimize the wash steps, a solvent should be identified that does not elute the analytes. A likely wash solvent is similar to the eluting solvent but is not strong enough to disrupt the analyte–sorbent bonds. An example would be an analyte forming hydrogen bonds with the sorbent that is eluted with a nonpolar organic solvent such as hexane. A mixture of hexane with slightly polar organic solvents should be suitable for washing the cartridge. Mixtures of various proportions should be prepared and applied to the cartridge. After confirming by analysis the least polar composition that will not elute the analytes, this composition should be used as the rinse solvent.

---

*Bed volume is defined as the amount of solvent required to fill all of the interstitial spaces and internal pores of the particles in the sorbent. For 40–60-μm sorbent particles, the bed volume is approximately 120 μL/100 mg.
†Defined as the volume of solvent required to fill the entire cartridge.

Table 1 lists sorbent phases and elution solvents for various applications. In addition to the sorbents listed in the table, specialty phases are commercially available for drug screening and environmental applications. These may be shipped with step-by-step procedures for specific applications.

Elution Solvents for GC Analysis

Special considerations apply when the sample is to be analyzed by GC:

- The solvent should be compatible with the GC detector.

    Electron–capture detectors (ECDs)—no halogenated solvents.

    Nitrogen phosphorus detectors (NPD's)—no nitrogen-containing solvents such as acetonitrile; halogenated solvent are also not recommended for these detectors.

- Buffers and other solvents containing nonvolatiles are not suitable for GC injection.
- The solvent should be compatible with the GC column phase; for example, a polar solvent such as water or methanol is not compatible with a nonpolar phase such as methyl silicone.
- Some compounds must be derivatized prior to GC analysis. Examples are metabolites of drugs such as cocaine and tetrahydrocannabinol and some herbicides that contain carboxylic acid functional groups. If derivatization is necessary, it is desirable to elute the compound in a dry and inert organic solvent such as hexane.

In some applications, it may be necessary to use an elution solvent that is not desirable for the subsequent GC analysis. Possible solutions include modification of the GC conditions or additional sample workup. An example of the former is illustrated when a nitrogen–phosphorus detector is to be used in a GC analysis for determining drugs and pesticides. If halogenated solvents are necessary to elute the analytes, it might be possible to inject the sample in the split mode if the concentration of the analytes is relatively high. If the sample is too dilute for split injection or if the elution solvent contains nitrogen, the detector bead current can be turned off while the solvent passes through the detector and turned on immediately after [5]. This procedure is feasible if the first analyte elutes from the GC column several minutes after the solvent. Another approach with incompatible solvents is to use solvent venting. If the solvent is too polar to be focused on a nonpolar GC column phase, possible solutions include injecting less than 1 µL or injecting in the split mode. If the sample is not sufficiently concentrated, another solution might be to install a retention gap [6] or precolumn. For samples that are eluted from the SPE cartridge in a buffer, an additional cleanup step might be required. This might be accomplished with solid-phase microextraction (SPME), described later in this chapter.

Table 1 Summary of Sorbents and SPE Procedures for GC

Sorbent interaction	Sorbent	Analyte functional group	Matrix	Typical elution solvents	Applications
Nonpolar	C_{18}–Octadecyl C_8–Octyl C_2–Ethyl CH–Cyclohexyl PH–Phenyl CN–End-capped cyanopropyl	*Hydrophobic Groups:* Aromatic rings Alkyl chains	*Aqueous:* Water Buffers Biological fluids	Methanol Acetonitrile Ethyl acetate Chloroform Hexane	Drugs of abuse Pesticides Therapeutic drug monitoring
Polar	CN–Cyanopropyl 2OH–Diol Si–Silica NH_2–Aminopropyl	*Hydrophilic Groups:* Hydroxyls Amines Heteroatoms (S, O, N)	*Nonpolar:* Hexane Oils Chloroform Lipids	Methanol Isopropanol Acetone	Lipid separations[a] Oil additives Carbohydrates[b] Phenols
Cation exchange	SCX–Benzenesulfonic acid (strong) PRS–Propylsulfonic acid (strong) CBA–Carboxylic acid (weak)	*Cations:* Amines Pyrimidines	*Aqueous:* Water Acidic buffers Biological fluids	Alkaline buffer[c] High ionic strength buffer[c]	Herbicides Pharmaceuticals Catecholamines[b]
Anion exchange	SAX–Quaternary amine (strong) PSA–Primary/secondary amine NH_2–Aminopropyl (weak) DEA–Diethylaminopropyl (weak)	*Anions:* Carboxylic acids Sulfonic acids Phosphates	*Aqueous:* Water Alkaline buffers Biological fluids	Acidic buffer[c] High ionic strength buffer[c]	Organic acids[b] Fatty acids[b] Phosphates

[a] Additional sample workup would be required for GC analysis of lipids, including possible hydrolysis to isolate the fatty acids.
[b] Derivatization would usually be necessary prior to GC injection.
[c] Analytes would normally be extracted into an organic solvent prior to GC analysis.

C. Batch Processing

In the early years of SPE technology, the ability to handle multiple samples was limited because of variable flow rates through the cartridges [2]. Improvements in the manufacturing of extraction tubes have made batch processing more feasible in recent years. These improvements include a narrow range of particle size, consistent cartridge-to-cartridge sorbent mass, and packing techniques that assure a uniform flow through the tube (Fig. 4).

Vacuum Manifolds for Manual Handling of Multiple Samples

Normally, liquid flow through SPE cartridges requires assistance by application of a vacuum. Manifold systems are available that allow the user to process up to 24 samples simultaneously. These systems may allow individual control of the flow rate through each tube and contain a variety of racks for collection of the eluants. Manifolds are also available for disk cartridges.

Automated Systems

In a recent survey of experts [2] on sample preparation and automation, the consensus was that there are only a handful of automated systems for SPE. It was stated that early attempts at automation were not successful because the performance did not meet users' expectations. Problems of lack of standardization of cartridge size and poor column-to-column reproducibility were cited; it was also

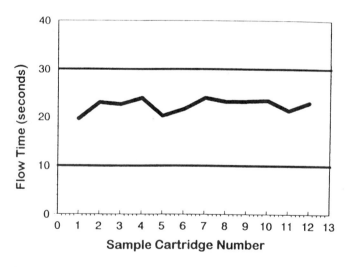

Fig. 4 Flow characteristics of 12 randomly selected SPE cartridges showing highly reproducible column-to-column flow characteristics. (From the *Varian Sample Preparations Products Catalog*, reprinted with permission.)

mentioned that many workers have too few samples to justify automation. An automated workstation that can be used to prepare environmental water samples for GC analysis is commercially available (Tekmar Co., Cincinnati, OH). This system, designed for up to six aqueous samples, can accommodate SPE cartridges and disks, but it does not offer online connection to the GC.

In Europe, online SPE-GC systems have been developed using column-switching techniques. Hankemeier et al. [7] described a fully online SPE–GC system for the determination of organophosphorous pesticides in water. This system used a stainless-steel precolumn packed with a styrene–divinylbenzene copolymer, rather than a conventional SPE cartridge. Systems such as these are not commercially available. The use of robotics has also been considered as an additional approach to automation of SPE, prior to GC analysis.

D. SPE/GC Applications

There are numerous reports in the literature of SPE cleanup in preparation for GC analysis. Some of these applications are summarized in this section.

Toxicology and Biomedical

Toxicology Including Drugs of Abuse. Veningerová and co-workers [8] published a procedure for determining chlorinated phenols and cresols in the urine of occupationally exposed workers. Urine samples were acidified and refluxed, and the pH was raised to 5, just prior to sample cleanup on a C_{18} cartridge. After derivatization and an additional SPE cleanup, the phenols were determined with GC–ECD. Recoveries of the various compounds of interest varied from 72% to 110%. Stevenson et al. [9] compared liquid–liquid extraction (LLE) with SPE for extraction of phencyclidine in urine prior to GC/MS analysis. Advantages of SPE over LLE were better removal of interferences resulting in lower detection limits. Dixit et al. [10] also compared LLE with SPE for determining fluoxipine and norfluoxipine and found better recoveries with SPE. Other researchers [11–15] have published SPE/GC procedures for several basic drugs in a variety of matrices. Recoveries in these procedures varied from 50% to 60% for cocaine and metabolites in the brain, up to 100% for benzodiazapines in plasma. Table 2 summarizes several recent publications on drug analysis in body fluids.

Biomedical Applications. A procedure for determining testosterone in urine was reported by Venturelli et al. [16]. In this method, two SPE cartridges were coupled, the first containing a nonpolar (C_{18}) sorbent and the second an anion exchanger. Urine was placed directly in the first cartridge. A final cleanup step with HPLC was required, prior to the analysis with GC/FID (flame-ionization detection). Saha and Giese [17] compared HPLC and SPE for the cleanup of a derivatized DNA-adduct prior to GC analysis. These adducts are used as biological markers for exposure to toxic chemicals. It was concluded that the SPE procedure surpassed the HPLC cleanup for removing interfering compounds.

Table 2 Procedures in the Literature for SPE of Drugs in Biological Matrices Followed by GC Analysis

Ref.	Drug	Matrix	LOD[a]	Detector
9	Phencyclidine	Urine	1 ng/mL	MS
11	Cocaine, benzoyleconine	Plasma	1 ng/mL 3 nf/ML	NPD
12	Cocaine, benzoyleconine cocaethylene	Brain	25 ng/g for all	MS
13	Benzodiazepines	Plasma	0.5–10 ng/mL	ECD, NPD
14	Oxazepam	Urine	50 ng/mL	MS
15	Basic drugs	Urine, plasma	100–200[b]	NPD
10	Fluoxetine Norfluoxetine	Serum	20–200[b]	ECD

[a]Limit of detection
[b]Range of concentration; limit of detection was not mentioned

Table 3 A Listing of Some Government-Regulated Procedures for Environmental Samples with SPE Cleanup and GC Analysis

	Analytes	Detector
EPA Method #		
506	Phthalates and adipates	ECD
508.1	Organochlorine/nitrogen pesticides	ECD
515.2	Chlorinated herbicides	ECD
525.2	Semivolatile organics	MS
548.1	Endothal	ECD
552.1	Haloacetic acids and dalapon	ECD
1631B	TCDD	ECD
German methods		
DIN 38407, section 6	Organophosphorous and nitrogen pesticides, including atrazine, simazine, cyanazine, trifluralin, and ethyl parathion	NPD
DIN 38407, section 14	Phenoxyalkyl carbonic acids including 2,4-D	MS

Environmental

Government-Approved Methods for SPE/GC. The United States Environmental Protection Agency (EPA) has approved SPE as a cleanup procedure for several GC methods. Two of the German standard methods for the examination of water, wastewater, and sludge include SPE for sample cleanup. Several government-approved methods as of early 1996, are listed in Table 3. The following subsections describe applications in the literature that are relevant to environmental monitoring.

Pesticides. Solid-phase extraction/gas chromatographic pesticide methods reported in the literature [7,18–23] include determination of triazine and urea herbicides and organophosphorous pesticides. These reports are summarized in Table 4. In all of these studies, the matrix was water.

Additional Compounds of Environmental Interest. Yook et al. [24] reported the analysis of a wide range of pollutants including phenols, polynucleararomatics, neutral aromatics, ethers, and nitrosamines. LLE was compared with SPE using 1-g C_{18} cartridges for 500-mL samples of water. Detection was with mass spectrometry (MS). In this case, recoveries were superior with LLE, but it was concluded that larger-capacity cartridges would have improved the results. Deans et al. [25] looked at relatively volatile compounds in acidic industrial effluents. These included toluene, xylene and mono-, di-, and trichlorobenzenes. SPE cartridges containing C_{18}, C_8, and phenyl-substituted cartridges were used. Recoveries varied from 70% to 90%; it was felt that losses were caused by analytes partitioning into the headspace due to storing the samples in bottles that were not completely full.

An unusual application of SPE/GC was reported by Johansson et al. [26]—the determination of selenium in natural waters. Selenium was derivatized with 1,2-diamino-3,5-dibromobenzene prior to SPE concentration and determined by

Table 4 Published SPE/GC Procedures for Pesticides

Ref.	Pesticide	Sorbent	Detector
7	Organophosphorous pesticides	Styrene–divinylbenzene	AED[a]
18	Atrazine	C_{18} cartridges	MS
19	Urea herbicides	C_{18} cartridges	NPD
20	Organophosphorous pesticides, triazines, carbamates	C_{18} cartridges	NPD
21	Organophosphorous and organochlorine pesticides	C_{18} cartridges	ECD, NPD
22	Organophosphorous and triazine pesticides	C_{18} cartridges	NPD, MS
23	Fungicides including captan, captafol, carbendazin, chlorothalonil and folpet	C_{18} disks	ECD, MS

[a]Atomic emission detector.

GC/ECD. Recoveries with SPE were similar to LLE, but the SPE procedure was less time-consuming and avoided the sample foaming problems that were observed with LLE.

Chemical Warfare Agents in Water. Häkkinen [27] reported the determination of a variety of chemical warfare agents in water. These included nerve agents (isopropyl methylphosphofluoridate and ethyl N,N-dimethylphosphoamidocyanidate) and 2,2′-dichlordiethylsulfide (Mustard gas). These compounds are very unstable and yields tended to be low. A C_{18} sorbent was used in this study; the author felt that a polar sorbent might have yielded better recoveries.

Slack et al. [28] coupled SPE with supercritical fluid extraction in an online GC determination of nitroglycerin and explosive nitrated aromatics. Supercritical CO_2 was used to elute the explosives from cartridges that contained a phenyl sorbent. The analytes were cryogenically trapped in the GC column. Prior to the GC analysis, the CO_2 was vented out of the GC by opening a valve in the carrier gas inlet line. Recoveries from spiked water were essentially 100%.

III. SOLID-PHASE MICROEXTRACTION

A. Background

With solid-phase microextraction (SPME) as with conventional SPE, analytes are concentrated by absorption into a solid phase; however, in practice, the two techniques are quite different. SPME was developed by Pawliszyn and associates at the University of Waterloo in Ontario, Canada [29]. Unlike conventional SPE, no solvents are used. Typical samples are in an aqueous or solid matrix.

The technique utilizes a short thin solid rod of fused silica (typically 1 cm long and 0.11 µm outer diameter), coated with an absorbent polymer. The coated fused silica (SPME fiber) is attached to a metal rod; the entire assembly (fiber holder) may be described as a modified syringe (Fig. 5). In the standby position, the fiber is withdrawn into a protective sheath. For sampling, an aqueous sample containing organic analytes or a solid containing organic volatiles is placed in a vial, and the vial is closed with a cap that contains a septum. The sheath is pushed through the septum and the plunger is lowered, forcing the fiber into the vial, where it is immersed directly into the aqueous sample or the headspace. Organic compounds in the sample are subsequently absorbed on the fiber. After a predetermined time, the fiber is withdrawn into the protective sheath and the sheath is pulled out of the sampling vial. Immediately after, the sheath is inserted into the septum of a GC injector, the plunger is pushed down, and the fiber is forced into the injector insert where the analytes are thermally desorbed and separated on the GC column. The desorption step is usually 1–2 min; afterward, the fiber is withdrawn into the protective sheath and the sheath is removed from the GC injector. Figure 6 is a schematic of this process.

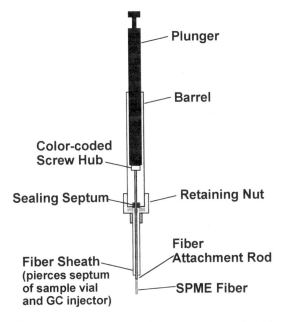

Fig. 5 SPME fiber holder for automated sampling. The manual version is similar.

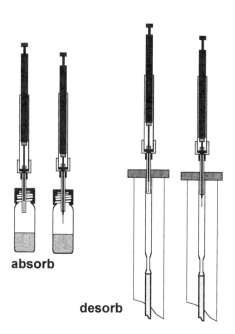

Fig. 6 Summary of the SPME sampling process showing headspace sampling over a liquid phase, followed by injection into the glass insert of a GC capillary injector. Immediately before removal of the fiber holder from the sampling vial and injector, the fiber is withdrawn into the sheath.

The fibers and the fiber holders are commercially available from Supelco Inc. The sampling process has been automated by Varian Associates for online operation on any gas chromatograph [30]. Several absorbent polymers, normally used to manufacture fused silica GC columns, are available on SPME fibers. Table 5 summarizes the phases that are available at the present time. Applications include volatiles and semivolatiles in foods [31–36], polymers [37], and environmental samples [38–58]. Some of these will be discussed in Section III.D. The technique is rugged, and fibers can usually be used for 100 or more samplings. SPME is relatively new and, at the time this was written (early 1996), was not yet approved by government regulatory agencies.

B. Principles of SPME

Solid-phase microextraction is an equilibrium technique; therefore, analytes are not completely extracted from the matrix. Nevertheless, the method is useful for quantitative work, and excellent precision and linearity have been demonstrated. Limits of detection and quantitation are often well below 1 ppb. Practically, if the sampling and desorption steps are precisely timed as is the case with automation, excellent precision can be attained without achieving equilibrium. The theory has been well documented by Arthur et al. [56] and Zhang and Pawliszyn [59] and it is recommended that the analyst study their publications for a com-

Table 5 Commercially Available SPME Fibers

Phase	Features
100 μm polydimethylsiloxane (PDMS)	Nonpolar phase with a high sample capacity, suitable for a wide variety of applications—volatile compounds such as dichloromethane to semivolatiles (some pesticides).
30 μm PDMS	Useful for semivolatile compounds such as pesticides. The thinner phase allows faster desorption, thus sample carryover is minimized.
7 μm PDMS	For semivolatiles. Bonded fiber which allows a higher desorption temperature than the above PDMS fibers (maximum operating temperature 320°C vs. 260°C for the 100 μm fiber). Shorter absorption and desorption times. Reduced sample capacity.
85 μm polyacrylate	Greater affinity for polar compounds (recommended for phenols). This phase is more of a solid than the PDMS phases; therefore, diffusion rates are slow and equilibration times are relatively long.
65 μm carbowax/divinylbenzene	Polar phase with a much stronger affinity for alcohols than the PDMS fibers.

plete understanding of the principles of SPME. Some concepts useful for method development will be discussed here.

With liquid samples, the recovery expected from SPME is dependent on the partitioning of the analytes among the three phases present in the sampling vial—the liquid (normally aqueous), the headspace above the liquid, and the fiber coating.* The equations governing this equilibrium process are

$$K_1 = \frac{C_L}{C_G}$$

$$K_2 = \frac{C_F}{C_L}$$

$$K_3 = \frac{C_F}{C_G}$$

where K_1 is the partition coefficient of an analyte between the liquid and headspace phases, K_2 is the partition coefficient of an analyte between the fiber and liquid phases, and K_3 is the partition coefficient of an analyte between the fiber and headspace phases; C_L, C_G, C_F are the concentrations of the analyte in these phases.

The distribution among the three phases after equilibrium is represented as

$$C_0 V_L = C_G V_G + C_L V_L + C_F V_F$$

where C_0 is the concentration of the analyte in the original liquid sample and V_L, V_G, and V_F are the volumes of the liquid, headspace, and fiber phases, respectively. The above equations can be rearranged to predict the effect of changing certain variables, such as changing the relative volumes of phrases or to calculate partition coefficients:

$$C_F = \frac{C_0 V_L K_1 K_2}{V_G + K_1 V_L + K_1 K_2 V_F}$$

For example, the above equation can be used to predict the concentration of an analyte in the fiber if the partition coefficients K_1 and K_2 have been previously determined. This will enable the analyst to determine recovery of the analyte from the matrix. These calculations are useful when equilibrium exists among the three phases.

Intuitively, it would appear that the techniques listed below should be signif-

---

*With solid samples, partitioning is also among three phases, but equilibrium is not achieved between the solid phase and the headspace. Sampling of solid phases is useful for qualitative identification of volatiles. Quantitative results are often achieved by preparing a slurry of the solid in a solvent prior to analysis.

icant in maximizing SPME sensitivity; the effectiveness of these measures will be discussed.

- Sampling the liquid rather than the headspace
- Mixing the sample during absorption
- Changing the partition coefficient between the liquid and headspace phases so that samples are driven into the headspace
- Changing the partition coefficient between the fiber and the other phases so that analytes are driven into the fiber
- Maximizing the ratio of liquid to headspace volumes in the vials
- Using larger sampling vials

Sampling the liquid rather than the headspace: After equilibrium is attained, the concentration of analytes in the fiber should be the same whether the fiber is immersed in the liquid or the headspace [59]. Figure 7 compares the rela-

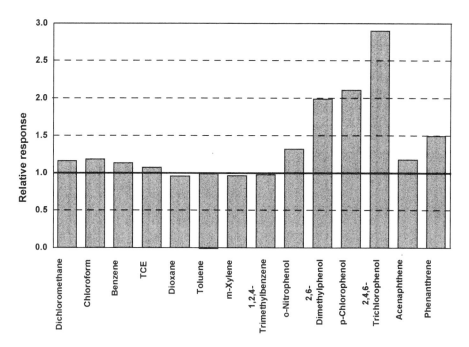

Fig. 7 Responses for several components (0.2–0.4 ppm in water at pH 2) were determined after SPME headspace and SPME liquid sampling. Absorption times were 10 min with 2 min desorption. The bars represent the detector response after liquid sampling, divided by the response from headspace sampling. Sample volumes were 1.2 for liquid sampling and 0.8 for headspace sampling—vial size 2.0 mL.

tive responses of compounds in a test mixture after SPME sampling of the liquid and the headspace. When the same mixture was sampled with a conventionally heated headspace, the phenols and polycyclic aromatic hydrocarbons (PAHs) were not detected. SPME headspace sampling is desirable if samples contain nonvolatile compounds such as salts and proteins. Practically, for compounds of very low volatility, liquid sampling is often preferable.

Mixing the sample during absorption: Mixing the sample is effective in increasing the response for compounds of low volatility, but is not necessary for volatiles. Figure 8 is a curve of mass absorbed by a fiber versus sampling time, with and without mixing. Note that with benzene, a volatile compound with a high diffusion rate, equilibrium and maximum response is achieved in less than 1 min, regardless of whether the sample is mixed. With o-xylene, mixing enhances the response approximately 10% after 2 min of sampling. With compounds of low volatility such as PAHs, mixing has a significant effect, whether the headspace or the liquid is sampled. With acenapthene, equilibrium was attained in 30 min with stirring, but the equilibrium time was reduced to 10 min when the rate of stirring was increased by one-third. Arthur et al. [57] explained

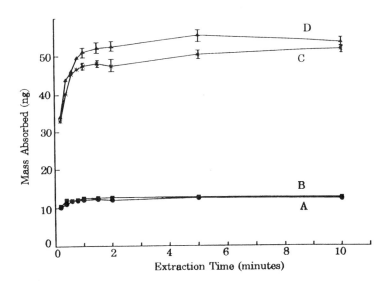

Fig. 8 Time profile of the mass absorbed by the fiber coating with a 1 ppm initial concentration of benzene and o-xylene aqueous solution: (A) benzene with a static aqueous phase; (B) benzene with a well-agitated aqueous phase; (C) o-xylene with a static aqueous phase; (D) o-xylene with a well-agitated aqueous phase. (Reprinted with permission from Ref. 59.)

this phenomenon by suggesting the existence of a depleted area around the fiber, due to slow diffusion of relatively large molecules in the liquid phase.

Changing the partition coefficient between the liquid and headspace phases so that samples are driven into the headspace and then into the fiber: Partitioning of organic analytes out of an aqueous phase can be effected by changing the composition of the liquid phase or by heating the sample. The addition of salt to aqueous samples is frequently used to drive polar compounds into the headspace [60]; it has a relatively insignificant effect on nonpolar compounds (Fig. 9).

Although heating the sample is often useful to enhance sensitivity in static headspace, it is less effective in SPME sampling. The difference arises from the fact that with SPME, there are three phases and heating alters the partitioning of the analyte between the headspace and the fiber to favor the headspace. Zhang and Pawliszyn [61] have shown that sensitivity can be enhanced significantly by heating the sample and simultaneously cooling the fiber with liquid CO_2. With wide fluctuations in temperature, thermostatting would be expected to improve precision; in such cases, an internal standard might have the same effect.

Changing the partition coefficient between the fiber and the other phases so that analytes are driven into the fiber: This would normally be achieved by changing to a fiber phase with better affinity for the analytes.

Maximizing the ratio of liquid to headspace volumes in the vials: Intuitively,

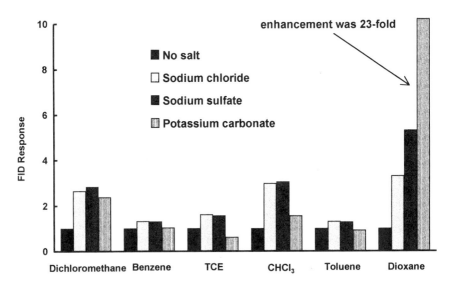

Fig. 9 Showing the effect of adding various salts to volatiles in water (1 ppm each) on the concentration in the headspace. "Salting out" had the most effect with dioxane, the most polar compound.

Table 6 Detector Area Counts after SPME Headspace Sampling of 100 ppm Methanol in Water in 2 ml Vials

	Sample volume	
	200 μL	600 μL
Vial 1	22,597	21,645
Vial 2	22,622	22,095
Vial 3	21,877	22,306

Note: Three vials were filled with 200 μL of sample and three vials were filled with 600 μL. Each vial was sampled once.

it might seem that, with a relatively more liquid phase, there would be a significant enhancement of sensitivity. It can be calculated that with polar compounds (where the partitioning favors the aqueous phase), there is virtually no increase in sensitivity when the ratio of the phases is altered. Table 6 demonstrates this with a sample of methanol in water. With nonpolar samples (low partition coefficients), sensitivity is enhanced when the proportion of liquid phase is increased; the magnitude of this effect varies with the partition coefficient.

Using larger sampling vials: As shown in Fig. 10, larger sampling vials are not effective in enhancing sensitivity if the relative volumes of headspace and liquid are the same; the advantage of larger vials is that they are easier to fill with solid and semisolid samples; furthermore, with solids, larger aliquots are likely to be more representative of the original sample.

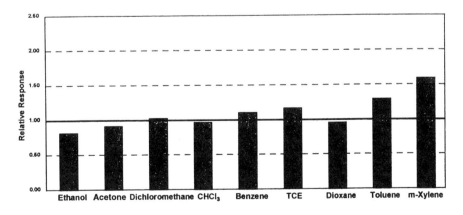

Fig. 10 Relative response with SPME sampling of the headspace over an aqueous mix in 16 mL vials versus 2 mL vials. In both cases, the liquid phase occupied 40% of the volume of the vial.

C. Optimizing SPME Sampling

Preparation of Standards and Samples

Experience has shown that poor precision and accuracy often result from improper sample handling prior to SPME sampling. Some guidelines are listed below:

- For standards and spiking solutions, it is a common practice to prepare a stock solution of analytes in methanol and then add a small aliquot to water. This method is acceptable with SPME and results are the same as adding organics directly to water if the total level of methanol is less than 1%.
- Saturation of standards and samples with salts such as sodium chloride or sodium sulfate is useful, not only to enhance sensitivity for polar analytes but also to minimize matrix differences when there are sample-to-sample variations in ionic strength.
- When determining acidic compounds such as phenols or basic compounds such as amines, the pH of the sample should be adjusted so that the analytes are in the nonionic state.
- When adding standards to aqueous samples, the concentrations of all the components should be low enough so that they remain in solution.
- The sampling vial should not be filled to the top. Immersion of the metal fiber-support rod in the liquid sample may result in sample carryover and the absorption and/or breakdown of analytes.

In general, extra care is required when handling standards containing volatiles. When preparing standards of volatiles in water, the liquid should fill the entire storage container without any headspace. Caution should be exercised during procedures such as making serial dilutions, adding salt, adjusting the pH, and transferring from the container to the sampling vial. Aqueous standards and samples remaining in the storage containers that were used to fill the sampling vials should not be used again.

Preparation of the GC Injector

To avoid breaking the fiber and to optimize the chromatography during the desorption step, an unpacked injector insert with an internal diameter of 0.75–0.80 mm should be used [62]. Figure 11 demonstrates the effect of the injector insert diameter on peak shape.

Users sometimes express concern that the sheath on the fiber rod might core the injector septum. This can be avoided by piercing a new septum prior to use, then removing and inspecting the new septum. If loose particles of septum material are present, they are removed and the septum is reinstalled. Figure 12 shows no change of retention time after 46 runs, indicating that the septum is intact. In

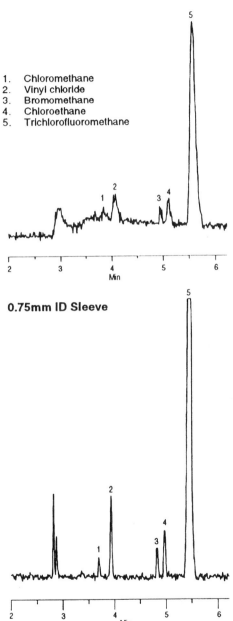

Fig. 11 Desorption of a SPME fiber in an injector with a 2 mm splitless insert versus a 0.75 mm splitless insert. (Reprinted with permission from Ref. 62.)

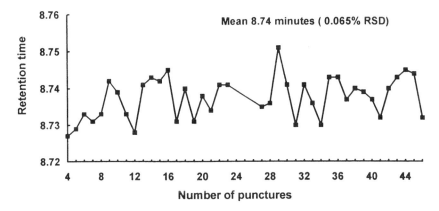

Fig. 12 Retention time stability over 46 injections into the same GC septum. Sample: nitrotoluene in water (1 ppm).

our laboratory, the practice is to change the septum and the fiber at the same time, after approximately 100 runs.

Achieving a Clean Blank Run

A new fiber must be conditioned by desorbing for a minimum of 5 min in an injector that is at least 10°C hotter than the temperature to be used during the analysis. The GC column should then be temperature programmed; this procedure should be repeated until there are no extraneous peaks. The amount of conditioning required is a function of the fiber phase and the detector. When interfering peaks persist after several conditioning runs, the problem can usually be traced to the septa installed in the sampling vials. It is sometimes necessary to bake the vial septa for several hours at 150°C to remove interfering compounds such as siloxanes and phthalates.

Selecting the Sampling Parameters

When a SPME method is being developed, a curve should be made by plotting various sampling times versus detector responses. The response with headspace sampling should also be compared with liquid sampling. Ideally, the sampling time should be no longer than the total GC cycle time minus the desorption time. If equilibrium is not achieved during this time, sample mixing might be required. Although good precision can be achieved without attaining equilibrium if timing is precise [62], for large molecules, mixing can have a dramatic effect on sensitivity.

Usually desorption is complete after 1 or 2 min, but this should be verified by inserting a blank after a sample to ascertain that there is no carryover.

Other Considerations

Use the guidelines in Table 5 when selecting a fiber. Fiber life will vary with experimental conditions, but, typically, there is no evident deterioration in chromatography even after 100 runs when desorbing into an injector heated to 220°C. This is often the case even when immersing the fiber into water that is saturated with salt and is at pH 2. One sign of an aging fiber is deterioration of precision. This might also be due to the aging of the septum. For this reason, it was recommended above that the septum and fiber be replaced at the same time.

We have observed that if a temperature-programmable injector is available, it is best to use it in the isothermal mode and inject at the temperature required to desorb the analytes. Cryogenics in the column oven are advantageous when the sample contains analytes with a wide range of volatility (e.g., vinyl chloride and PAHs); this is generally true and is not a characteristic of SPME sampling. If the sample contains only volatile compounds, it is possible to avoid cooling the column to subambient temperatures by selecting a column with a film thickness of 1 μm or greater.

D. Applications

Industrial

Solid-phase microextraction sampling has been useful in a number of industrial applications, including the determination of residual solvents in pharmaceuticals [63], monomers in polystyrene polymers [37], and volatiles in a corrosive industrial formulation [64]. Among the most promising SPME applications, however, is flavor and fragrance analysis (Fig. 13). Penton has shown that SPME has advantages over static headspace (SHS) for determining volatiles in a fruit beverage [32]. Table 7 compares the responses and precision for various volatiles in a fruit beverage when sampling with SPME at ambient temperatures versus sampling with SHS at 60°C.

In another study [36], terpene alcohols were detected at the ppb level in wine samples that contained 20% ethanol. With SHS, the terpene alcohols could not be detected below the ppm level because of absorption in the sample path.

Environmental

Recently, several articles have appeared in the literature in which SPME has been used in the analysis of environmental samples. A wide variety of pollutants have been determined in groundwater and drinking water, including pesticides [39,46], semivolatile nitrated aromatics [44], phenols and nitrophenols [40,43], BTEX's and petroleum compounds [65], PAHs and polychlorinated biphenyls [48], and volatiles [47,53]. Macgillivray et al. [45] compared SPME with purge and trap for substituted benzene compounds in water and Penton [47] compared

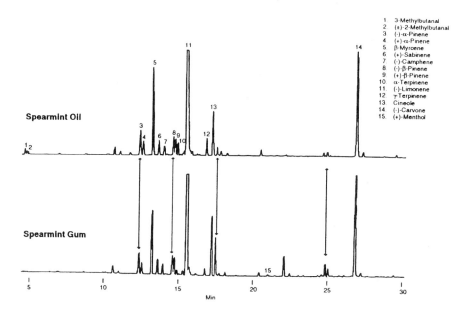

Fig. 13 Spearmint flavor components determined with SPME. [From *Supelco Bulletin* 869 (1995), reprinted with permission.]

Table 7 Detector Response of Volatiles in a Fruit Beverage When Sampled with SHS at 60°C and with SPME at Ambient Temperatures

	FID area counts	
Compound	SHS	SPME headspace
Ethyl acetate	64,257	11,732
Ethyl butyrate	14,690	13,085
Ethyl isovalerate	5,051	5,504
Isoamyl acetate	3,846	6,667
Ethyl valerate	742,651	1,364,563
Limonene	12,890	15,952
Benzaldehyde	285,258	709,229

SPME with static headspace, purge and trap, and direct injection, for several volatile analytes in water (Table 8).

In their study on phenols, Buchholz and Pawliszyn [43] used an in situ derivatization technique in which acetate derivatives were formed by bonding the derivatization reagent to the SPME fiber. Detection limits for the phenols varied from 0.01 to 1.6 ppb with GC/MS.

Solid-phase microextrusion has also been used for determining pollutants in air. Chai and Pawliszyn [38] measured several halogenated hydrocarbons plus benzene and toluene in air samples, including laboratory air and air collected in canisters. Samples were collected on a 100 μm polydimethylsiloxane (PDMS) fiber. The fiber with the absorbed volatile analytes could be stored at −70°C for up to 2 days.

Toxicology and Forensic Applications

Several volatile compounds are of interest in toxicology. Butler [66] reported the analysis of volatiles and ethylene glycol in urine. SPME appears to give results comparable to SHS for determining ethanol in blood [67] (Fig. 14). SPME fibers have been used for sampling in arson investigations. Future applications of SPME will probably include determination of drugs in biological fluids.

Table 8 Minimum Detection Limits (S/N=4) with Different Sample Introduction Methods

	Minimum detectable quantities (ppb)				
	SPME (ambient)	SHS (ambient)	SHS (heated[a])	Purge and trap	Direct injection
Dichloromethane	12	10	0.78	0.05	80
Chloroform	8.6	20	1.5	0.04	240
Benzene	0.3	1.4	0.1	0.003	17
Trichloroethylene	1.2	8.5	0.8	0.01	108
Dioxane	45	900	5.9	0.6	94
Toluene	0.18	2.2	0.2	0.003	20
m-Xylene	0.13	3.3	0.2	0.003	26
1,2,4-Trimethylbenzene	0.12	3.6	0.2	0.005	29

Note: SPME liquid and SPME headspace values were similar. For SPME sampling, absorption times were 15 min; desorption times were 1 min.
[a]Samples were analyzed by conventional SHS at 75°C.

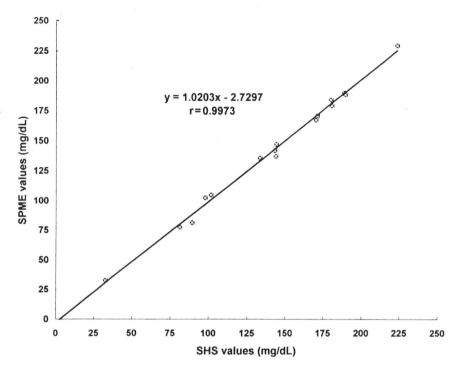

Fig. 14 Correlation between blood alcohol measured in blood and urine samples from drivers. Samples were split and measured with SPME headspace and with conventional static headspace.

IV. CONCLUSIONS

Both conventional SPE and SPME are helping laboratories to achieve the goals of minimizing the use of solvents for sample preparation and freeing workers from the tedium of sample cleanup. SPE can replace liquid–liquid extraction in many instances; SPME can be used instead of liquid–liquid extraction and/or static headspace. For some dirty samples, if may be of value to combine SPME and SPE. When a SPE procedure requires the elution of analytes into a buffer, rather than extracting into an organic solvent prior to injection into a GC, it would be convenient to use a SPME fiber to concentrate the analytes. Table 9 summarizes some of the main features of SPE and SPME.

Table 9 Summary of SPE and SPME Features

	SPE	SPME
Sample matrices	Liquids—organic or aqueous	Aqueous liquids, solids, gas (rarely)
Sample pretreatment	Blood, tissue, and some environmental samples usually require pretreatment to remove solids or proteins; many liquid samples do not	For determining volatiles, sample pretreatment usually not required. Analytes not amenable to headspace sampling may require some matrix cleanup
Analytes	Semivolatiles and slightly volatile compounds; recovery a problem with volatiles	Volatiles and semivolatiles
Use of organic solvents	Minimal—much less than with liquid–liquid extraction	None
Recovery of analytes	Generally, recoveries close to 100% expected	Equilibrium method—quantitation usually by comparing to spiked blank matrix or by standard additions.
Automation	Limited online for GC	Online available
Government regulatory agency approval	Several approved methods	Relatively new technique (no approved methods at the present time)

REFERENCES

1. J. V. Hinshaw, *LC–GC*, *13*: 536 (1995).
2. R. E. Majors, *LC–GC*, *13*: 742 (1995).
3. R. Hansson, C. Wiklund, E. Ponten, and K. Irgum, *Seventh Symposium on Handling of Environmental and Biological Samples in Chromatography*, Lund, Sweden, May 7–10, 1995.
4. N. Simpson and K. C. Van Horne (eds.), *Sorbent Extraction Technology Handbook*, 2nd ed., Varian Sample Preparation Products, Harbor City, CA, 1993.
5. Z. Penton, *GC Application Note No. 30*, Varian Chromatography Systems, Walnut Creek, CA, 1990.
6. K. Grob and B. Schilling, *J. Chromatogr.*, *391*: 3 (1987).
7. T. Hankemeir, A. J. H. Louter, F. D. Rinkema, and U. A. T. Brinkman, *Chromatographia*, *40*: 119 (1995).
8. M. Veringerová, V. Prachar, J. Uhnák, M. Lukácsová, and T. Trnovec, *J. Chromatogr. B*, *657*: 103 (1994).
9. C. C. Stevenson, D. L. Cibull, G. E. Platoff, D. M. Bush, and J. A. Gere, *J. Anal. Toxicol.* *16*: 337 (1992).

10. V. Dixit, H. Nguyen, and V. M. Dixit, *J. Chromatogr.*, *563*: 379 (1991).
11. L. Virag, S. Jamdar, C. R. Chao, and H. O. Morishima, *J. Chromatogr. B*, *658*: 135 (1994).
12. A. Hernandez, W. Andollo, and W. L. Hearn, *Forensic Sci. Int.*, *65*: 149 (1994).
13. Y. Gaillard, J. P. Gay-Montchamp, and M. Ollagnier, *J. Chromatogr.*, *622*: 197 (1993).
14. J. G. Langner, B. K. Gan, R. H. Liu, L. D. Baugh, P. Chand, J.-L. Weng, C. Edwards, and A. S. Walia, *Clin. Chem.*, *37*: 1595 (1991).
15. X.-H. Chen, J.-P. Franke, J. Wijsbeek, and R. A. de Zeeuw, *J. Anal. Toxicol.*, *18*: 150 (1994).
16. E. Venturelli, A. Manzari, A. Cavalleri, M. Benzo, G. Secreto, and E. Marubini, *J. Chromatogr.*, *582*: 7 (1992).
17. M. Saha and R. W. Giese, *J. Chromatogr.*, *629*: 35 (1993).
18. Z. Cai, V. M. Sadagopa Ramanujam, D. E. Giblin, M. L. Gross, and R. F. Spalding, *Anal. Chem.*, *65*: 21 (1993).
19. S. Scott, *Analyst*, *118*: 1117 (1993).
20. C. de la Colina, A. Peña Heras, G. Dios Cancela, and F. Sánchez Rasero, *J. Chromatogr. A.*, *655*: 127 (1993).
21. G. E. Miliadis, *Bull. Environ. Contam. Toxicol.*, *52*: 25 (1994).
22. M. Psathaki, E. Manoussaridou, and E. G. Stephanou, *J. Chromatogr. A.*, *667*: 241 (1994).
23. J. S. Salau, R. Alonso, G. Batlló, and D. Barceló, *Anal. Chim. Acta*, *293*: 109 (1994).
24. K. S. Yook, S.-M. Hong, and J.-H. Kim, *Anal. Sci. Technol.*, *7*: 441 (1994).
25. I. S. Deans, C. M. Davidson, D. Littlejohn, and I. Brown, *Analyst*, *118*: 1375 (1993).
26. K. Johansson, U. Örnemark, and Å. Olin, *Anal. Chim. Acta*, *274*: 129 (1993).
27. V. Häkkinen, *J. High Resolut. Chromatogr.*, *14*: 811 (1991).
28. G. C. Slack, H. M. McNair, S. B. Hawthorne, and D. J. Miller, *J. High Resolut. Chromatogr.*, *16*: 473 (1993).
29. C. L. Arthur and J. Pawliszyn, *Anal. Chem.*, *62*: 2145 (1990).
30. C. L. Arthur, I. M. Killam, K. D. Buchholz, J. Pawliszyn, and J. R. Berg, *Anal. Chem.*, *64*: 1960 (1992).
31. F. Pelusio, T. Nilsson, L. Montanarella, R. Tilio, B. Larsen, S. Facchetti, and J. Madsen, *J. Agric. Food Chem.*, *43*: 2138 (1995).
32. Z. Penton, *Food Testing Anal. 2*: 16 (1996).
33. X. Yang and T. Peppard, *J. Agric. Food Chem.*, *42*: 1925 (1994).
34. S. B. Hawthorne, D. J. Miller, J. Pawliszyn, and C. L. Arthur, *J. Chromatogr.*, *603*: 185 (1992).
35. C. Wooley and V. Mani, *Supelco Reporter*, *13*: 9 (1994).

36. Z. Penton, *SPME Application Note 6*, Varian Chromatography Systems, Walnut Creek, CA, 1995.
37. Z. Penton, *SPME Application Note 7*, Varian Chromatography Systems, Walnut Creek, CA, 1995.
38. M. Chai and J. Pawliszyn, *Environ. Sci. Technol.*, *29*: 693 (1995).
39. R. Eisert and K. Levsen, *J. Am. Soc. Mass Spectrom.*, *6*: 1119 (1995).
40. B. Schaefer and W. Engewald, *Fresenius' J. Anal. Chem.*, *352*: 535 (1995).
41. B. L. Wittkamp and D. C. Tilotta, *Anal. Chem.*, *67*: 600 (1995).
42. G. Brand, *Proc. Water Qual. Technol. Conf., Part I*: 273, San Francisco, CA, 1994.
43. K. D. Buchholz and J. Pawliszyn, *Anal. Chem.*, *66*: 160 (1994).
44. J. Horng and S. Huang, *J. Chromatogr. A*, *678*: 313 (1994).
45. B. Macgillivray, J. Pawliszyn, P. Fowlie, and C. Sagara, *J. Chromatogr. Sci.*, *32*: 317 (1994).
46. L. Nolan, R. Shirey, and R. Mindrup, *Proc. Water Qual. Technol. Conf., Part II*: 1761, San Francisco, CA, 1994.
47. Z. Penton, *Proc. Water Qual. Technol. Conf., Part I*: 1027, San Francisco, CA, 1994.
48. D. W. Potter and J. Pawliszyn, *Environ. Sci. Technol. 28*: 298 (1994).
49. L. P. Sarna, G. R. B. Webster, M. R. Friesen-Fischer, and R. S. Ranjan, *J. Chromatogr. A*, *677*: 201 (1994).
50. R. E. Shirey, *Kankyo Kagaku*, *4*: 496 (1994).
51. C. L. Arthur, K. D. Buchholz, D. W. Potter, S. Motlagh, L. Killam, and J. Pawliszyn, *Proc. Water Qual. Technol. Conf., Part II*: 1315, Miami Beach, Fla, 1993.
52. K. D. Buchholz and J. Pawliszyn, *Environ. Sci. Technol.*, *27*: 2844 (1993).
53. M. Chai, C. L., Arthur, J. Pawliszyn, R. P. Belardi, and K. F. Pratt, *Analyst*, *118*: 1501 (1993).
54. R. Mindrup and R. Shirey, *Proc. Water Qual. Technol. Conf. Part II*: 1545, Miami Beach, Fla, 1993.
55. Z. Zhang and J. Pawliszyn, *J. High Resolut. Chromatogr.*, *16*: 689 (1993).
56. C. L. Arthur, D. W. Potter, K. D. Buchholz, S. Motlagh, and J. Pawliszyn, *LC–GC*, *10*: 656 (1992).
57. C. L. Arthur, L. M. Killam, S. Motlagh, M. Lim, D. W. Potter, and J. Pawliszyn, *Environ. Sci. Technol.*, *26*: 979 (1992).
58. T. Nilsson, F. Pelusio, L. Montanarella, B. Larsen, S. Facchetti, and J. Madsen, *J. High Resolut. Chromatogr.*, *18*: 617 (1995).
59. Z. Zhang and J. Pawliszyn, *Anal. Chem.*, *65*: 1843 (1993).
60. B. V. Ioffe and A. G. Vitenberg (Eds.), *Headspace Analysis & Related Methods in Gas Chromatography*, John Wiley & Sons, New York, 1984, pp. 61–62.
61. Z. Zhang and J. Pawliszyn, *Anal. Chem.*, *67*: 341 (1995).

62. R. E. Shirey, *Supelco Reporter*, *13*: 2 (1994).
63. L. C. Nelson, A. Smith, S. Scypinski and S. R. Shaw, *Eastern Analytical Symposium*, Nov. 1994.
64. Z. Penton, *SPME Application Note 8*, Varian Chromatography Systems, Walnut Creek, CA, 1995.
65. T. Górecki and J. Pawliszyn, *J. High Resolut. Chromatogr.*, *18*: 161 (1995).
66. M. G. Butler, *Eastern Analytical Symposium*, Nov. 1995.
67. Z. Penton, *SPME Application Note 9*, Varian Chromatography Systems, Walnut Creek, CA, 1995.

6
Capillary Electrophoresis of Proteins

Tim Wehr, Roberto Rodriguez-Diaz, and Cheng-Ming Liu
Bio-Rad Laboratories, Hercules, California

I.	INTRODUCTION	238
II.	STRATEGIES FOR REDUCING PROTEIN–WALL INTERACTIONS	240
	A. Operation at pH Extremes	240
	B. Use of Buffer Additives	241
	C. Capillary Coatings and Other Surface Modifications	243
III.	CAPILLARY ZONE ELECTROPHORESIS	248
	A. Sample Preconcentration Techniques	249
	B. Enzyme Assays	258
	C. Affinity Capillary Electrophoresis	261
	D. Chiral Separations	265
	E. Analysis of Protein Folding	269
	F. Analysis of Milk and Dairy Products	270
	G. Metalloproteins	276
	H. Glycoproteins	279
	I. Cereal Proteins	280
IV.	CAPILLARY ISOELECTRIC FOCUSING	282
	A. The CIEF Process	284
	B. Capillary IEF Optimization	295
	C. Applications	303

V.	SIEVING SEPARATIONS	310
	A. Analysis of Native Proteins	311
	B. Analysis of SDS–Protein Complexes	312
VI.	CLINICAL APPLICATIONS	323
	A. Serum Proteins	323
	B. Urine Proteins	335
	C. Cerebrospinal Fluid	339
	D. Hemoglobin	340
	E. Enzymes or Isoenzymes	345
	F. Peptide Hormones	350
	G. Soluble Mediators from Immune Cells	350
	H. Lipoproteins	351
	REFERENCES	352

I. INTRODUCTION

Separation technology has been central to the elucidation of protein structure and function. Both chromatography and electrophoresis have been used for decades to isolate and characterize proteins and their components. As powerful as these protein separation techniques are, they are not without limitations. Chromatographic separations are based on the interaction of an analyte with the surface of the stationary phase. However, proteins are by nature very surface-active molecules. They also possess low diffusion constants and display poor mass-transfer kinetics during the chromatographic process. As a consequence, resolution is often less than desired, protein recovery may be low, and native proteins may be denatured during the separation. Gel electrophoresis, on the other hand, is a laborious and time-consuming technique requiring preparation of the gel, separation of the sample, staining and destaining, and gel-drying. The gel must be treated with a dye or stain to visualize the separated proteins, and because the uptake of stain many occur in a nonlinear fashion, the intensity of the stained bands may be poorly correlated with amount of protein. For this reason, gel electrophoresis is, at best, a semiquantitative technique.

Capillary electrophoresis (CE) is a relatively new separation technology which combines aspects of both electrophoresis and high-performance liquid chromatography (HPLC). Like electrophoresis, the separation depends on differential migration in an electrical field. Since its first description in the late 1960s, capillary electrophoresis techniques analogous to most conventional electrophoretic methods have been demonstrated: zone electrophoresis, displacement electrophoresis, isoelectric focusing, and sieving separations. Unlike conventional electrophoresis, however, most separations are performed in free solution without the requirement for casting a gel. As in HPLC, detection is accomplished as the separation progresses, with resolved zones producing an electronic signal as they migrate past the monitor point of a concentration-sensitive (e.g., UV absorbance or fluorescence) detector. Therefore, the need for staining and destaining is eliminated. Data presentation and interpretation are also similar to HPLC; the output (peaks on a baseline) can be displayed as an electropherogram and integrated to produce quantitative information in the form of peak area or height. Capillary electrophoresis and HPLC share similar sample introduction methods: A single sample is injected at the inlet of the capillary and multiple samples are analyzed in serial fashion. This contrasts to conventional electrophoresis in which multiple samples are frequently run in parallel as lanes on the gel. This limitation of CE in sample throughput is compensated by the ability to process samples automatically using an autosampler. Compared to its elder cousins, CE is characterized by high resolving power, sometimes higher than electrophoresis or HPLC. The high separation efficiencies (plate numbers) obtained in CE result from the absence of contributions to band-broadening other than diffusion. The use of narrow-bore capillaries with excellent heat-dissipation properties enables the use of very high field strengths (sometimes in excess of 1000 V/cm), which decreases analysis time and minimizes band diffusion. When separations are performed in the presence of electroosmotic flow (EOF), the plug-flow characteristics of EOF also contribute to high efficiency. In contrast, the laminar flow properties of liquid chromatography increase resistance to mass transfer, reducing separation efficiency.

Because of these many advantages, CE shows great promise as an analytical tool in protein chemistry. In some cases it may replace HPLC and electrophoresis, but more often it is used in conjunction with existing techniques, providing a different separation selectivity, improved quantitation, or automated analysis. An anticipated benefit of performing protein separations in open-tubular capillaries was the reduced potential for surface interactions. In fact, this proved not to be the case; the high surface-to-volume ratio of the capillaries and the high surface activity of the fused silica capillary wall has proven to be a major problem in applying CE to protein separations. Much of the research in separation chemistries and capillary wall modifications has been directed toward improving CE performance in protein separations.

Many general reviews of capillary electrophoresis to have been published [1–3], as well as specific reviews of application of CE protein analysis [4–7]. This review will focus primarily on new developments. Because the literature base in this field is large and expanding rapidly, it is not possible to cover every reported application, and we have elected to concentrate on contributions of wider interest. The review is organized by separation mode with selected applications in each mode discussed in detail. As noted above, protein–wall interactions are a central problem in all applications of CE to protein analysis; it is therefore appropriate to begin with a discussion of approaches to minimizing wall effects.

II. STRATEGIES FOR REDUCING PROTEIN-WALL INTERACTIONS

Interactions of protein with the inner wall of fused silica capillaries has been the major obstacle to successful application of capillary electrophoresis to protein separations. At pH values above about 2, the weakly acidic silanol groups on the capillary surface become ionized, and the charge density on the wall increases with pH to a maximum of about 10, at which point the silanol groups are fully dissociated. This characteristic of fused silica has two consequences for protein analysis. First, proteins with basic amino acid residues positioned on the protein surface can participate in electrostatic interactions with ionized silanols. Protein adsorption at the capillary wall can result in band-broadening, tailing, and, in the case of strong interaction, reduced detector response or complete absence of peaks. Second, changes in the state of the wall during an analysis or from run to run can alter the magnitude of EOF, resulting in changes in analyte migration times and peak areas. Protein adsorption can alter the zeta potential of the capillary wall, changing EOF and degrading reproducibility. Three strategies have been employed to minimize protein–wall interactions: operation at extremes of pH, use of buffer additives, and use of wall-coated capillaries.

A. Operation at pH Extremes

The simplest approach to minimizing protein–wall interaction is to use a buffer pH at which interactions do not occur. Neutral and acidic proteins (pI $\leq$ 7) will carry net negative charges; under alkaline conditions, the capillary surface will also be anionic due to silanol ionization, and protein–wall adsorption will be prevented by columbic repulsions. The same approach can be used for basic proteins (pI 7–10) if very alkaline buffers are used (e.g., pH 11–12). This approach has been used successfully in some cases [8], but inadequate selectivity and the risk of protein degradation under very alkaline conditions has prevented this from being a general strategy for protein separations. An alternative strategy is operation at very low pH (<3). Under these conditions, the degree of silanol ionization and the magnitude of EOF are very low. At such low pH values, most proteins will bear a net positive charge and (in the absence of significant EOF) migrate electrophoretically toward the cathode. Phosphate buffers have proven to be very effective for operation in this pH range because of their high buffering capacity and low UV absorbance. In addition, McCormick [9] has suggested that complexation of phosphate groups with surface silanols contributes to reduced EOF and reduced polypeptide adsorption. However, although use of low-pH conditions has been remarkably successful for separations of complex mixtures of peptides, the approach has been less successful for proteins due to limited selectivity.

Clearly, successful application of capillary electrophoresis to most protein separation problems requires the ability to operate under conditions where protein charge differences are greatest and, preferably, where the proteins are in their native states. Consequently, since the introduction of CE over a decade ago, there have been tremendous efforts to develop conditions for protein analysis at physiological pH in which protein adsorption is minimized and EOF is either minimized or stably controlled.

B. Use of Buffer Additives

The use of fused silica capillaries has advantages in simplicity, low cost, and good capillary lifetime. For this reason, many investigators have searched for buffer components and buffer additives which would permit protein separations to be achieved with uncoated capillaries.

Green and Jorgenson[10] proposed the use of high concentrations of monovalent alkali metal salts to reduce protein adsorption and determined the order of effectiveness to be $Cs^+ > K^+ > Na^+ > Li^+$. The counterion had little effect on adsorption but, in many cases, contributed unacceptably high UV absorbance. Chen et al. [11] applied this approach to the separation of milk proteins. Although effective in reducing protein–wall interactions, the use of high ionic strength electrolytes generates excessive joule heat. This can be circumvented by the use of capillaries with a small internal diameter or by operation at lower field strengths, approaches which compromise detection sensitivity and analysis time.

A variety of amine-containing organic bases has been used as additives to reduce protein adsorption (see Table 1). Alkyl amines added to the electrophoresis buffer under conditions where they are protonated will participate in electrostatic interactions with silanols on the capillary wall, stabilizing the state of the double layer and the level of EOF [21,22]. Protein–wall interactions are reduced, resulting in improved peak shape and separation efficiencies. At very low pH, alkylamine additives reverse the direction of EOF toward the anode [23,24]. Use of amine additives increases operating current, and protein-additive interactions may affect the separation. The latter phenomenon may be advantageous in some cases; Landers et al. [19,30] have demonstrated that the use of alkyl diamines and bis-quaternary ammonium alkanes provided improved separation of protein glycoforms, which was suggested to arise in part from additive-protein interactions.

The use of zwitterions as buffer additives for capillary electrophoresis was first proposed by Bushey and Jorgenson [12]. Zwitterions reduce protein–wall interactions without increasing the conductivity of the electrophoresis buffer. Ideally, the additive should be zwitterionic over a wide pH range, have good solubility to enable use at high concentrations, produce a stable electroosmotic flow, and exhibit low UV absorbance. These authors demonstrated that zwitterions such as betaine and sarcosine could be used to improve protein separations, although best results required use of these zwitterions in combination with salts.

Table 1 Additives Used in CZE

Additive	Application	Ref.
Betaine	Basic proteins	12
Cadaverine	Acidic and basic proteins	13
Ethylene glycol (20%)	Acidic and basic proteins, serum proteins	14
Cationic and zwitterionic fluorosurfactants	Basic proteins	15–17
2-(N-cyclohexylamino)ethanesulphonic acid (CHES)	Insulins	18
N,N-bis(2-hydroxyethyl)-2-aminoethanesulphonic acid (BES)	Insulins	18
3-[(1,1-dimethyl-2-hydroxyethyl)amino]-2-hydroxypropanesulphonic acid (AMPSO)	Insulins	18
3-(cyclohexylamino)-2-hydroxy-1-propanesulphonic acid (CAPSO)	Insulins	18
1,4-Diaminobutane	Protein glycoforms	19–21
Triethylamine	Basic proteins	22, 24
Triethanolamine	Basic proteins	24
Galactosamine	Basic proteins	23
Glucosamine	Basic proteins	23
Trimethylammonium propylsulfonate (TMAPS), trimethylammonium chloride (TMAC)	Monoclonal antibody	25
Trimethylammonium propylsulfonate (TMAPS)	Acidic, neutral, and basic proteins	26
(Trimethyl)ammonium butylsulfonate (TMABS)	Acidic, neutral, and basic proteins	26
2-hydroxyl-3-trimethylammonium propylsulfonate (HTMAPS)	Acid, neutral, and basic proteins	26
3-(Dimethyldodecylammonio)propanesulfonate	Basic proteins	27
Cetyltrimethylammonium bromide	Acidic and basic proteins	28
Chitosan	Basic proteins	29
Amino acids	Basic proteins	22
Hexamethonium bromide, hexamethonium chloride	Protein glycoforms	30
Decamethonium bromide	Protein glycoforms	30
Polydimethyldiallylammonium chloride	Basic proteins	31
Phytic acid	Acidic and basic proteins	32
Ethylenediamine	Basic proteins	21
1,3-diaminopropane	Basic proteins	21
N,N-Diethylethanolamine	Basic proteins	22
N-Ethyldiethylamine	Basic proteins	22
Triethanolamine	Basic proteins	22

Since this work was published, a wide variety of zwitterionic compounds have been used for capillary electrophoresis separations of proteins, and these are summarized in Table 1.

C. Capillary Coatings and Other Surface Modifications

Surface modifications to the capillary wall may be grouped into two categories: covalent and dynamic coatings. Covalent coatings are attached to the wall through a chemical bond, whereas dynamic coatings physically interact with the silica or with a first layer attached to the silica. Use of dynamic coatings generally requires the incorporation of the wall-interacting compound in the run or conditioning buffer.

Capillary coatings operate by physically blocking the access of the sample molecules to the capillary wall. Coatings and other inner wall modifications are an area of intense research in CE. Much controversy has been generated over which approach is more convenient and produces a more stable modification, thus yielding better and more reproducible results. Perhaps a missing ingredient in the discussion of coatings is the reproducibility and effectiveness of available protocols (e.g. methods fully optimized and standardized for particular coating procedures).

Coatings usually consist of one or two layers. One-layer coatings use a compound that can be reacted with the capillary wall to mask the charges typical of fused silica and/or to change the characteristics (sometimes dramatically) of the capillary surface. In two-layered coatings, the first layer (e.g., bifunctional moieties, hydrophobic layers) is used to anchor the second layer (often polymeric), which typically defines the characteristics of the surface. The main differences among coatings are the type of first layer, and/or type of anchorage bond used, and the composition of the second layer. To be useful for routine analysis, a coating must be stable for prolonged periods of time under typical analysis conditions. Loss of coating and changes due to interaction with sample or other buffer components leads to degradation of efficiency and poor reproducibility.

The diversity of chemistries described in the literature reflects the continuous search for more stable coatings and perhaps the inadequacy of any single approach to provide satisfactory results for all applications. To the CE practitioner who prepares his/her own capillaries, a particular problem arises because of the high number of nonoptimized coating "recipes" that are continuously being published and because once optimized, the originally explicit chemistry may be lost behind a commercial trade name.

Coating comparisons do not always lead the user to better chemistries because optimization is achieved for only one of the methods compared. Hjertén noted that even the same coating process does not produce the same quality coating from column to column, and this may reflect variations in the quality of the fused silica capillary. Thus, he suggested that capillary pretreatment and coupling chemistries may have to be optimized for different sources of capillary or even for different batches of silica.

Coating chemistries have been extensively reviewed previously [33–35] and a description of each coating is beyond the scope of this chapter. Nevertheless, some historically important and new developments can be addressed.

One-Layer Wall Modifications

Silanized capillaries with carbon moieties ranging from 1 to more than 18 have been used to render the capillary wall more hydrophobic. In most cases, buffer additives are used to avoid sample adsorption and/or to control EOF.

Polybrene is a positively charged polymer that absorbs tenaciously to the silica wall and has been used to reverse the charge of the capillary wall [36], thus allowing the analysis of basic proteins at pH values below their pI (Fig. 1). This coating has also been employed for the analysis of small molecules (e.g., inorganic ions).

Several compounds have been used to physically cover the capillary wall by filling the column with a solution containing the material, flushing it out (usually with an inert gas), leaving a thin layer of the compound deposited on the wall when the solvent evaporates. Typically, the solute used is not easily dissolved by the buffer system used for electrophoresis. Polyvinylalcohol [46,47] and cellulose acetate coatings are examples employing this technique [42].

Two-Layer Wall Modifications

Polyacrylamide (PA) Coatings. One of the most popular wall coverage materials is polyacrylamide, and in this section we describe techniques that maintain the second layer (polyacrylamide) constant but differ in the way it is anchored to the wall.

Hjertén [52]) introduced a new method of coating capillaries shortly after successfully adapting the IEF process to the capillary format. In his approach, linear hydrophilic polymers are chemically attached to the wall via a bifunctional group (siloxane chemistry), which also provides a vinyl group used to anchor the polyacrylamide chains polymerized *in situ*. Hjertén not only published a procedure to synthesize coated capillaries but also described the theoretical basis for elimination of nonspecific adsorption and EOF. His discovery that EOF is reduced as the viscosity near the ionic double layer (Stern layer) is increased has been exploited using other ligands and bonded chemistries and is also behind the rationale of some dynamic coatings. Typically, a well-coated linear PA capillary exhibits over a 50-fold EOF reduction as compared with uncoated columns.

Polyacrylamide has been attached to a vinyl group directly anchored to a chlorinated silica by reaction with a Grignard reagent [45]. This reaction produced a silicon-carbon bond. The vinyl group was then reacted with acrylamide. Through this procedure, electroosmosis is virtually eliminated. The silicon–carbon bond provided high stability, even when exposed to pH 10.5 for 5 days.

Linear polyacrylamide can be cross-linked after attachment to a capillary

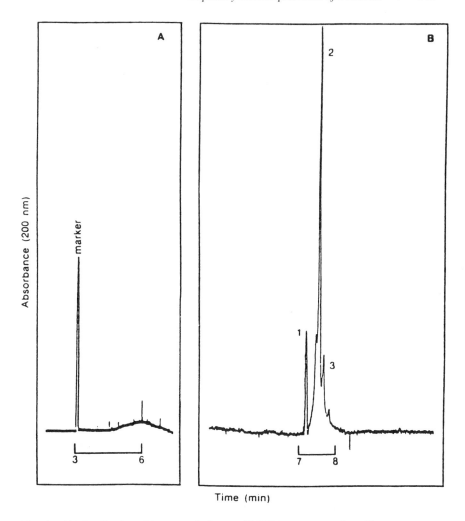

Fig. 1 (A) Application of isoenzymic forms of LDH in an uncoated capillary. Conditions: field strength, 400 V/cm; current, 4 µA; buffer, 5 mM sodium Phosphate, pH 7.0; capillary length, 90 cm (68 cm effective length); capillary diameter, 50 µm; separation temperature, 30°; polarity normal. (B) Separation of LDH isoenzymes under conditions of capillary charge reversal. Peaks 1, 2, and 3 are isoenzymes of pI 8.3, 8.4, and 8.55, respectively. Conditions same as (A) except that polarity is reversed. Horizontal scale in minutes. (Reproduced from Ref. 36 with permission.)

first treated with 7-oct-1-enyltrimethoxysilane [50]. This silane is known to cross-link extensively and, according to the authors, renders a more stable first layer to which the polymer is anchored.

Karger et al. [40] developed a coating that used vinyl siloxanediol to anchor linear or cross-linked acrylamide. The hydroxyl groups of the siloxanediol were used as cross-linkers in the first layer. Acrylamide was reacted with the vinyl group of the deposited siloxanediol; then the resulting polyacrylamide chains were cross-linked by filling the column with a 37% solution of formaldehyde (pH 10, adjusted with NaOH) and incubating for 3 h. Linear PA reduced EOF more efficiently than cross-linked PA, but the latter showed a lower degree of EOF change as a function of operating time at a pH of 8.8.

Other Polymeric Coatings. Polymeric coatings can also be used to produce capillaries with a desired level of EOF [41]. In one case, the column was covered with a mixture of monomers of acrylamide and 2-acrylamido-2-methyl-1-propanesulfonic acid (AMPSA). The tube was treated with a bifunctional silane before polymerization of the final layer. The resulting capillaries exhibited a remarkably constant EOF value over a pH range of 2–9 as compared with bare silica capillaries. Uses of constant EOF capillaries include the performance of MECC at acidic pH values which generate low EOF and therefore exhibit prolonged analysis time.

Some natural polymers exhibit very high hydrolytic stability even at extreme pH, and for that reason Hjertén and Kubo [43] attached dextrans and cellulose derivatives, and Mechref and El Rassi [44] attached high-molecular-weight dextrans to the capillary wall. The method described by El Rassi is composed of three steps: first the silica was treated with an oligomeric epoxysilane by incubating at 96°C; then the dextran was anchored to the first layer in the presence of boron trifluoride; finally, the dextran was cross-linked with diepoxypolyethylene glycol in the presence of boron trifluoride. Capillaries prepared by the above method were exposed to various harsh conditions (e.g., dilute NaOH) to test for stability, and a group of selected proteins was analyzed to demonstrate the performance of the columns. EOF reduction was found to be affected by the size of the dextran used, with the lowest EOF achieved with a 150 kD dextran (7–10 times the EOF reduction as compared with uncoated capillaries). Model basic proteins analyzed at pH 5 yielded up to 750,000 plates/m (Fig. 2). Acidic proteins showed a decreased number of plates per meter (111,000).

Cellulose derivatives have also been attached to the capillary wall using several procedures [49,50]. Cellulose-coated capillaries were found to withstand 30 days of exposure to 10mM NaOH (pH 12) [42].

A problem encountered in our laboratory working with natural polymers is that they tend to provide a less efficient wall coverage as judged by the presence of higher EOF values compared to compounds polymerized *in situ*. One possible

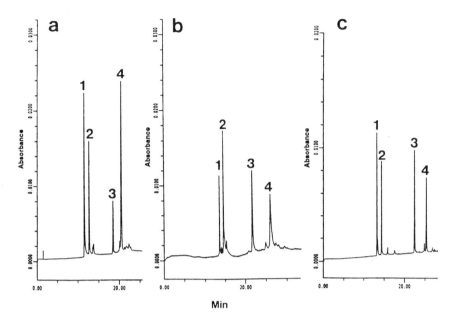

Fig. 2 Typical electropherograms of four standard basic proteins obtained with capillaries coated with dextran 45 kDa (a), dextran 71 kDa (b), and dextran 150 kDa (c). Running electrolyte: 100 mM phosphate, pH 5.0; pressure injection 1 s; applied voltage, 15 kV. Samples: 1, lysozyme; 2, cytochrome c; 3, ribonuclease A; 4, α-chymotrypsinogen A. (Reproduced from Ref. 44 with permission.)

explanation is the steric hindrance of the attached molecules with the incoming ones to unoccupied available patches on the column surface. A good coverage would require a perfect ordering side by side of such bulky molecules.

Righetti [51] developed N-acryloylaminoethoxyethanol (AAEE) monomers which can be used to form a polymeric coating. He demonstrated that this compound exhibits a much higher hydrolytic stability than acrylamide and is also more hydrophilic. Coatings prepared with this material show improved performance and higher stability (the authors' laboratory, unpublished data).

Surfactant-Coated Capillaries. Towns and Regnier [48] derivatized the capillary surface with octadecyltrichlorosilane to form a covalently bonded hydrophobic phase. When an aqueous solution of a nonionic oxyethylene-based surfactant (e.g., Brij 35) at a concentration above its critical micelle concentration was introduced into the capillary, the hydrophobic portion of the surfactant formed an absorbed layer on top of the covalently bonded (silane) hydrophobic layer. The polar head group of the surfactant was exposed to the aqueous solu-

Table 2 Commercially Available Coated Capillaries

Coating type	Source	Characteristic	Applications
Linear polyacrylamide	Bio-Rad Laboratories	Linear; hydrophilic	Polypeptides peptides, nucleic acids, small molecules
Proprietary	Beckman Instruments	Hydrophilic	Polypeptides, nucleic acids
CElect-P	Supelco, Inc.	Hydrophilic	Polypeptides
CElect-H	Supelco, Inc.	Weakly hydrophobic (C_1)	CIEF
CElect-H1	Supelco, Inc.	Moderately hydrophobic (C_8)	
CElect-H2	Supelco, Inc.	Highly hydrophobic (C_{18})	
Micro-Coat	Applied Biosystems	Dynamic, cationic	Polypeptides
CE-100-C18	Isco, Inc.	Hydrophobic (for use with dynamic coatings)	Polypeptides
CE-200-glycerol	Isco, Inc.		Polypeptides
CE-300-sulfonic	Isco, Inc.		Nucleotides
DB1	J & W Scientific	Hydrophobic	CIEF
Proprietary	Dionex		Nucleic acids
MicroSolv CE	MicroSolveCE	Deactivated	Polypeptides
PEG coated	MicroSolveCE	Weakly hydrophilic	Polypeptides

tion filling the column, forming a hydrophilic layer that masked the hydrophobic layer and the unreacted silanol groups, thus effectively reducing protein interaction with the surface of the capillary.

A list of commercially available capillaries, their characteristics, and suggested applications are presented in Table 2.

III. CAPILLARY ZONE ELECTROPHORESIS

Capillary zone electrophoresis (CZE) offers several advantages in comparison to other CE separation modes. In most cases, CZE is simple and straightforward to perform: A single buffer is used throughout the capillary and electrode vessels, and the sample is introduced as a zone or plug at one end. Capillary preparation often involves only filling the capillary with the separation buffer, although uncoated capillaries generally require prior washing or conditioning steps. The technique is inexpensive in comparison to capillary gel electrophoresis or isoelectric focusing, as low-cost buffers and salts are used. Separation selectivity can be easily varied by manipulation of buffer pH or use of additives. For these reasons, CZE is the most widely used CE mode for protein separations.

The development of a CZE separation is typically a three-step process: determination of optimum buffer composition, selection of the appropriate capillary type, and determination of high-voltage power-supply parameters.

Buffer selection will be dictated by the pH value required to achieve satisfactory analyte mobility and resolution. The buffer should have good buffering capacity at this pH, should exhibit low UV absorbance at the detection wavelength (typically 190–200 nm for CZE) for good sensitivity, and low conductivity to minimize joule heating. Additives may be incorporated into the buffer to suppress protein–wall interactions or to control EOF (e.g., neutral salts, organic bases, zwitterions, neutral or charged polymers), to solubilize the analyte or reduce protein aggregation (e.g., surfactants, chaotropes, organic modifiers), or to modulate protein charge or shape (e.g., metal ions or chelating agents for metal-binding proteins, borate ion for glycoproteins).

Capillary selection includes choice of internal diameter (small i.d. to reduce joule heat and improve resolution, large i.d. for higher sensitivity or greater capacity for micropreparative applications) and length (longer lengths for improved resolution, particularly with EOF-driven separations, or shorter lengths for rapid separation, particularly with wall-coated capillaries). Capillary coatings, described in detail earlier, can be neutral to suppress protein adsorption and EOF or charged to reverse the direction of EOF and reduce adsorption of basic proteins.

Power-supply parameters include high-voltage setting (high field strength for rapid separations, low field strength to minimize joule heat) and polarity ("normal" polarity with detector at the cathodic end for EOF-driven separations, "reversed" polarity with detector at the anodic end for reversed EOF using cationic capillary coatings, or either polarity with neutral capillary coatings). The majority of protein separations are carried out under constant-voltage conditions, although constant current may be used in systems with poor temperature control.

The very small sample volumes injected into the capillary in CZE (typically a few nanoliters) limits detectivity, particularly for samples in which the analyte is at low concentration. Therefore, methods for preconcentrating the sample prior to or during injection can be highly desirable when analyzing very dilute protein samples.

A. Sample Preconcentration Techniques

Sample preconcentration techniques are used with two purposes: (1) to increase concentration in order to achieve detections; and (2) to eliminate disturbances of the electrophoretic system when the conductivity of the sample is significantly different than the background electrolyte (during hydraulic or electrokinetic [electrically driven injection in the presence of EOF] sample introduction).

Preconcentration to Improve Detection

There are many advantages to performing electrophoresis in narrow-bore capillaries, and theoretically decreasing the internal diameter would potentially provide higher resolution. The main problem associated with the use of capillaries is detection. Because the most widely used methods of detection in CE are based on optical absorptivity, the signal generated depends of the path length the light has to travel (along with the concentration and absorption coefficient of the sample). In CE, the optical path is usually only 25–100 μm (the i.d. of the capillary for online detection), so the signal generated by the detector is weak. For this reason, CE requires relatively high initial sample concentration.

For sample components present at low concentrations, several strategies have been described to preconcentrate. These techniques include zone sharpening [56,57], online packed columns [58], and transient CITP [59,60].

Principles of Concentration. Most concentration procedures utilize some form of transport phenomena to achieve concentration. The main idea is to mobilize (transport) sample molecules in space, slowing the velocity of the leading particles and/or accelerating the trailing ones (i.e., creating conditions for a differential velocity as a factor of position within the system), thus allowing the trailing molecules to catch up with the ones originally moving up front. Differential transport can be achieved by using temporary physical barriers and/or uneven transport force distribution. Chromatographic methods can also be used to trap the molecules present in low concentration but large volumes, and then desorb them with a reduced volume of solvent.

The principle of using physical barriers to concentrate is better illustrated by using as an example a classical technique such as ultrafiltration. In this method, the leading particles are stopped by a dialysis membrane, allowing the molecules in solution to eventually reach the same physical space at the membrane, thus increasing concentration. Because the membrane is semipermeable, desalting occurs concomitantly with concentration.

Electrophoretic techniques can be used to concentrate by manipulating the voltage distribution along the capillary axis or by changing the analysis conditions [59] (e.g., pH) after the concentration step has been achieved. Two concentration methods based on voltage distribution are described in the following sections: zone sharpening and isotachophoresis.

Methods for Preconcentration

Physical Barriers. For low volumes of solutes, Hjertén et al. [62] developed physical barriers (gels, dialysis membranes) directly attached to one end of the capillary. In this approach, the capillary is filled with the sample diluted in the leading buffer (ITP is later used as a "mobilization" step) and the tip of the column is pressed against a polymerized acrylamide gel ($T=40\%$, $C=3\%$) containing buffer, and then drawn slowly off the gel. This process leaves a gel plug equal to the depth to which the capillary is pushed into the gel. An electric field is applied with the polarity set so that migration of the proteins proceeds in the direction of the gel. Since the pores of the gel are very small, the proteins accumulate on its surface. Once the concentration process is completed, a mobilization step (e.g., ITP) using a terminating buffer for a short period of time is used to avoid peak-broadening. After mobilization, the separation proceeds (as CZE) by replacing the trailing buffer for a vessel containing leading electrolyte (Fig. 3). When the physical barrier is a dialysis membrane, the proteins may be detached from the membrane using a hydrodynamic force (e.g., by raising the electrode vessel closest to the membrane).

Electrical Force Distribution and/or Mobility Manipulation. The first and simplest method for online sample preconcentration is zone sharpening by stacking [66]. In this procedure, the sample is dissolved in a buffer of lower conductivity (most commonly the diluted run buffer) than the run buffer. Upon applying

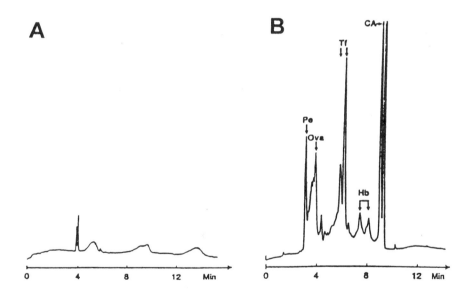

Fig. 3 High-performance zone electrophoresis of model proteins. (A) Prior to concentration (concentration of each protein: 20 μg/mL); (B) following concentration toward a small-pore polyacrylamide gel and a short mobilization by displacement electrophoresis. The width of the applied sample zone was in (A) 3-4 mm and in (B) 140 mm (= the length of the capillary). Following concentration, the zone width in (B) was about 0.2 mm. (Reproduced from Ref. 63 with permission.)

an electric field, the sample components become concentrated at the interface of the sample and run buffers. The key element of zone sharpening is the lower conductivity of the sample solution as compared to the run buffer. Maximum concentration would, theoretically, be achieved in complete absence of any salt ions (including buffer ions) in the sample solution. Under these conditions, the sample molecules must carry all the electric current in the portion of the system occupied by the sample solution. Because (most) sample molecules have relatively low mobilities, a high voltage drop occurs in this region, and the sample ions are accelerated. The next zone (the sample–buffer interface) contains buffer ions (higher conductivity) and thus lower voltage. When the sample molecules reach this area, they are decelerated and the trailing ions, still moving fast toward this zone, eventually catch up with them, and thus the sample becomes concentrated. When the differential conductivity is removed (by replacing the sample solution with run buffer in electrophoretic injection or by ion diffusion during hydraulic injection), the voltage is again distributed evenly along the capillary, and electrophoresis proceeds normally. Besides increasing sample concentration, zone sharpening also improves resolution by narrowing the starting zones.

Zone sharpening is less effective as the concentration of buffer ions (or any other competing ions) in the sample is increased. For this reason, sample desalting is often necessary in CE. Special desalting methods are required when the characteristics of the sample (e.g., very low volumes) do not allow the use of classical techniques (e.g., ultrafiltration, dialysis).

Hjertén et al. [53] described an off-line method for desalting and concentrating microliter volumes of protein samples using small-pore polyacrylamide gels. The gels can be prepared in several formats: tube gels with a fused silica "piston," small gel-filled test tubes with wells, or gel-filled petri dishes containing wells. The tube gel format is used as a syringe, allowing an individual sample to be introduced by drawing the piston. The other two formats are useful for processing multiple samples. Gels were cast in a low-ionic-strength buffer (typically 10-fold dilution of the electrophoresis buffer) at a polyacrylamide concentration providing a pore size appropriate for the application ($T=22\%$, $C=8\%$ for desalting or $T=18\%$, $C=5\%$ for desalting with concentration). The pore size of the gel allows small molecules to diffuse away from the sample, and this was proven to be advantageous when analyzing complex matrices (e.g., amniotic fluid) in which the pattern was greatly simplified (Fig. 4).

Isotachophoresis. A typical setup for isotachophoresis includes a leading electrolyte and a trailing electrolyte with high and low mobility, respectively, than any of the sample components. Another important characteristic of the ITP system is the presence of a counterion (with charge opposite to the sample, leading and trailing electrolytes) that typically possesses buffering capacity (and, thus, is the molecule that maintains the pH at the desired value). Upon the appli-

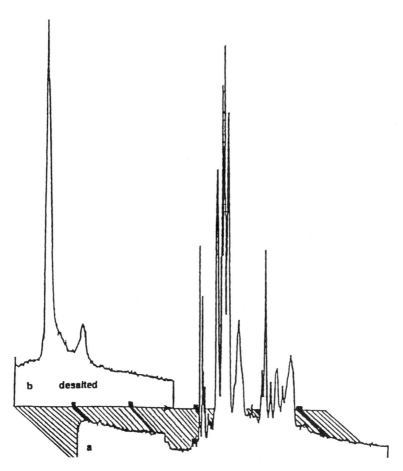

Fig. 4 (a) Free-zone electrophoresis of nondesalted amniotic fluid. Buffer: 100 mM Tris-acetate, pH 8.6. (b) Free-zone electrophoresis of desalted amniotic fluid. Following desalting only two protein peaks appear in the electropherogram [as compared to (a)], which facilitates clinical diagnostics. (Reproduced from Ref. 53 with permission.)

cation of an electric field, the sample components are distributed according to mobility (with the sample of the highest mobility just behind the leading electrolyte, and the sample ion of the lowest mobility just in front of the trailing electrolyte). The voltage gradient is unevenly distributed along the capillary, with the voltage being "concentrated" as the mobility of the ions decrease. Once the system reaches a steady state, the sample components migrate as contiguous zones, because there is no background electrolyte in the sample zones. For this reason, a CZE step is required to achieve total resolution.

The uneven distribution of the voltage eliminates diffusion, because an ion that diffuses into a zone of higher mobility encounters a lower voltage and slows down. If it diffuses into a trailing zone, the voltage is higher, and thus it speeds up. The key phenomenon exploited for ITP preconcentration is that the concentration of the sample zones always reaches a value that is determined by the composition of the leading electrolyte [61]. Obviously, the composition of the leading electrolyte can be manipulated to achieve concentration. It is important to note that often it is necessary to destack the samples before proceeding with the CZE separation. Some destacking procedures (also referred as "mobilization" by some authors) are described throughout the text. The main advantage of ITP preconcentration is that it can be used in the presence of salts [54] (unlike zone sharpening by stacking).

Preconcentration by ITP can be performed online or in coupled columns. The former uses a transient ITP step and gradually changes into a CZE mode to achieve separation. ITP preconcentration in coupled columns uses one column for ITP concentration and another column for CZE separation. An advantage of transient ITP is the simplicity of instrumentation, as it can be performed in most commercially available CE instruments. Coupled-column ITP CZE offers a greater potential to concentrate samples (because the injection volume can be increased as compared with transient ITP, where a major portion of the capillary has to be reserved for the CZE stage of the analysis), but it requires building (and coupling) a homemade system.

Foret et al. [54] described general strategies to perform sample preconcentration by ITP. Transient as well as coupled-column ITP are amply illustrated through the use of model proteins. In a subsequent article [64], the authors emphasized the need to optimize the preconcentration step and provided equations to determine optimal preconcentration as well as destacking times, which depend on the composition and the length of capillary originally occupied by the sample. Because the ionic strength of the sample matrix causes variability in migration time, sample desalting was found to increase reproducibility. To produce quantitative results, constant current was used, because under this condition, the sample zones move at a constant speed when they pass the detection point. When constant voltage is used, quantitation may suffer from the transient conductivity states of the sample components. Two methods were studied by the authors. The first was a typical ITP setup (trailing electrolyte–sample–leading electrolyte), and after concentration, the trailing electrolyte was replaced by leading electrolyte to transform the run into a CZE analysis. Through the use of ITP, the sample (model proteins) introduced into a 50 cm by 75 μm column (internally coated with polyacrylamide for most analysis) was increased ninefold (450 nL) compared to a typical injection (50 nL). Likewise, the gain in detection limit was increased almost 10 times. Migration time reproducibility was about 1% relative standard deviation (RSD), whereas the area was about 5% RSD. The second method employed fast-moving ions (leading electrolyte) in the sample and the background electrolyte (BE) was a slow-moving ion (trailing electrolyte). In this case, the BE was the one eventually used for CZE, whereas in typical ITP systems, the leading electrolyte becomes the BE after replacing the trailing solution. Upon the application of an electric field, the ions automatically formed characteristic ITP zones, but eventually the concentration of the leading electrolyte (present in quantities determined only by its level in the original sample plug) decreases by diffusion, the electric field eventually becomes homogeneous, and the process is transformed into CZE without the need to replace any of the initial solutions. Coated capillaries were required for the analysis of basic proteins, because samples containing low concentrations of proteins (less than 10^{-8} M) were lost to adsorption to the capillary wall. Detection limits for the protein mixture after ITP preconcentration were estimated to be around 10^{-9} M.

A simpler system [65] (peptidelike substances) was used to study the effect of parameters such as injection volume, ITP times, and electrolyte composition in the performance of preconcentration by ITP. It was found that if destacking is initiated before completion of preconcentration, the process is not efficient (low signal-to-noise ratio). If destacking is prolonged, resolution is diminished (because both processes occur in the same capillary, when stacking is prolonged, the length of capillary available for CE is short, resulting in incomplete separation).

Conductivity Gradients. An uneven electric field can be obtained by manipulating the conductivity of one or more segments of the electrophoresis system. In one approach [62,63], an electrode vessel was filled with a buffer of very high ionic strength (high conductivity results in low voltage across the reservoir). After concentration the sample was moved into a dialysis tubing (attached to the capillary) and then back into the capillary. By reversing the polarity of the electric field, a zone-sharpening effect occurred (because the sample had been dialyzed).

Sample Focusing. Sample focusing [62,63] is very similar to isoelectric focusing (IEF) except for the lack of the ampholytes (other than the polypeptides) used to create a pH gradient in the IEF process (Fig. 5). In this method, the capillary is completely filled with sample (in high-pH buffer), the cathode reservoir contains a high-pH buffer (10 mM Tris/HCl, pH 8.5) and the anode reservoir a low-pH buffer (500 mM Tris/HCl, pH 2.5). Upon the application of an electric field, the sample components migrate toward the anode. Because the pH in the anode is 2.5, the proteins are unable to exit the capillary (they become protonated and thus are repelled by the anode). Strong electrolytes (salt ions) are free to leave the capillary, thus the sample is desalted as it is being concentrated. Once concentration has been achieved, a "mobilization" step (transient ITP) is performed by replacing the acidic buffer by a terminating electrolyte (30 mM

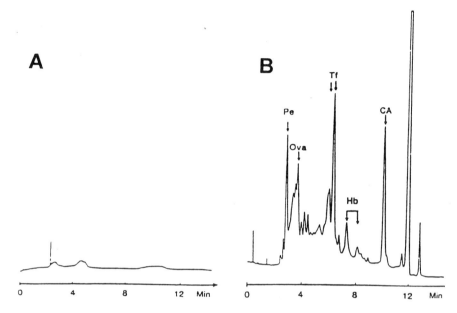

Fig. 5 High-performance capillary zone electrophoresis of model proteins (A) Prior to concentration and (B) following concentration toward a steep, nonbuffering pH gradient and a short mobilization by displacement electrophoresis. The large peak at 12 min may correspond to a moving boundary. (Reproduced from Ref. 63 with permission.)

glycine/NaOH, pH 10), reversing the polarity by applying voltage for a brief period of time. Finally, the trailing electrolyte is replaced with a high-pH buffer similar to the one contained inside the capillary and in the cathode reservoir. At this point, the separation continues as regular zone electrophoresis.

Preconcentration Using ITP with Hydraulic Counterflow. A combination of ITP and hydraulic forces was described by Hjertén et al. [62,63]. First, the capillary was filled with the sample dissolved in buffer (containing the leading electrolyte, 15 mM HCl/Tris, pH 8.5), and a terminating buffer (100 mM glycine/NaOH, pH 8.5) was placed in one of the electrode reservoirs. The other reservoir was filled with leading buffer. A hydrodynamic flow is generated by elevating one of the buffer vessels (the flow direction must be opposite that of the electrophoretic migration of the sample). When the electric field was applied, the sample was sandwiched between the leading and trailing buffers, and its migration was countered by the hydrodynamic flow, thus making the sample zones stationary. It is critical to match the counterflow velocity to the sample velocity to avoid loss of sample by migration into the electrode vessels. If the sample ions have mobilities intermediary to the trailing and leading electrolytes, they become concentrated at the boundary, and diffusion is eliminated (for reasons described earlier for concentration by ITP). Upon completion of the concentration stage (about 15 min), the trailing buffer was replaced by the leading electrolyte solution and the separation performed by CZE. This method can increase protein concentration 400–1000-fold when the proper conditions are selected. Precautions should be taken to avoid loss of the most rapid and slowest of the sample components during preconcentration. Quantitative differences for rapid and slow proteins were observed when the sample was diluted in the trailing instead of the leading buffer.

Preconcentration Through Adsorption to a Chromatographic Support. Chromatographic supports, and specially affinity supports, can also be used to preconcentrate samples to be analyzed by CE [67]. This method has been demonstrated to work off-line as well as online. The main advantage of online preconcentration is instrumental simplicity, whereas off-line preconcentrators offer more flexibility (and also the potential for higher concentration levels). Typical concentration factors are several hundred-fold. Online solid-phase concentration [68] has been used for the analysis of proteins such as metallothionein. The concentrator was constructed using a 5 mm length of polyethylene tubing (280 µm i.d.) fitted over the inlet end of a fused silica capillary. A 0.5 mm frit was made from fragments of a 0.45 µm nitrocellulose membrane packed into the concentration column using a small wire. Octadecyl packing material suspended in ethanol was then drawn into the concentrator by suction through the free end of the capillary. Another frit was placed at the end of the concentrator, and a piece of capillary (sample end) was fitted after the frit. After a brief conditioning step, the sample was injected into the column (up to 5 min) using low (0.5 psi) or high pressure (18 psi). The system was then flushed with buffer to remove any unbound protein and to prepare the capillary for electrophoresis. The sample was then eluted with 33–50% acetonitrile using a small low-pressure pulse. Optimized conditions improved the sensitivity over 700-fold.

Chromatographic supports such as beads derivatized with C_{18} are nonselective and thus concentrate all proteins present in the sample. In instances when this is undesirable, selective preconcentration can be used, e.g., by immunoaffinity using chromatographic supports derivatized with protein G and loaded with antibodies [69]. In this application, the antibodies used had affinity for insulin and formed non-covalent interactions with protein G, thus the antibodies had to be replenished between analyses. The concentration column was a 30 cm by 150 µm capillary equipped with frit made out of 40–60 µm glass beads and immobilized by a short exposure to an open flame. To reduce protein adsorption, the capillary wall was treated with polybrene. Online experiments used a flow-gated interface to couple the columns, whereas for off-line analysis, the samples were first collected from the concentrator and then injected into the electrophoresis column. A 1000-fold concentration of insulin was achieved using this affinity chromatographic step.

Preconcentration to Regulate Sample Conductivity

The introduction of a zone with different conductivity into a capillary produces a heterogeneous distribution of the electric field. When the introduced zone is small, the electrical current usually "recuperates" to the level normally seen with the capillary filled with background electrolyte. But when the zone is of significant length (overloading), the migration time can be severely affected [55] (and in extreme cases, the current drops to zero). Caution should be taken when using large-bore and short columns, as they can be easily overloaded. For this reason it is necessary to preconcentrate diluted samples (obviously diluting the concentrated ones is not a problem) to increase their conductivity, thus improving reproducibility (and often resolution).

Hirokawa et al. [55] constructed a coupled device that allowed the concentration of diluted samples. The first section of the apparatus (ITP section) consisted of two electrode reservoirs filled with leading (10 mM HCl + β-alanine + 0.1% hydroxymethylcellulose, pH 3.5) and trailing (10 mM succinic acid + β-alanine + 0.1% hydroxymethylcellulose, pH 3.9) electrolytes. A separation column (fused silica capillary, 60 cm × 0.66 mm o.d. × 0.53 mm i.d.) was also filled with leading electrolyte and was equipped with online tandem UV and potential gradient detectors. An electric field was applied (constant current, 100 µA) until the sample components reached a second column (about 1 h), which was used as an interface to the CZE section of the system. The sample was transported to a separation CZE column through the use of two valves that switched the electrical circuit to the CZE section of the apparatus. The capillary and electrode reservoirs were filled with a homogeneous buffer. To prove the usefulness of this system, the authors injected samples that differed widely in their conductivity and in the amount injected. Without the ITP preconcentration, the migration times varied (nonlinearly) with sample strength and amount injected. With the ITP/CZE apparatus, the migration times were remarkably constant.

A summary of buffer systems used for ITP preconcentration of proteins is provided in Table 3.

B. Enzyme Assays

Application of capillary electrophoresis to the determination of enzyme activity typically involves off-line incubation of enzyme and substrate with timed injections of the reaction mixture into the capillary to separate and quantitate the low-molecular-weight substrate, intermediates, and product. An approach for in-tube enzyme assay using CZE has been developed by Regnier's group [70–72]. This system, termed electrophoretically mediated microassay (EMMA), is based on electrophoretic mixing of enzyme and substrate under conditions where mobilities of enzyme and product are different. (Fig. 6). The capillary was prefilled with all of the required components for the assay (buffer, substrate), and the en-

Table 3 Buffer Systems Used for ITP Preconcentration of Proteins

Method	Leading	Trailing	Ref.
ITP	120 mM HCl/Tris pH 8.9	190 mM Taurine/Tris pH 8.1	59
Transient ITP	20 mM Triethylamine-acetic acid, pH 4.4	10 mM Acetic Acid	54
Coupled ITP	10 mM Ammonium acetate–acetic acid, pH 4.8 + 1% Triton	20 mM ε-aminocaproic acid–acetic acid, pH 4.4	54
Coupled ITP	10 mM HCl/β-alanine, pH 3.5 + 0.1% MC	10 mM succinic acid/β-alanine, pH 3.9 + 0.1% MC	55
Counterflow-ITP	15 mM HCl/Tris pH 8.5	100 mM succinic acid/β-alanine, pH	62, 63
Transient ITP	10 mM ammonium acetate, pH 3.6	50 mM acetic acid, pH 3.1	65
Transient ITP	20 mM Triethylamine–acetic acid, pH 4.3	10 mM Acetic acid	64
Transient ITP	Na–acetate (added to the sample)	20 mM β-Alanine/acetic acid, pH 4.3 (also used as BE)	64

zyme was introduced at the capillary inlet. Both product and enzyme were transported to the UV-vis-detector, and product was detected at a selective wavelength where the enzyme did not interfere. The relative effective mobilities of enzyme and product were adjusted by manipulating the rate of EOF using covalent (e.g., epoxy polymer) or dynamic (e.g., nonionic surfactant adsorbed onto an octadecyl layer) coatings. The coatings also served to reduce adsorption of enzyme to the capillary wall. The analysis could be carried out in constant potential mode, in which enzyme was injected and transported down the capillary under constant voltage for the course of the analysis; sensitivity in the constant-potential mode could be increased by reducing the potential to increase incubation time. The highest sensitivity was achieved by operation in the zero-potential mode; in this mode; the enzyme was first injected and mixed with the substrate, then voltage was turned off for a fixed incubation period to accumulate product. In the final step, potential was reapplied to transport product to the detection point. A limitation of the zero-potential mode was band spreading caused by diffusion of the product. This problem was solved by using a polyacrylamide gel-filled capillary cast under conditions such that the porosity of the gel reduced product diffusion without introducing protein sieving effects. In this case, incubation times of up to 2 h could be used in the zero-potential mode. Electrophoretically medicated microassay was demonstrated for glucose-6-phosphate dehydrogenase (Fig. 7) with glucose-6-phosphate + NADP/6-phosphogluconate + NADPH substrate/product (NADPH product detected at 340 nm),

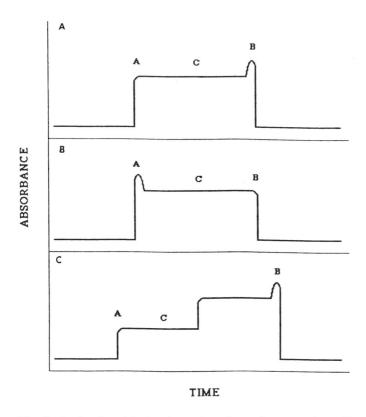

Fig. 6 Predicted models showing various electropherograms in capillary electrophoretic enzyme assay. The moving velocities for enzyme–substrate complex (ES) and product (P) are (A) ES > P and (B) P > ES. A multiple isoenzyme form is shown in (C) with the moving velocity of their common product smaller than those of these isoenzymes. (Reproduced from Ref. 70 with permission.)

for alcohol dehydrogenase with ethanol + NAD/acetaldehyde + NADH substrate/product (NADH product detected at 340 nm), for β-galactosidase using o-nitrophenyl β-galactoside/o-nitrophenol substrate/product (product detected at 405 nm), and for alkaline phosphatase using p-nitrophenyl phosphate/p-nitrophenol substrate/product (product detected at 405 nm). Alkaline phosphatase was also assayed by EMMA using p-aminophenylphosphate as the substrate and electrochemical detection with a prototype electrochemical detector installed in a modified commercial CE system [66]. Electrochemical detection provided an enzyme detection limit 10-fold lower in concentration than UV, but at the expense of greater complexity and limitations on separation conditions imposed by the detection system.

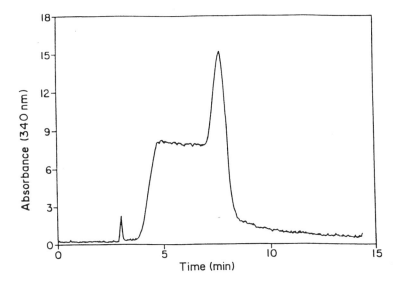

Fig. 7 Electropherogram showing the accumulated product peak resulting from on-column assay of NADPH formed by the reaction of glucose-6-phosphate dehydrogenase with NADP. The running buffer contained all the reagents necessary to assay G-6-PDH; several minutes were allowed to elapse following injection of the G-6-PDH sample and the start of electrophoresis. (Reproduced from Ref. 70 with permission.)

Recently Sun and Hartwick [74] have described a different approach for in-tube monitoring of enzyme activity. A multipoint detection scheme was used which employed a capillary with four windows positioned along the capillary length and mounted in a common detector cell. The electrophoresis buffer contained the enzyme, and conversion of substrate to product could be monitored at four time points during migration; an internal standard was included to compensate for varying internal capillary diameters at the window positions. This system was used to monitor the adenosine deaminase conversion of adenosine to inosine.

C. Affinity Capillary Electrophoresis

Affinity capillary electrophoresis (ACE) is a method for studying receptor–ligand binding in free solution using capillary electrophoresis. The technique depends on a shift in the electrophoretic mobility of the receptor upon complexation with a charged ligand; it has been applied to study of protein–drug, protein–protein, protein–carbohydrate, and protein–nucleic acid interactions. Conventional methods of studying receptor–ligand interactions include equilibrium dialysis using radiolabeled species, UV or fluorescence spectroscopy, nuclear magnetic resonance, and differential scanning calorimetry. Compared to these methods, ACE offers several advantages, including the requirement for very small amounts of analytes, speed, automation, and the ability to determine ligand interactions with multiple receptors. In addition, pure receptor preparations or accurate concentration values are not required, as only migration times are measured. In a typical ACE experiment, the receptor is injected into a capillary containing free ligand at a variety of concentrations; depending on the kinetics of the on and off processes, incremental shifts in migration times are observed and Scatchard analysis of migration time shifts (usually normalized to a neutral EOF marker or a nonbinding reference protein) in response to ligand concentration is used to estimate the ligand–receptor binding or dissociation constant. A model system for measuring protein–ligand interactions using carbonic anhydrase as the receptor and arylsulfonamides as ligands (Fig. 8) has been well characterized [75,76] and the values for binding constants determined

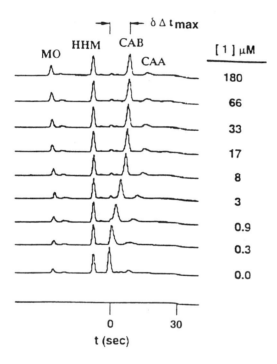

Fig. 8 Affinity capillary electrophoresis of bovine carbonic anhydrase B in Trisglycine buffer (pH 8.4) containing various concentrations of a charged aryl sulfonamide as affinity ligand. Horse heart myoglobin (HHM) and mesityl oxide (MO) were used as internal standards. (Reproduced from Ref. 75 with permission.)

by ACE were in good agreement with those determined by other techniques. In cases where the receptor–ligand dissociation time is of the same magnitude as its migration time, the analysis of the extent of peak-broadening can be used to estimate K_{on} and K_{off} values. Binding constants for neutral ligands can be estimated from shifts in the migration time of charged ligand–proteins complexes in the presence of increasing amounts of a competing neutral ligand. Several limitations must be addressed when developing an ACE system: Wall interactions can interfere with the separation process and affect the receptor–ligand complex equilibrium, whereas the use of additives to reduce wall effects may also alter receptor–ligand interaction.

The estimation of binding constants by changes in protein mobility due to ligand complexation can be problematic if the magnitude of EOF is affected by varying ligand concentration. The ligand may affect EOF by interacting with the wall (especially cationic ligands) or by changing the viscosity or thermal conductivity of the electrophoresis buffer. Gomez et al. [77] demonstrated that mobility corrections using a neutral marker (mesityl oxide) could compensate for a variable EOF to yield accurate estimates of binding constants.

High-performance size-exclusion chromatographic techniques for determining protein–drug binding have been adapted for capillary zone electrophoresis [78]. Using warfarin binding to bovine serum albumin as a model system, the Hummel–Dreyer method, vacancy peak method, and frontal analysis were compared. Frontal analysis was found to be the most reproducible and yielded binding constant estimates in closest agreement with previously reported values.

Affinity capillary electrophoresis has been used to characterize protein-sugar interactions [79]. Lactobionic acid was used as the charged ligand to determine association constants for three β-galactosidase-specific lectins. These studies were performed near the isoelectric points of the lectins which, in the absence of ligand, migrated with a neutral EOF marker. In the presence of the acidic disaccharide ligand, migration of the sugar–protein complex was retarded and Scatchard analysis of the data yield association constants essentially identical to those obtained from equilibrium dialysis.

Application of ACE to determination of protein–ligand binding stoichiometry for charged ligands was described by Chu et al. [80]. Determinations for weak-binding systems (fast off rate) were performed by including the ligand in the electrophoresis buffer as well as the sample at concentrations well above K_d. At a fixed receptor concentration, the point at which the free-ligand response in the electropherogram passed through zero was used to estimate the concentration of bound ligand. Aryl sulfonamide binding by carbonic anhydrase ($n = 1$) was used as a model weak-binding system. Determinations for strong-binding (slow off rate) systems were performed simply by varying the ligand concentration at which the free ligand was detectable in the electropherogram; partially bound intermediates could be detected as peaks migrating between unbound receptor and fully bound ligand–receptor complexes. Examples of strong-binding systems were human serum albumin/anti-HSA ($n = 2$) and streptavidin/biotin ($n = 4$). In the latter case, partially bound intermediates could not be resolved due to insufficient charge differences. By substituting a biotinylated 15mer oligonucleotide for biotin, enhanced charge differences permitted resolution of all intermediates.

The possibility of using CZE to monitor antibody–antigen complexation was described several years ago [81]. Recently, Reif et al. [82] reported the use of fluorescein isothiocyanate-labeled protein G as an affinity ligand in conjunction with laser-induced fluorescence detection for quantitation of FITC-protein G-antibody complexes. The technique was applied to determination of IgG in serum using FITC–protein G or FITC–protein G tagged with anti-IgG. Very rapid separation conditions were used to minimize dissociation of antibody–antigen complexes. One drawback of this approach was the broad, multicomponent FITC–protein G peak resulting from antigen tagged with variable numbers of fluorophores, which interfered with quantitation of the antibody–antigen complex at low IgG concentrations.

Affinity capillary electrophoresis has been used to assess the binding of the immunosuppressant deoxyspergualin (DSG) and DSG analogs to the heat-shock proteins Hsc70 and Hsp90, which function as molecular chaperons. Nadeau et al. [83] employed uncoated capillaries using Tris-glycine buffer at pH 8.3 to determine K_d values for DSG binding to murine, bovine, and trypanosomal Hsc70 proteins. Liu et al. [84] compared DSG binding to human T-cell Hsc70 using uncoated and coated capillaries and demonstrated that K_d values obtained with coated capillaries at pH 2.8 were eightfold to ninefold lower than those obtained with uncoated capillaries at pH 5.3 or 6.95.

Separation of human and bovine serum albumin was shown to be improved by addition of antisteroidal inflammatory drugs as affinity ligands to the electrophoresis buffer, and this system was used to estimate the binding constants of HSA and BSA for ibuprofen and flurbiprofen [85]. This study employed coated capillaries so no correction for EOF was required; α-lactoglobulin (which did not bind the ligands) was used as an internal reference standard.

Chadwick et al. [86] studied the irreversible binding of N-oxidized metabolites of the antiarrhythmic drug procainamide to hemoglobin using CZE with uncoated capillaries and a 0.1 M sodium tetraborate buffer at pH 8.5. Estimates of the number of protein binding sites and the protein–ligand association constant were obtained by determinations of amount of free ligand as a function of total ligand concentration, and the values obtained were in close agreement with those determined using flow injection analysis with electrochemical detection of the free ligand.

Capillary zone electrophoresis has been used to monitor competitive adsorption of human serum albumin onto polyisobutylcyanoacrylate nanoparticles coated with human orosomucoid [87]. In this study, CZE was used to detect the amount of HSA and orosomucoid in the supernatant solution following incubation of particles with the competitive ligand, and to quantitate both proteins following desorption from the particles with detergent. Improved resolution of HSA and orosomucoid was obtained by the addition of 25mM SDS to the electrophoresis buffer; the detergent bound strongly to HSA (increasing its effective mobility) but bound inefficiently to the highly glycosylated orosomucoid. Complexation with SDS also improved HSA peak shape, presumably by reducing electrostatic interaction with the negatively charged capillary wall.

The use of CZE to separate bound from free antigen in a competitive immunoassay for angiotensin II was described by Pritchett et al. [88]. The antigen was labeled with cyanine dye and detected by laser-induced fluorescence using a helium–neon laser. Separations were performed in an uncoated capillary using a borate buffer at pH 10.25. Angiotensin was analyzed in serum samples with a concentration limit of detection of about 10^{-10} M.

D. Chiral Separations

Proteins have been widely used as stationary phases for separation of chiral compounds in HPLC, and their application as chiral selectors in capillary electrophoresis has been extensively investigated (see Refs. 89–92 for recent reviews and Table 4 for the characteristics of proteins used as chiral selectors). The use of capillary electrophoresis with protein chiral selectors offers many advantages, including simplicity (no requirement to attach the protein to a support), flexibility (any soluble protein can potentially be used and a given protein may interact with a variety of analytes), and capacity (the protein is in a free solution). Because separation is achieved by the interaction of the analyte with a protein pseudophase, the technique is therefore a mode of electrokinetic chromatography. To achieve adequate resolution, the mobility of the protein-bound complex must be significantly different from that of free analyte. Resolution of an enantiomeric pair arises from differential migration of enantiomer–protein complexes (usually in the presence of EOF), and because kinetics of complex formation and protein mobility can be highly sensitive to pH, adjustment and control of the buffer pH is key to optimization of a method. Proteins used as chiral selectors in CZE are described in Table 4.

The use of proteins as pseudophases for chiral separations has two major limitations. As in applications of CE for separation of protein analytes, the use of proteins as chiral selectors is limited by protein–wall interactions, which can effect EOF and reduce analytical reproducibility. Proteins also exhibit strong absorbance in the low UV, and detection sensitivity is extremely limited due to high background absorbance. Detection can be performed at longer wavelengths, but this limits the application of the technique to compounds with absorbance maxima in the high-UV range. For these reasons, several investigators have employed columns packed with proteins coupled to gels or particles in which the

Table 4 Characteristic of Proteins Used as Chiral Selectors

Protein	M_r	pI	% Carbohydrate
α_1-Acid glycoprotein	44,000	2.9–3.2	45
Avidin	70,000	10–10.5	20.5
Bovine serum albumin	67,000	4.7–4.9	–
Cellobiohydrolase I	60,000–70,000	3.9	6
Conalbumin	77,000	6.1–6.6	–
Fungal cellulase	60,000–70,000	3.9	6
Human serum albumin	68,000	4.7	–
Ovomucoid	28,000	3.8–4.5	20

detection segment is free of packing. However, the magnitude of EOF flow in packed columns is low compared to open-tubular columns, and packed columns must be operated at low ionic strengths or low field strengths to avoid problems in bubble formation.

Birbaum and Nilsson [93] immobilized bovine serum albumin by crosslinking the protein with glutaraldehyde in situ within the capillary to form a gel; the gel terminated before the detection point which enabled detection of analytes at 214 nm; the separation of D- and L-tryptophan was demonstrated.

Barker et al. [94] employed bovine serum albumin (BSA) as an enantioselective buffer additive for separation of stereoisomers of leucovorin, a folate derivative used in cancer therapy. Poor reproducibility and capillary lifetime were observed with uncoated capillaries, presumably due to protein–wall interactions. Use of a polyethylene glycol-coated capillary resulted in almost a 10-fold improvement in reproducibility and capillary lifetime. In cases where BSA and the analytes had similar mobilities or where the analytes exhibited weak binding to the protein, resolution of enantiomers was poor. These authors [95] demonstrated that the addition of a UV-transparent polymer additive (e.g., 5% 2×10^6 M_r dextran) selectively retarded migration of the BSA additive and allowed separation of enantiomers of several drugs and amino acid derivatives which could not be resolved with BSA in the absence of the polymer. In a later refinement of this approach, these authors [96] immobilized BSA on the dextran polymer using cyanogen bromide. The mobility of the BSA–dextran polymer was very low relative to that of the analytes, permitting rapid separations of enantiomers in short capillaries. An additional advantage of this approach was the ability to vary the selector phase ratio simply by dilution with underivatized dextran. Vespalec et al. [97] investigated human serum albumin as a buffer additive for chiral separations of amino and carboxylic acids. They observed that slow changes in the enantioselectivity of the buffer system could be avoided by heating of the albumin solution prior to use. At high albumin concentrations, adsorption of protein to the wall of fused silica capillaries caused strong decreases in EOF and, consequently, long analysis times; this was remedied by using linear polyacrylamide-coated capillaries. Evaluation of serum albumins from seven different animal sources for the separation of enantiomers of ofloxacin (an antibacterial quinoline) demonstrated that enantioselectivity was variable for the different proteins [98,99]. Interestingly, successful resolution was achieved with BSA although protein-drug binding was observed with human serum albumin (HSA). Chemical modification with palmitic, glucosamide, or acetyl groups reduced or eliminated BSA enantioselectivity; the addition of other drugs to the electrophoresis buffer provided information about the ofloxacin binding sites on native BSA. Lloyd et al. [92] compared the use of HSA as a chiral selector immobilized on packed beds or in free solution for electrochromatographic separations of a variety of enantiomers. Capillaries with HSA phases on 7 μm dp silica exhibited low EOF and lower efficiency compared to open-tubular systems. Free-solution studies employing dextran as an additive to increase the mobility differences of the free versus bound ligand suggested that the polymer additive affected protein-ligand binding.

A widely used protein for chiral separations in HPLC is α1-acid glycoprotein (ACP, orosomucoid), and its use in CE separations has been investigated [100]. AGP is a strongly acidic glycoprotein (pI 2.7) containing 47% carbohydrate; both the amino acids of the primary sequence and the sugars of the carbohydrate moieties can participate in chiral interactions. Capillaries packed with AGP bonded to 5 μm silica particles were used for chiral separation of a group of β-blockers, barbiturates, and nonsteroidal anti-inflammatory drugs. The effects of buffer pH and ionic strength, organic modifier type and concentration, and field strength on EOF and enantioselectivity were studied. Separation efficiencies were generally higher than those obtained with HPLC columns packed with AGP phases but lower than those observed in CE separations using protein additives in free solution.

Ovomucoid is an acidic glycoprotein obtained from egg white which has been used as a chiral stationary phase in HPLC. Ishihama et al. [101] investigated the use of ovomucoid in free solution for electrokinetic chromatographic separation of several drugs. Poor results were obtained using uncoated capillaries due to protein–wall interactions, but the use of PEG-coated capillaries, hydroxypropylcellulose as a buffer additive, or 2-propanol as an organic modifier improved reproducibility and efficiency.

Avidin has been used as a chiral selector for a variety of acidic compounds [102]. Avidin is a very basic protein (pI ~ 10) and required the use of coated capillaries to reduce protein–wall interactions. In the absence of electroosmotic flow, avidin migrated in a direction opposite to that of the analytes, and optimization of the buffer pH was required to achieve enantiomeric resolution with reasonable analysis times. As is often observed when using proteins as chiral selectors, the slower-migrating member of an enantiomeric pair exhibited significant peak asymmetry (perhaps due to slow binding kinetics); increasing the capillary temperature improved peak shape. In some cases where very strong analyte-protein interaction reduced analyte mobility, interaction could be reduced by the addition of an organic modifier (e.g., 2-propanol) to the electrophoresis buffer.

The enantioselectivities of four proteins (BSA, ovomucoid, orosomucoid, and fungal cellulase) used as buffer additives were compared by Busch et al. [103]; five pairs of enantiomers (tryptophan, benzoin, warfarin, pindolol, promethazine, and disopyramide) were used as analytes. All four proteins except ovomucoid were able to resolve enantiomers of some but not all of the analytes, and selectivity was affected by buffer pH and the type and concentration of organic modifier used (1-propanol, dimethyloctylamine). Wistuba et al. [104] compared the effectiveness of BSA, AGP, ovomucoid, and a mixture of α-, β-, and γ-casein as chiral selectors for six 2,4-dinitrophenyl amino acid derivatives. BSA provided best resolution of D- and L-glutamic acid.

The use of a soluble protein as a chiral selector under conditions which permitted low-UV detection was described by Valtcheva et al. [105]. The capillary was partially filled with buffer containing cellobiohydrolase, leaving the detection window of the capillary protein-free; separations were carried out at a buffer pH such that the injected analytes (a series of β-blockers) migrated toward the detector (cathodic end) while the enzyme migrated toward the capillary inlet (anodic end). Electroosmotic flow was eliminated by using a linear polyacrylamide-coated capillary, and hydrodynamic flow was eliminated by introducing an agarose plug at the capillary end. Enantiomeric resolution required high buffer ionic strength (0.4 M phosphate), necessitating the use of low field strengths to prevent excessive heating. In addition, the use of high concentrations of an organic modifier (up to 30% isopropanol) was necessary to prevent peak-broadening due to hydrophobic interactions of the analytes with the enzyme. Involvement of the enzyme active site in chiral recognition was evidenced by impairment of enantioselectivity by the enzyme inhibitor cellobiose and inhibition of enzyme activity by one of the analytes (propanolol). Tanaka and Terabe [106], using a similar approach, developed a method for chiral separations which did not require use of an agarose plug in the capillary and which could be fully automated using a commercial CE system. In this method, the capillary was prefilled with the electrophoresis buffer; then buffer containing the chiral selector was introduced at low pressure to form a separation zone occupying only the segment of capillary before the detection point. The separation conditions (pH and polarity) were adjusted so that the analytes injected at the capillary inlet migrated through the segment of chiral selector, whereas the protein had low mobility and did not migrate significantly during the analysis. This approach offered several advantages: High protein concentrations could be used to effect chiral separations without compromising sensitivity, only small volumes of protein are needed to fill the separation segment of the capillary, and automation of the method provided excellent reproducibility. Bovine serum albumin, α_1-acid glycoprotein, ovomucoid, and conalbumin were used as chiral selectors in this study.

E. Analysis of Protein Folding

Capillary electrophoresis has been shown to be a valuable tool for analysis of protein folding. Unlike techniques such as chromatography and gel electrophoresis, capillary zone electrophoresis is performed in free solution and migration is a function of the intrinsic properties of the molecule. The ability of CZE to distinguish different folding states of a protein depends on changes in solvent-accessible charge, and the migration rate thus reflects a cross section of the conformational states. Moreover, peak shape can provide information on the distribution of the protein among folding states. Capillary electrophoresis separations are performed in a short time frame, permitting detection of short-lived unfolding intermediates which might not be observed using chromatography or gel electrophoresis. Rush et al. [107] first described the effect of thermally induced conformational changes on migration behavior of α-lactalbumin, observing a sigmoidal dependency of the viscosity-corrected mobility on temperature, with a transition temperature which agreed closely with that determined by intrinsic fluorescence measurements. Strege and Lagu [108] used CZE to monitor reformation and interchange of disulfide bonds during reoxidation of reduced trypsinogen. In this study, capillary electrophoresis was performed under low-pH conditions to minimize protein–wall interactions for this basic protein. A population of refolding intermediates distributed between native and unfolded trypsinogen was resolved, and resolution was improved by the addition of ethylene glycol and sieving polymers. These authors also monitored the transition of bovine serum albumin from the native folded state to the unfolded state using CZE in the presence of increasing amounts of urea [109], and the resulting plot of EOF-corrected migration versus urea concentration was similar to urea denaturation profiles obtained with other techniques. Using capillary electrophoresis in the presence of 0–8M urea, Kilár and Hjertén [110] detected intermediate unfolding states of transferrin as distinct peaks and were able to resolve unfolding intermediates for each of the five transferrin glycoforms (2-, 3-, 4-, 5-, and 6-sialotransferrin). Hilser et al. [111] monitored the migration behavior of lysozyme as a function of capillary temperature and observed a sigmoidal behavior of a protein-unfolding transition. Calculation by van't Hoff analysis of the transition temperature, entropy, and enthalpy of unfolding yielded values in close agreement with those determined by differential scanning colorimetry and confirmed that the temperature-dependent decrease in electrophoretic mobility represented a two-state thermal denaturation. As observed previously for α-lactalbumin [107], lysozyme exhibited sharp peaks at low and high temperatures but migrated as a broad peak at the intermediate temperatures with maximum broadening at the transition midpoint (Fig. 9). This behavior was consistent with fast interconversion of folded and unfolded states during transition, with the mobility of the peak determined by the equilibrium population of molecules in each state. The appearance of a second minor peak in the transition region was interpreted as a slow-refolding subpopulation of molecules, perhaps arising from protein interaction with the uncoated capillary surface. This interpretation was supported by the disappearance of the minor peak when the experiments were repeated using polyacrylamide-coated capillaries. In a recent review, Hilser and Freire [112] simulated protein migration behavior in response to thermally induced unfolding to demonstrate the utility of CZE for quantitative analysis of protein folding. Two cases were treated: a slow-time regime in which interconversion is much slower than electrophoresis time (yielding peaks for folded and unfolded species) and a fast-time regime in which interconversion is much faster than electrophoresis time, and the peak represents a time-averaged combination of component species. In the fast-time regime, apparent thermodynamic parameters can be calculated for two-state unfolding and kinetic rate constants obtained from an analysis of the peak width. In the slow-time regime, true thermodynamic parameters can be obtained, and characterization of unfolding transitions is not limited to the two-state case.

F. Analysis of Milk and Dairy Products

The analysis of milk and dairy products is important because of the high level of consumption of these products by humans. The nutritional value of these food sources stems from their significant content of proteins, sugars, vitamins, and fatty acids. Most analyses are thus aimed at the characterization of milk composition to assure quality and consistency in these products. Milk components greatly influence the resulting dairy products after processing, and the quality of these products is, to a certain degree, unpredictable without accurately knowing the composition of milk (both qualitatively and quantitatively). Other areas of research include adulteration, processing stability, storage, genetic selection, and, lately, genetic engineering (by introducing genes that express their products in the milk of mammals, where they can be easily isolated later).

Milk and dairy product polypeptide analyses are frequently performed by HPLC and/or slab gel electrophoresis. HPLC in many cases lacks the selectivity necessary to resolve sample components. Slab gel electrophoresis is labor intensive and does not provide quantitative information. Selectivity and quantitation are of crucial importance to the applications mentioned above. Despite the drawbacks of slab gel electrophoresis, most of the protein nomenclature is derived from gel analyses and, therefore, the performance of electrophoresis in capillaries as a simplified and yet powerful alternative to gels is very attractive.

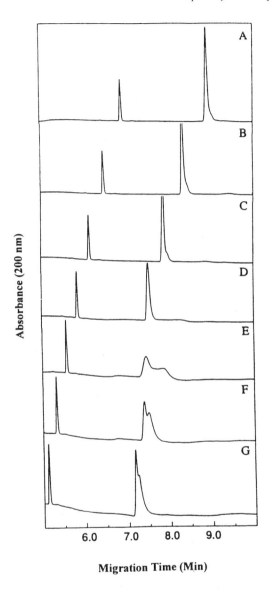

Fig. 9 Capillary electrophoresis of lysozyme at (A) 37°C, (B) 42°C, (C) 47°C, (D) 52°C, (E) 57°C, (F) 62°C, and (G) 67°C. The first peak corresponds to a seven-residue peptide mobility standard and the second peak is lysozyme. Although a plot of mobility versus temperature was linear over the temperature range for the peptide standard, the plot for lysozyme was sigmoidal in the transition region of 50–60°C and peak-broadening was observed. The more slowly migrating peak was attributed to interactions of the protein with the fused silica capillary. (Reprinted from Ref. 11 with permission.)

Analysis can be performed in whole milk or in fractions isolated from milk (e.g., whey and casein analysis).

Whey Analysis

Whey is the fraction of milk that remains after precipitation and removal of caseins. A common method to precipitate caseins is to acidify milk to a pH of 4.6 using acetate buffer or other acidic solutions (e.g., hydrochloric acid). β-lactoglobulins (composed of about 162 amino acids) are the main protein components of whey. There are as many as eight variants of β-lactoglobulin, with variants A and B being the most abundant. Lactoglobulins A and B differ by two amino acids (substitution of Asp for Gly at position 64, and substitution of Val for Ala at position 118), and this provides them with a charge differential that can be exploited, especially in the neutral-to-basic range of the pH scale.

In the authors' lab [115], the separation of β-lactoglobulins, BSA and α-lactalbumin is routinely achieved using a 300 mM Na–borate buffer, pH 8.5, and linear polyacrylamide-coated capillaries (Fig. 10). BSA can be easily resolved from all other whey components after denaturation with SDS and running under sieving conditions. This method does not resolve the β-lactoglobulins [115,120].

These proteins plus β-lactoglobulin C have also been analyzed using a 2-(*N*-morpholino)ethanesulphonic acid (MES) buffer pH 8.0 + 0.1% Tween 20 + 0.1% ethanolamine in uncoated capillaries [113]. Although good resolution was achieved with this solution, it was recognized that the buffer capacity of the solution may be compromised (pK_a of mES is 5.3–7.3). Tween and ethanolamine were added to suppress interaction of the polypeptides with the capillary wall. Another buffer system used in conjunction with uncoated capillaries is 100 mM borate, pH 8.2 containing 30 mM sodium sulfate [114]. The use of an internal standard and a rinsing protocol that included water, 5 mM sodium hydroxide, water, nitrogen, and run buffer yielded a reproducible method that was used to analyze whey from milk exposed to different treatments and also milk from several species.

In summary, methods for whey protein analysis employ high pH, coated capillaries, or additives to suppress protein adsorption.

Casein Analysis

The precipitated fraction after milk acidification contains a group of proteins collectively referred to as caseins. This fraction yields a number of bands when analyzed at pH 6.0 using a 100 mM MES buffer in the presence of 8 M urea [115]. Under these conditions, the milk of different cows under controlled diets produced two sets of patterns, but because the individual casein standards produced more than one peak (especially κ-casein), identification of the sample components was not possible. Other approaches for casein characterization are described below in the section of whole milk analysis.

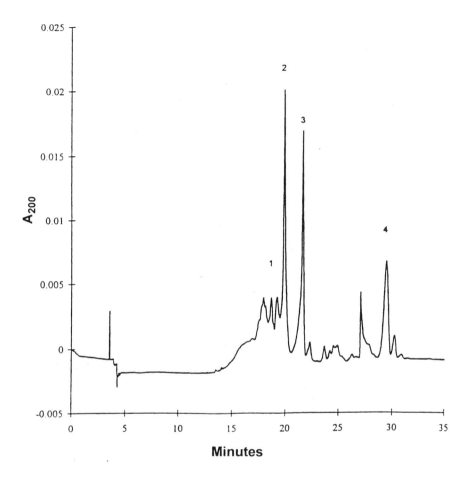

Fig. 10 Analysis of whey using a borate buffer at pH 8.5 and a capillary coated with linear polyacrylamide. Peak identities: (1) bovine serum albumin, (2) β-lactoglobulin A, (3) β-lactoglobulin B, (4), α-lactalbumin. Because most proteins in whey have a pI lower than 8, the analysis was performed with injection at the anode.

Peptide Analysis

Peptide analysis is usually used on enzymatic digests of proteins or to detect naturally occurring peptides (e.g., hormones) and protein degradation products [115,117,118,121]. To assure that all peptides present are electrically charged, electrophoresis is performed at pH extremes. Most commonly, acidic pH is preferred to alkaline pH (at low pH there is little or no EOF, thus results tend to be more reproducible). Variation of pH between 1.8 and 3.0 greatly influences resolution. When uncoated capillaries are used, adsorption can be high at pH > 2.0 (negative silanols adsorbing positive peptides).

A correlation of peptide concentration (as a parameter to measure protein integrity) and curd texture was established using capillary electrophoresis [117]. The study also aimed to correlate peptide concentration with bacterial count. Protein standards and the resulting peptides of the hydrolyzates of pure proteins (by the action of several enzymes) were first characterized. A peptide (presumably λ-CN) produced by hydrolysis of α_{s1}-casein was selected as a marker because it was present even after extensive plasmin hydrolysis of the parent protein. This peptide was shown to increase in concentration if measured at different time intervals spanning 6 days of milk storage under refrigeration at 34°F. After 6 days, the concentration of the peptide had doubled. At the same time, the hardness of the curd formed by the same milk also decreased, with its lowest point at the sixth day of the analysis. Interestingly, the number of bacterial colonies forming per milliliter of milk followed a similar pattern as the peptide, reaching a maximum on the sixth day of the experiment.

Whole Milk Analysis

Milk polypeptides [116] can be analyzed by CE using a citrate buffer, pH 2.5, containing 6 M urea after dilution of the sample 1 : 5 with a "reduction buffer" (73 mg trisodium citrate dihydrate + 38 mg DTT + 6 M urea dissolved in 50 mL of water, then titrated to pH 8.0 with NaOH). The separation was carried out in a hydrophilic coated capillary at 45°C. The pH of the buffer was varied between 2.5 and 3.35 to optimize between resolution and analysis time (lower pH produced best resolution at the expense of time). This buffer system was shown to produce different patterns for milk of various species (including human). Of all methods reviewed, this protocol produced the most impressive results.

All major milk polypeptides can also be analyzed using a 100 mM phosphate buffer, pH 2.5, containing 8 M urea [115]. The sample is simply prepared by adding solid 8 M urea to the milk, incubating for 5 min at 40°C (to dissolve the urea), and centrifuging for 5 min using an microcentrifuge to remove fat and precipitates that may clog the capillary. The remaining clear (yellowish) solution can be injected directly into the column after adjusting the pH with a 1 : 10 dilution of the run buffer (dilution of the run buffer is recommended to achieve zone sharpening, thus increasing resolution and sensitivity). Using this method, electropherograms with different patterns were obtained for bovine and human milk (Fig. 11).

Whole milk can also be analyzed using a 250 mM borate buffer, pH 10.0 in uncoated capillaries [119]. Using this system, β-casein and α-lactalbumin were resolved from all other sample components, but casein and β-lactoglobulin migrated very close to each other (without complete resolution) in fresh nonfat milk. A dramatic difference was observed when analyzing powdered milk: β-

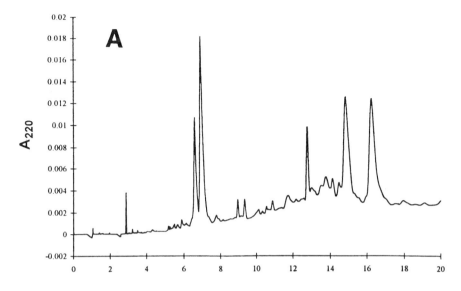

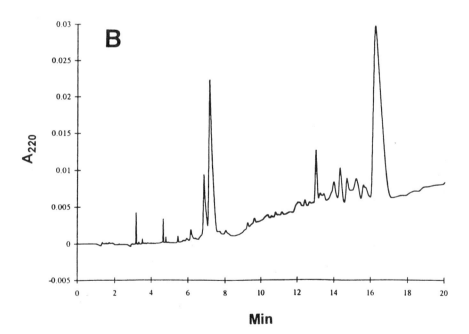

Fig. 11 Analysis of bovine milk by CZE at pH 6.0. The two panels (A and B) represent milk from two individual cows fed under the same controlled diet. All analyses were performed using 0.1 M MES buffer containing 8 M urea in a linear polyacrylamide-coated capillary, injection at the anode, and detection at 220 nm.

Lactoglobulin was totally absent and both caseins appeared as broad peaks, possibly due to denaturation during the drying process.

G. Metalloproteins

Capillary zone electrophoresis has been widely applied to the separation of metalloproteins and metal-binding proteins, and this subject has been recently reviewed by Richards and Beattie [122]. Applications include the determination of purity, structural studies, metal-binding studies, stability studies, and the determination of enzymatic activity.

The oxidation state of protein-bound metal ions can affect migration time, and CZE has been used to identify redox changes in proteins. In a study on the influence of capillary temperature in CZE of proteins, Rush et al. [123] noted that electrophoresis of the heme-containing protein myoglobin at elevated temperature under constant current operation resulted in the appearance of a second slower-migrating species. Further investigation demonstrated that this was not due to protein conformational change but to reduction of the heme-associated iron from the ferric to ferrous state. It was speculated that on-column reduction might be due to a reducing impurity in the electrophoresis buffer or to autoreduction by residues in the polypeptide; in any case, this phenomenon should be considered in the development of separation methods for metalloproteins. Sun and Hartwick [74] employed a multipoint detection scheme to monitor the conversion of myglobin Fe^{3+} to Fe^{2+} during the separation, and used this system to determine the effect of temperature on the myoglobin degradation rate constant. Lee and Yeung [124] analyzed the proteins in a population of 29 human red blood cells by introducing single erythrocytes into a capillary; each cell was lysed after injection, and proteins were separated electrophoretically and detected by native laser-induced fluorescence. Variations in the relative ratios of hemoglobin A and methemoglobin were observed and attributed to variations in in-vitro oxidation, reflecting the age distribution of the sampled erythrocytes.

The effect of metal binding on electrophoretic mobility for three metal-binding proteins was investigated by Kajiwara [125]. Using Tris-tricine or Tris-glycine buffers, it was observed that addition of calcium ions to the electrophoresis caused an increase in the mobility of the calcium-binding proteins calmodulin and parvalbumin. This change in mobility was attributed mainly to the change in net charge rather than changes in Stokes radii of the proteins. Carbonic anhydrase, a zinc-binding protein, did not display a shift in migration time with a zinc-containing electrophoresis buffer, perhaps due to the weak affinity of the active site for Zn^{2+} ions. Thermolysin is a metalloprotease which binds four Ca^{2+} ions for conformational stability and one Zn^{2+} ion at the active site. In the presence of EDTA thermolysin migrated as a broad peak, suggesting conformational instability; the addition of Zn ions reduced mobility but did not improve peak shape. Upon the addition of calcium ions or both calcium and zinc ions, thermolysin appeared as a very sharp peak with increased mobility. Metal-dependent migration shifts were also observed for calmodulin, parvalbumin, and thermolysin using MEKC, indicating that metal binding reduces the hydrophobic character of the protein.

Metallothioneins are low-molecular-weight, cysteine-rich metal-binding proteins which in many species exist in a variety of isoforms due to genetic polymorphism. Application of CZE to the separation of metallothionein isoforms was first described by Beattie et al. [126] using a Tris-HCl buffer at pH 9.1 in an uncoated capillary. These conditions could easily resolve MT-1 and MT-2, the two major metallothionein classes, and the method was used for determination of metallothioneins in liver extracts from Zn-induced rats. Removal of interfering protein by precipitation or acetonitrile was necessary for analysis of tissue samples. Liu et al. [127] described a Tris-sodium tetraborate buffer system for separation of MT-1 and MT-2 using an uncoated capillary, but the method was unsuitable for metallothionein analysis in tissue samples due to protein adsorption to the capillary. Separations of metallothionein isoforms under a variety of CZE conditions has been reported by Richards et al. [128–130]. In addition to the resolution of MT-1 and MT-2, separations also revealed heterogeneity within the major groups. The authors compared separations achieved using uncoated fused silica capillaries, capillaries dynamically coated with a polyamine polymer, and capillaries covalently coated with polyacrylamide. Separations in uncoated capillaries were performed under alkaline conditions using a borate buffer and neutral or acidic conditions using phosphate buffers. The use of a borate buffer at pH 8.4 separated MT-1 and MT-2 but did not resolve isoforms within the major classes. The use of a phosphate buffer at pH 7 reduced EOF and allowed the resolution of additional isoforms. The phosphate buffer also provided better sensitivity due to its reduced absorbance in the low UV; this was advantageous due to the lack of aromatic residues in metallothioneins. Under acidic conditions, zinc and cadmium ions dissociate from the polypeptides, and the use of phosphate buffer at pH 2.0 permitted resolution of apothionein isoforms. The absence of metals was confirmed by use of UV scanning detection which yielded spectra characteristic of the metal-free proteins. Scanning detection could also be used to distinguish the type of metal bound to the metallothioneins under neutral conditions. The use of very-high-ionic-strength buffers (0.5 M) provided improved resolution and rapid analysis due to reduction of EOF but required the use of small-bore (20 μm i.d.) capillaries to minimize joule heat. Further improvement in separations of metallothionein isoforms were obtained using the amine-coated capillary with reversed EOF at pH 7, but resolution was highly dependent on the composition of the electrophoresis buffer. The polyacrylamide-coated capillary, which exhibited negligible EOF, permitted separations to be performed with either normal or reversed polarity over a wide pH range to achieve additional selectivity, and provided improved resolution at pH 7 compared to either uncoated or polyamine-coated capillaries. These authors also employed MECC [131] for successful separation of isoforms from a variety of species using a borate buffer containing 75 mM SDS. Interestingly, metallothioneins typically exhibit broad bands in SDS–PAGE due to the formation of intermolecular and intramolecular linkages via cysteine groups but produced sharp peaks in the MECC separations.

The application of capillary zone electrophoresis for the characterization of ferritin subunit composition and associated ferritins was reported by Zhao et al. [132]. Ferritins are multisubunit proteins functioning in the concentration and storage of iron. The spherical holoferritin complex consists of 24 subunits organized around a mineral core containing up to 4500 iron atoms and variable amounts of inorganic phosphate. These authors have used CZE in coated capillaries with a variety of buffer systems to study the purity of ferritin, the subunit composition of ferritin preparations, and assembly of the ferritin complex from individual subunits. The latter study derived from the fact that ferritin exists as subunits at low pH but as the native complex at physiological pH. Results indicated that apo, holo, and reconstituted ferritins have the same mobility, suggesting that the iron core does not change the overall charge density. However, during iron deposition in the assembly process, significant mobility shifts were observed, suggesting that CE may be a useful tool for studying the process of core formation in both mammalian and bacterial ferritins. One caution in these studies was the observation that migration behavior in the presence of variable amounts of iron was strongly affected by the buffer composition, indicating that the electrophoretic conditions could markedly affect metal binding.

The application of CZE to the characterization of a fibrinolytic metalloproteinase was described by Markland et al. [133]. Snake venom fibrolase, a 23 kDa zinc-containing enzyme purified from southern copperhead venom, was resolved into two isoforms using an uncoated capillary with 100 mM Tris-HCl at pH 7.6. Crystalline fibrolase was also resolved into two isoforms, whereas recombinant fibrolase could be resolved into three components. The authors were also able to use CZE in the presence of urea to monitor the inactivation of the metalloenzyme by incubation with EDTA, because the apoenzyme was well separated from the Zn-associated protein.

Improved sensitivity for detection of metal-binding proteins using an online preconcentration scheme was described by Cai and El Rassi [134]. A preconcentration capillary containing a metal chelating coating was used in tandem with the separation capillary. The preconcentration capillary was prepared by first etching the interior of the silica tube to increase the surface area, and then attaching a covalent hydrophilic coating with activated epoxide groups. Finally, iminodiacetic acid (IDA) was attached to the activated coating, and this moiety, after complexation with zinc ion, served as an affinity ligand for selectively concentrating metal-binding proteins. The preconcentration column was directly coupled to the separation capillary with a PTFE union; the separation capillary was also coated with a covalently linked polyether to minimize protein adsorption. In the initial binding step, a dilute solution of the metal-binding protein was introduced into the capillary by gravity; interaction of certain amino acid side chains (primarily histidine and cysteine) with the IDA affinity group enabled concentration of the protein on the surface. In a subsequent debinding step, a strongly competing ligand (EDTA) was introduced to elute the concentrated protein into the separation capillary as a narrow band. This system was able to increase detection sensitivity 25-fold for carbonic anhydrase; optimum binding was obtained at neutral pH, whereas optimum debinding was achieved with 50 mM EDTA at pH 3.5.

H. Glycoproteins

Glycoproteins often exist as multiple glycoforms sharing a common amino acid primary sequence but differing in the number, location, and structure of carbohydrate groups attached to the polypeptide chain. The importance of glycosylation patterns in the biological activity of glycoproteins has generated strong interest in methods for separation of glycoforms, particularly in the case of therapeutic glycoproteins. A variation in the number of sialic acid residues confers charge microheterogeneity to glycoproteins, and gel electrophoresis and isoelectric focusing have been used successfully for their characterization. Therefore, CZE and CIEF (capillary isoelectric focusing) are obvious candidates for automated analysis of protein glycoforms. A typical strategy for optimizing CZE separations of glycoforms starts with determination of conditions which yield the best resolution of glycoforms, followed by enzymatic cleavage (e.g., with neuraminidase) of carbohydrate moieties. The disappearance of peaks in the electropherogram following enzyme treatment confirms their identity as glycoproteins.

The separation of protein glycoforms was first described by Kilár and Hjertén [135], who used CZE with a Tris-borate + EDTA (pH 8.4) buffer and a coated capillary to separate the di-, tri-, penta- and hexasialo isoforms of iron-free human serum transferrin. Yim [136] used CZE to resolve glycoforms of recombinant tissue plasminogen activator (rtPA), a 60 kDa glycoprotein containing complex N-linked oligosaccharides attached to the polypeptide chain at two (type II) or three (type I) sites. Using an ammonium phosphate buffer at pH 4.6 containing 0.01% reduced Triton X-100 + 0.2 M ε-aminocaproic acid (added to stabilize solubility of the protein) in a linear polyacrylamide-coated capillary, approximately 15 glycoforms were partially resolved.

Erythropoeitin (EPO) is an acidic (pI 4.5–5.0) 35 kDa glycoprotein with three N-linked and one O-linked polysaccharide chains comprising 40% of the protein mass. Tran et al. [137] investigated the effects of buffer pH and organic modifiers and achieved partial resolution of five glycoforms using an acetate-phosphate buffer at pH 4.0. However, column equilibration times of up to 11 h were required for good reproducibility. Watson and Yao [138] were able to achieve complete separation of six EPO glycoforms by adding 1,4-diaminobutane (DAB) and 7 M urea to a tricine-NaCl buffer at pH 6.2. The alkylamine additive reduced EOF and allowed glycoforms to be resolved in order of increasing sialic acid content. Landers et al. [19,30] also used DAB as well as hexa- and decamethonium salts as additives for separation of glycoforms of ovalbumin, pepsin, and human chorionic gonadotropin. Morbeck et al. [139] used a 25 mM borate buffer (pH 8.8) containing 5 mM 1,3-diaminopropane to resolve up to eight glycoforms of human chorionic gonadotropin, a 38 kDa glycoprotein consisting of an α subunit with two N-linked glycosylation sties, and a β subunit with the two N-linked sites plus four O-linked glycosylation sites.

Bovine pancreatic ribonuclease is a mixture of unglycosylated Rnase A and a family of five glycoforms bearing variable-length oligomannoses at a single N-linked site (Rnase B Man-5 to Man-9). Using a buffer consisting of 20 mM sodium phosphate + 5 mM sodium tetraborate + 50 mM SDS (pH 7.2), Rudd et al. [140] were able to separate all five glycoforms, enabling the use of capillary electrophoresis as a tool in determining the importance of glycosylation in the stability and functional properties of the enzyme.

Recombinant human bone morphogenetic protein (rhBMP) is a basic dimeric protein consisting of two identical 114-residue subunits with single glycosylation sites at Asn56 carrying two N-acetylglucosamines and five to nine mannose units. Using a pH 2.5 phosphate buffer with coated capillaries, Yim et al. [141] were able to resolve the 15 possible glycoforms into 9 peaks, each of which differed by only 1 mannose unit (Fig 12). The authors demonstrated that mobilities of the glycoforms decreased in proportion to the number of mannose units present and concluded that frictional drag of the large mannose residues had a strong effect on migration.

I. Cereal Proteins

The quality of food products made from wheat and other cereals is often dependent on the type and content of storage proteins in the grain endosperm. Grain-quality differences among different varieties has been characterized by extraction of the endosperm proteins and comparison of the separation patterns obtained with native PAGE under acidic conditions, SDS–PAGE, and reversed-phase HPLC. Several groups have investigated CZE as an alternative method for characterization of cereal proteins.

Bietz, and Bietz and Schmalzried [142,143] compared CZE in uncoated capillaries under basic and acidic conditions for the separation of gliadins extracted with 30% ethanol from wheat seeds and flour. Basic conditions (60 mM borate buffer at pH 9 containing 20% acetonitrile and 1% SDS) at 40°C yielded resolution comparable to reversed-phase HPLC, but reproducibility was poor without extensive capillary conditioning and between-run wash protocols. The use of acidic conditions (100 mM sodium phosphate buffer at pH 2.5 containing

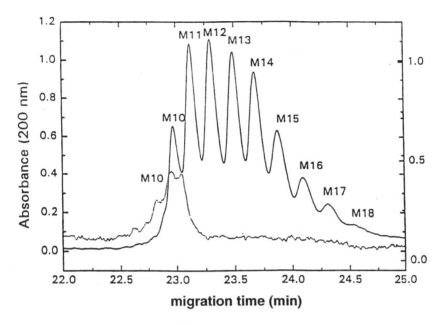

Fig. 12 Overlay of the CZE profiles of intact (solid trace) and α(1–2) mannosidase digested (dotted trace) recombinant human bone morphogenetic protein. The labeled peaks indicate individual glycoforms of the dimeric protein which differ by one mannose residue at the single glycosylation site on each monomer. Glycoforms which possess the same number of sugar residues were not resolved under these conditions, resulting in separation of the 15 glycoforms into 9 peaks. (Reproduced from Ref. 141 with permission.)

hydroxypropylmethycellulose, HPMC) provided improved reproducibility and significantly higher resolution. The patterns obtained under these conditions for extracts from six wheat varieties indicated CZE could be successfully used to differentiate genetically related varieties.

Werner et al. [144] employed CZE under acidic conditions for separation of gliadins extracted from wheat with 70% ethanol. In this study, the capillary was dynamically coated with a cationic polymer which reversed the direction of EOF and minimized protein adsorption. Four gliadin fractions (α, β, γ, ω) were purified by gel filtration and analyzed by acidic PAGE and CZE; the four fractions exhibited the same relative mobility in both techniques. Application of this method to separation of 12 cultivars yielded distinguishable patterns for each cultivar.

Lookhart and Bean [145] developed a rapid method for wheat cultivar differentiation using an acidic phosphate buffer containing HPMC with 20 μm uncoated capillaries at 45°C. The migration order of the major endosperm protein classes (albumins, globulins, gliadins, and glutenins) was established, and the migration order of the gliadin subclasses was compared with acid–PAGE, SDS–PAGE and HPLC [146]. This method was able to differentiate closely related cultivars which were not differentiable by acidic PAGE. These authors [147] used the same analysis conditions to characterize endosperm proteins from other grains as well. Avenins were extracted from oats using 70% ethanol, and prolamins were extracted from rice using 60% n-propanol. CZE separation patterns could be used to differentiate U.S. oat cultivars and to distinguish both U.S. and Philippine rice cultivars. Differences in CZE protein profiles were observed for cultivars which exhibited identical profiles with reversed-phase HPLC or acidic PAGE. This method was later refined [148] by using a single 1 M phosphoric acid wash between injections (which improved reproducibility) and by the additions of 20% acetonitrile to the electrophoresis buffer (which improved resolution). These optimized conditions were applied to the identification of proteins associated with the presence of wheat–rye chromosomal translocations in several wheat cultivars [149]. The novel translocation-associated proteins migrated in the ω-gliadin region, as expected for proteins (secalins) of rye origin (Fig. 13). Capillary zone electrophoresis has also been used to characterize soluble proteins remaining after coagulation of wheat albumin and globulin fractions at elevated temperatures [150].

Wong et al. [151] resolved the 7S-rich (conglycinin) and 11S-rich (glycinin) fractions of soy protein using an uncoated capillary with a 20 mM borate buffer at pH 8.5 and used these conditions to monitor protein hydrolysis with several proteases.

IV. CAPILLARY ISOELECTRIC FOCUSING

Capillary isoelectric focusing (CIEF) is a technique that resolves amphoteric sample mixtures according to isoelectric point (pI) differences. This method is mainly employed to analyze proteins and, to a lesser extent, peptides, amino acids, and small organic zwitterions. The process can be divided in three stages: injection, focusing, and mobilization (the latter two are sometimes combined). Injection consists of introducing the sample+ampholyte mixture into the capillary. Because injection is a common step for all modes of CIEF, we consider CIEF methods as single-step or two-step processes, referring to the occurrence of focusing and mobilization as separate events (two step) or simultaneously (single step). In contrast to other modes of CE, the injection volumes in CIEF usually occupy from several centimeters to the whole length of the capillary.

The key element in CIEF is the formation of a pH gradient during focusing by synthetic compounds collectively termed ampholytes (*ampho*teric electro*lytes*). In CIEF, focusing of the ampholytes occurs concurrently with the focus-

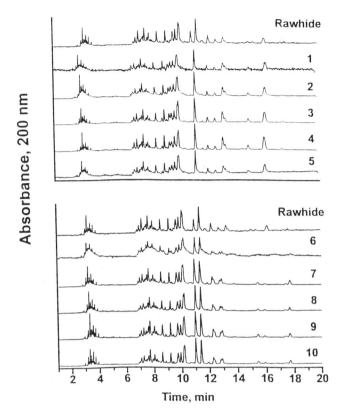

Fig. 13 Comparison of CZE protein profiles of sister lines of the wheat cultivar Rawhide which do carry (traces 1–5) or do not carry (traces 6–10) the 1BI.1RS rye protein chromosomal translocations. Protein peaks characteristic of the translocation migrate as a doublet around 13 min. (Reproduced from Ref. 149 with permission.)

ing of the sample components. Focusing begins once the sample has been introduced; one end of the capillary is immersed in an acid (e.g., 20 mM phosphoric acid), the other end in a base (e.g., 30 mM NaOH), and an electric field is applied. Sample molecules initially charged migrate along the pH gradient until they reach their pI. At that point, they stop moving and form a usually very sharp band. The resolving power of CIEF is very high, in part due to the antidiffusion character of the method: If an isoelectric molecule diffuses away from the point where pH=pI, it acquires a net charge that brings it back to its pI zone.

Several CE reviews have been published [2,152-158,161], but generally, not much emphasis has been placed on CIEF. A comparison between CIEF and slab gel IEF (isoelectric focusing) was published by Righetti and Gelfi [204].

Capillary isoelectric focusing was introduced in 1985 by Hjertén and Zhu

[159]. In their early work, the need for mobilization was recognized (see below), and they offered two alternatives to transport focused protein zones toward the detector: hydraulic pressure and ion addition.

Hydraulic mobilization was achieved by pumping phosphoric acid into the capillary tube a rate of 0.05 µL/min. During hydraulic mobilization, the electric field was maintained to reduce distortion of the focused zones. Chemical mobilization toward the anolyte was accomplished by replacing the phosphoric acid with sodium hydroxide after focusing. Chemical mobilization toward the catholyte was performed by replacing the sodium hydroxide with phosphoric acid after focusing. In a subsequent article, Hjertén et al. [169] described the theoretical basis of chemical mobilization. Because capillaries with an i.d. of 200 µm were used, the voltage applied was limited to 1000-3000 v. Although CIEF was performed in free solution and gel-filled capillaries, only the former has become an accepted technique. To eliminate EOF, the capillaries used were internally treated with methylcellulose. Later, various strategies to perform CIEF in uncoated capillaries were developed [165,168,170,172].

A. The CIEF Process

The performance of CIEF includes an injection, the focusing process, and transport of the focused proteins toward the detection point. As stated before, injection is a common step to all modes of CIEF, whereas focusing and mobilization can be carried out as one or two separate stages. First, we will deal with focusing and two-step CIEF, followed by single-step CIEF techniques.

Focusing

After sample + ampholyte injection into the capillary, the next step in CIEF is focusing, which begins with the immersion of one end of the capillary in anolyte and the other in catholyte, followed by application of high voltage. Typically, the catholyte solution is 20–40 mM NaOH, and the anolyte is 10–20 mM phosphoric acid. For narrow-bore capillaries (e.g., 25 µm i.d.), field strengths of 300–900 V/cm are commonly used. Upon application of high voltage, the charged ampholytes and proteins migrate under the influence of the electric field. Because ampholytes possess higher mobilities, a pH gradient begins to develop with low pH toward the anode (+) and high pH toward the cathode (−); the range of the pH gradient in the capillary is defined by the composition of the ampholyte mixture. At the same time, protein components in the sample migrate until, at steady state, each protein becomes focused in a narrow zone where pH equals its isoelectric point (Fig. 14). Focusing is achieved rapidly (typically 2–5 min for short capillaries) and is accompanied by an exponential drop in current (Fig. 15). When focusing is complete (as determined by attainment of a minimum current value or minimum rate of current decrease), the final step (mobilization) begins with the substitution of the

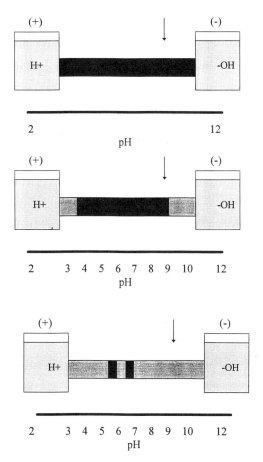

Fig. 14 Focusing process during CIEF. Top panel: The capillary is filled with a protein–ampholyte mixture. The ends of the capillary are then immersed in anolyte (H$^+$) and catholyte (OH$^-$). Middle panel: upon the application of an electric field the ampholytes and protein migrate toward their isoelectric points. Because the ampholytes have a higher mobility, they rapidly focus, creating a pH gradient. Protein zones are formed at both ends of the gradient. Bottom panel: The protein zones at both ends of the gradient converge at the point where the pH equals their pI.

anolyte or catholyte solutions with a suitable mobilization solution or by applying a hydraulic force.

Mobilization

Because most CE instruments use online detection at a fixed point along the capillary, CIEF must include a means of transporting the focused zones past

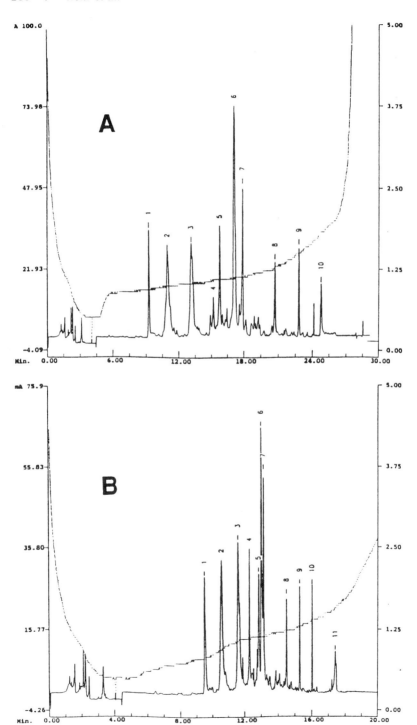

the detection point. Three approaches have been used to mobilize focused zones:

- In **chemical mobilization** (ion addition), changing the chemical composition of the anolyte or catholyte causes a shift in the pH gradient, resulting in electrophoretic migration of focused zones past the detection point [159,169].
- In **hydraulic mobilization**, focused zones are transported past the detection point by applying pressure [159,207,208] or vacuum [162] at one end of the capillary.
- In **electroosmotic mobilization**, focused zones are transported past the detection point by electroosmotic pumping [166,170,172]. Mobilization using electroosmotic flow (EOF) is used in single-step CIEF only.

Chemical Mobilization. At the completion of the focusing step, high voltage is turned off and the anolyte or catholyte is replaced by the mobilization reagent. High voltage is again applied to begin mobilization. As in focusing, field strengths of 300–900 V/cm can be used for mobilization, with optimum separations achieved in small-i.d. capillaries using a field strength of about 600 V/cm. The choice of anodic versus cathodic mobilization and the composition of the mobilizing reagent depend on isoelectric points of the protein analytes and the goals of the separation. As the majority of proteins have isoelectric points between 5 and 9, cathodic mobilization (mobilization toward the cathode) is used most often.

To date, the most common chemical mobilization method is the addition of a neutral salt such as sodium chloride to the anolyte or catholyte (Fig. 16); sodium serves as the nonproton cation in anodic mobilization and chloride functions as the nonhydroxyl anion in cathodic mobilization. A suggested cathodic mobilization reagent is 80 mM NaCl in 40 mM NaOH. At the beginning of mobilization, current initially remains at the low value observed at the termination of focusing but gradually begins to rise as the chloride ions enter the capillary. Later in mobilization, when chloride is present throughout the tube, a rapid rise in current signals the completion of mobilization. Ideally, mobilization should cause focused zones to maintain their relative position during migration (i.e., zones should be mobilized as a train past the monitor point). In practice, move-

Fig. 15 Monitoring current during CIEF. The solid line depicts detection of the protein zones; the dotted line traces the current behavior during the analysis. The initial current depends on concentration of salts present in the sample solution, the concentration of ampholytes, and the concentration of sample components. During focusing, the current drops rapidly (first 4 min in both panels). (A) During ion-addition mobilization with NaCl, the current rises slowly at the beginning, and then sharply at the completion of mobilization. (B) During ion-addition mobilization using a zwitterion, the final current does not increase as much as with the use of NaCl.

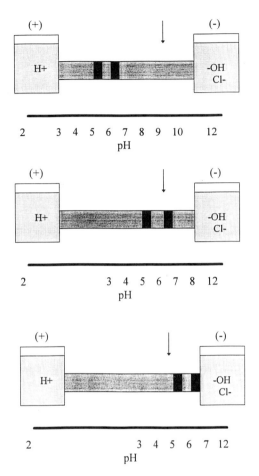

Fig. 16 Mobilization by ion addition (also referred as chemical mobilization). A competing ion (in this example Cl−) is added to one of the focusing reagents (in this example to the catholyte), disrupting the equilibrium attained during focusing. The addition of the competing ion results in a pH shift that sweeps the entire length of the capillary, effectively mobilizing the focused protein zones.

ment of ions into the capillary causes a pH change at the capillary end which progresses deeper into the tube. The rate of change depends on the amount of co-ion moving into the capillary, the mobility of the co-ion, and the buffering capacity of the carrier ampholytes. The actual slope of the pH gradient changes across the capillary, becoming shallower in the direction opposite to mobilization (Fig. 17). Neutral and basic proteins are efficiently mobilized toward the cathode with

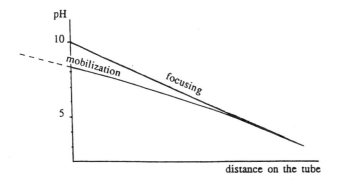

Fig. 17 Slope of the pH gradient during CIEF. Note that during focusing, the gradient is linear along the axis of the capillary, but during mobilization (in this case by ion addition using NaCl), the "modified" gradient is shallow in the side opposite to mobilization. (Reproduced from Ref. 201 with permission.)

sodium chloride, and mobilization times correlate well with pI. However, acidic proteins at the far end of the capillary are mobilized with lower efficiency and may exhibit zone-broadening or be undetected. Use of zwitterions is an alternative approach which provides more effective mobilization of protein zones across a wide pH gradient [201]. For example, cathodic mobilization with a low-pI zwitterion enables efficient mobilization of proteins with pIs ranging from 4.65 to 9.60. The proposed mechanism for zwitterion mobilization couples a pH shift at the proximal end of the tube with a displacement affect at the distal end as the zwitterion forms an expanding zone within the gradient at its isoelectric point. Effective zwitterion mobilization depends on selection of the appropriate mobilization reagent. For example, cathodic mobilization requires a zwitterion with an isoelectric point between the pH of the anolyte and the pI of the most acidic analyte protein. The current level increment during zwitterion mobilization is less than that observed in salt mobilization.

Principles of Chemical Mobilization: The theoretical bases of chemical mobilization were described by Hjertén et al. [169]. At steady state. the electroneutrality condition in the capillary during focusing can be expressed as

$$C_{H^+} + \sum C_{NH_3^+} = C_{OH^-} + \sum C_{COO^-} \tag{1}$$

where C_{H^+}, C_{OH^-}, $C_{NH_3^+}$, and C_{COO^-} are the concentrations of protons, hydroxyl ions, and positive and negative groups in the ampholytes, respectively. In anodic mobilization, the addition of a nonproton cation X^{n+} to the anolyte introduces another term to the left side of the equation:

$$C_{X^{n+}} + C_{H^+} + \sum C_{NH_3^+} = C_{OH^-} + \sum C_{COO^-}$$

Migration of the nonproton cation into the capillary will result in a reduction in proton concentration (i.e., an increase in pH). Similarly, the addition of a nonhydroxyl anion Y^{m-} to the catholyte in cathodic mobilization yields a similar expression,

$$C_{H^+} + \sum C_{NH_3^+} = C_{OH^-} + \sum C_{COO^-} + C_{Y^{m-}}$$

indicating that migration of a nonhydroxyl anion into the capillary results in a reduction in hydroxyl concentration (i.e., a decrease in pH). Progressive flow of nonproton cations (anodic mobilization) or nonhydroxyl anions (cathodic mobilization) will therefore cause a progressive pH shift down the capillary, resulting in mobilization of proteins in sequence past the detector point. The pH shift occurs first near the end of the capillary, but because small ions, like OH^-, H^+, Na^+, Cl^-, and so forth, have a very high mobility, the pH transition is propagated quickly to the rest of the gradient. If snapshots could be used, the pH changes would be seen as pulses that sweep the capillary from one end to the other, altering the pH gradient as they move along. In reality, the changes are not pulses but a continuous process. Ion-addition mobilization is generally performed in capillaries with the internal surface treated to eliminate EOF [52,163,167].

Hydraulic Mobilization. Another approach to transport the focused protein zones toward the detector is hydraulic mobilization [159]. Three hydraulic mobilization approaches have been described: pressure, vacuum, and gravity (siphoning).

Capillary isoelectric focusing with hydraulic mobilization was first described by Hjertén and Zhu [159]. Mobilization was accomplished by displacing focused zones from the capillary by pumping anolyte solution into the capillary using an HPLC pump equipped with a T-connection to deliver a flow rate into the capillary of 0.05 µL/min. Voltage was applied during mobilization to maintain the protein zones focused, and on-tube detection using a UV detector was employed.

During gravity mobilization [153,207], the proteins are transported toward the detection point using a difference in the levels of anolyte and catholyte contained in the reservoirs (Fig. 18). The force generated by the liquid height difference can be manipulated to be extremely small compared with pressure or vacuum, and there is no requirement for high-viscosity polymers in the sample. Furthermore, gravity can be applied from the beginning of the analysis (injecting a plug of sample instead of filling up the capillary) to resolve the sample components by dynamic (single step) CIEF; that is, the focusing and mobilization step are combined in one.

Mobilization by vacuum using on-tube detection [162] was performed using a dimethylpolysiloxane-coated capillary combined with the addition of methylcellulose to the catholyte and ampholyte solutions in order to suppress EOF. In

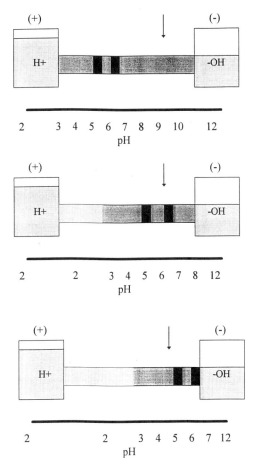

Fig. 18 General principle of hydraulic mobilization. After focusing has been completed, the protein zones are transported toward the detector by applying a hydraulic force. Pressure, vacuum, and siphoning (gravity) have all been used to achieve mobilization. Gravity mobilization is illustrated in this schematic figure.

this approach, a four-step vacuum-loading procedure was used sequentially to introduce segments of catholyte (20 mM NaOH + 0.4% methylcellulose), ampholytes + methylcellulose, sample solution, and a final segment of ampholytes + methylcellulose from the anodic end of the capillary (Fig. 19). Focusing was carried out for 6 min at a field strength of 400 V/cm, then mobilization of focused zones toward the cathode was performed by applying vacuum at the capillary outlet, with voltage simultaneously applied to maintain focused protein zones. Relative mobility values for proteins were calculated by normalizing zone

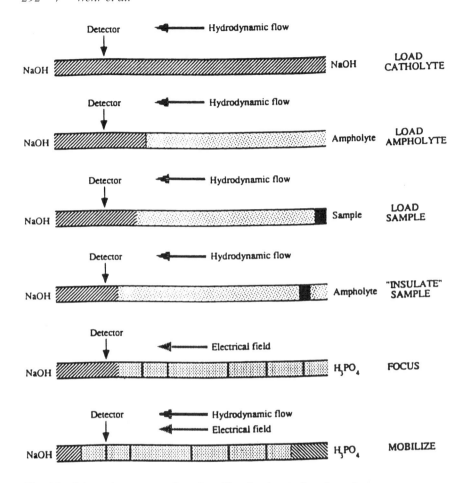

Fig. 19 Schematic representation of capillary isoelectric focusing using vacuum mobilization. (Reproduced from Ref. 162 with permission.)

migration times to the migration times of the catholyte-ampholyte and ampholyte-anolyte interfaces. A plot of relative mobility values versus pI values of protein standards was linear over a pH range of 2.75–9.5. Using this approach, Pedersen et al. [192] successfully analyzed nuclease isoforms by CIEF which were not resolved satisfactorily by other modes of CE (CZE and emulsion MECC).

Single-Step CIEF

Single-step or dynamic CIEF is a variation of CIEF in which focusing and transport of the sample matrix occurs simultaneously. The advantage of single-step CIEF is the simplification of the pattern, but because the capillary is only partially filled with sample–ampholytes solution some resolution and sensitivity are sacrificed. Single-step CIEF can be performed in the presence of EOF or any suitable hydraulic force.

Single-step CIEF was performed in a capillary first derivatized with octadecyltrichlorosilane, and then dynamically coated with various surfactants [164,165] such as Brij 35, pH-108, methycellulose, polyvinyl alcohol, or polyvinyl pyrrolidone (Fig. 20). Copolymers of hydrophilic and hydrophobic monomers interact with the hydrophobic octadecyl layer through the hydrophobic portion of the chain, leaving the hydrophilic segment exposed to the aqueous phase. Several volumes of a polymer–surfactant solution are introduced into the column to cover the capillary wall and then added to the sample+ampholyte solution to maintain a dynamic equilibrium between the wall and the solution. Polymers with higher molecular weights reduce electroosmosis more effectively than surfactants with smaller chains.

Capillary IEF can also be performed in uncoated capillaries [166] through the use of polymers to slow down the EOF. Employing methycellulose at a concentration of 0.4%, deoxyhemoglobin and methemoglobin, which differ by 0.1 pI unit, were resolved (Fig. 21). Myoglobin, carbonic anhydrase, and carbonic anhydrase II were analyzed using 0.05% methylcellulose. Because EOF is not eliminated under these conditions, a separate mobilization step was not necessary because the protein zones were transported by the EOF. The authors reported observing broad peaks when focusing acidic proteins (e.g., pI below 4.8). The problem was later traced to anodic drift [171] and was corrected by increasing the anolyte concentration. Total analysis time was reduced by employing the shorted segment of the capillary from the detection point, and resolution was maintained by manipulating the TEMED (tetramethylethylenediamine) concentration used. Optimal separation of proteins spanning the whole pH range is difficult with EOF-driven CIEF. Salt concentration higher than 10 mM was shown to greatly diminish resolution. A similar approach for performing CIEF in uncoated capillaries was independently reported [170,172]. The main difference between the two methods is the distribution of the reagents inside the capillary: in the first approach, the injection filled most of the capillary with the sample + ampholyte solution relying on TEMED to focus all proteins in the segment of the capillary before the detection point, whereas the second method injected only a small plug of sample + ampholytes. The latter added no polymer to the sample but included small quantities (0.06%) in the catholyte, which was used to fill the capillary before sample introduction.

Molteni and Thormann [202] studied the factors that influence the performance of CIEF with EOF displacement. The amount of methylcellulose (MC) or hydroxymethylcellulose (HMC) was found to be optimal in a 0.06–1% range. The polymer improved resolution when added to the catholyte and sample solutions but did not show any benefit if added to the anolyte. Lower polymer concentration produced better resolution of transferrin isoforms (Fig. 22). The

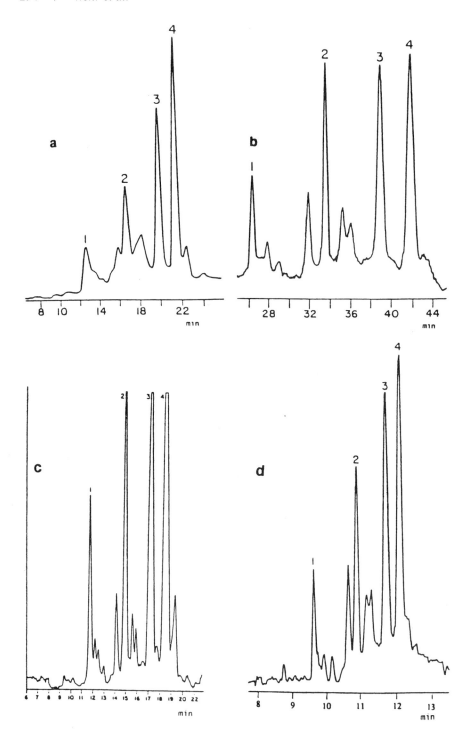

concentration of NaOH (catholyte) had a significant effect on migration times of the protein zones, mainly by affecting the rate of electroosmosis. Longer analysis times were observed at higher concentrations of NaOH with an improvement in resolution. On the other hand, higher concentrations of anolyte (phosphoric acid) shortened the analysis time and diminished resolution. Migration was also affected by the concentration of the ampholytes used (1%, 2.5%, and 5%), with slower mobilization at lower concentrations (1%). Resolution of isoforms was decreased when increasing the voltage from ~ 150 V/cm to 285 V/cm. The initial length of the capillary occupied by the sample was also found to affect migration times and resolution. Longer sample zones provided better resolution at the expense of analysis time. Because electroosmosis plays a key role in this approach, reproducibility was achieved only through extensive capillary conditioning. A rinse sequence of 0.1 M NaOH, water, and catholyte (5 min each) was recommended. Huang et al. [206] compared the linearity of elution time and pI of model polypeptides in coated and uncoated capillaries. Better linearity was obtained with pressure mobilization in coated capillaries than in uncoated capillaries with EOF-driven transport.

B. Capillary IEF Optimization

Synthetic pI Markers

Due to the large number of variables that need to be controlled when performing CIEF in order to achieve high reproducibility, the use of internal standards has been explored extensively. Difficulties such as the presence of impurities, questionable protein stability (especially under isoelectric conditions), and sample–ampholyte interaction has made model proteins unsatisfactory for use as internal markers for CIEF.

Slais and Friedl [203] described a number of synthetic compounds that possess specific pI values. Originally intended for ion-exchange chromatography with pH gradient elution [195], their potential as internal pI markers in CIEF was soon explored. Caslavska et al. [195] evaluated six synthetic markers using CIEF

Fig. 20 Electropherograms of Hb variants separated by (a) a methyl cellulose (15 cP)-coated octadecylsilane-derivatized (MC-15 + C_{18}) capillary (25 µm i.d., 20 cm total length, 16 cm effective length); (b) a methyl cellulose (25 cP)-coated octadecylsilane derivatized (MC-25 + C_{18}) capillary (25 µm i.d., 20 cm total length, 16 cm effective length); (c) a methyl cellulose (4000 cP)-coated octadecylsilane-derivatized capillary (25 µm i.d., 16 total length, 12 cm effective length); (d) a 124,000 MW poly(vinylalcohol)-coated octadecylsilane-derivatized (PVA-124 000 + C_{18}) capillary (25 µm i.d., 16 cm total length, 11 cm effective length). Sample solutions: 1 mg/mL Hb in 1–2% ampholyte with a polymer additive. Voltage: 500 V/cm. EOF mobilizations were employed without interrupting the experiments. The major peaks 1, 2, 3, and 4 correspond to Hb C, Hb S, Hb F, and Hb A, respectively. (Reproduced from Ref. 164 with permission.)

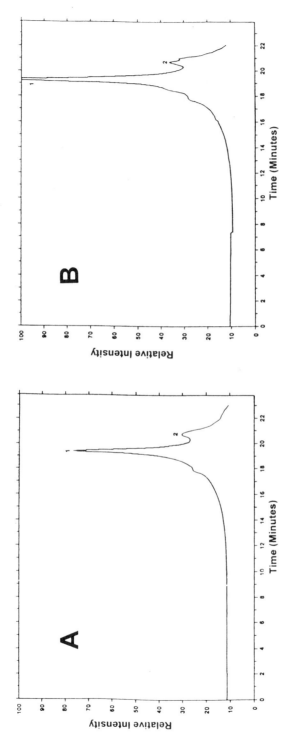

Fig. 21 (A) Capillary IEF of hemoglobin in a polyethylene glycol-coated capillary. Capillary dimensions: 50 μm i.d.; 0.1 μm film thickness; 60 cm total length, 40 cm effective length. Anolyte: 40 mM glutamic acid. Catholyte: 40 mM arginine. Applied voltage: 30 kV. Sample: hemoglobin, 3 mg/mL (consisting of 75% methemoglobin, pI 7.2, and 25% deoxyhemoglobin, pI 7.1) in 5% pharmalyte 6.0-8.0 + 0.1% methyl cellulose. (B) Capillary IEF of hemoglobin in an uncoated capillary. All conditions as on (A) except that the capillary was uncoated and the concentration of methyl cellulose was increased to 0.4%. Peak 1, methemoglobin; peak 2, deoxyhemoglobin. (Reproduced from Ref. 166 with permission.)

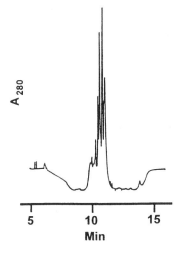

Fig. 22 Capillary IEF of human serum transferring (0.4 mg/mL) using 2.5% ampholine 4–6, 5–7, and 7–9 (3 : 6 : 1, respectively), 0.03% methyl cellulose added to the catholyte and sample. Capillary dimensions: 75 μm i.d., 50 cm effective length, 70 cm total length. Initial injection zone occupied about 50% of the effective capillary length. Applied voltage: 20 kV. (Reproduced from Ref. 202 with permission.)

in the presence of EOF (Fig. 23). They observed that all of the markers but one exhibit sharp bands that did not interfere with the focusing process utilized. The addition of proteins, however, slowed down EOF and thus increased migration times. Absorption spectra obtained during CIEF were presented. Through the use of the markers, the pIs of cytochrome c and myoglobin were estimated, and the values obtained were in accordance with those found in the literature.

Ampholyte Optimization
The pH gradient created by the ampholytes determines the resolving power of the IEF system used. Shallow gradients produce better-resolved protein bands. The slope of the gradient is determined by the number of chemical species present in such gradient. For that reason narrow-range ampholytes (richer in the number of species for the covered range) produce higher resolution. Ampholyte optimization usually consists of preparing experimental blends of ampholyte cuts (and/or ampholytes from different vendors) and, in some cases, varying the different concentrations used.

Desalting Strategies
The deleterious effects of high concentration of salt ions in the sample matrix have been extensively reported [201]. In addition to conventional techniques such

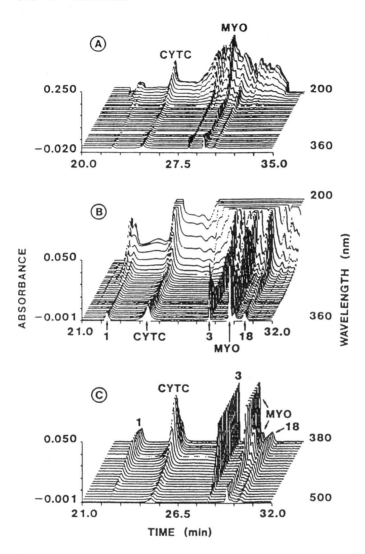

Fig. 23 Three-dimensional CIEF data for six synthetic pI markers. (A) UV absorbance (200–360 nm) showing the complete gradient; (B) a section of the UV data; and (C) a section of the visible absorbance (380–500 nm) obtained in a second experiment. The pH gradient was formed using 1% ampholine 3.5–10 and 2% ampholine 5–8.0. (Reproduced from Ref. 195 with permission.)

as dilution, dialysis, gel permeation, and ultrafiltration, ingenious new methods designed specifically for CIEF have been reported.

Online desalting by ion-ampholyte replacement was described by Liao and Zhang [208]. In this approach, a desalting step is added to the CIEF process following injection of the sample and ampholytes into the capillary but prior to sample focusing. A solution of ampholytes titrated with HCl to a pH of 4.0 is used as anolyte. Another solution titrated to pH 11.0 with NaOH is used as catholyte. An electric field is applied, employing constant current (to minimize heating) (e.g., 10–40 µA). Under this condition, the ampholytes placed in the reservoirs possess a net charge, and salt ions exiting the capillary can be replaced by a combination of ampholytes and protons (anode) or ampholytes and hydroxyls (cathode), resulting in a reduction of pH gradient compression. The electric field is maintained until the voltage reaches a determined value (e.g., 3–10 kV), then it is turned off. The CIEF process then continues as a regular CIEF analysis using conventional anolyte, catholyte, and mobilization reagents. This routine has been used to desalt samples containing up to 0.5 M NaCl (Fig. 24).

This method is advantageous when only a small quantity of sample is available. The main drawbacks are the increase in analysis time and variations introduced by the continuous addition of ampholytes (samples containing different amounts of salt will be focused with correspondingly different final concentrations of ampholytes). This procedure requires the suppression of electroosmosis. To remove salts from the sample before the performance of CIEF [173], a cartridge with an attached dialysis fiber has been used. The dialysis fiber is immersed in a beaker containing a 4% ampholyte solution. One end of the fiber is attached to the electrophoresis column, whereas the other is used to introduce the sample. Because molecular-weight cutoff of the fiber can be manipulated (e.g., 9 kD), it serves a dual purpose: while salt ions diffuse out, it allows ampholytes to diffuse in, thus reducing the total amount of sample mixed with the ampholytes. Due to the narrow diameter of the fiber (75 µm), desalting is accomplished in about 1 min.

Capillary Selection

To obtain good resolution when performing CIEF with chemical mobilization, it is essential that electroosmotic flow (EOF) be reduced to a very low level. In the presence of significant levels of EOF, attainment of stable focused zones is prevented, resulting in band-broadening, and in some instances multiple peaks ("duplicates") from the nascent zones formed at both capillary ends. Therefore, the use of coated capillaries is necessary for this technique. A viscous polymeric coating is recommended for greatest reduction in EOF, and the use of neutral, hydrophilic coating materials reduces protein interactions. Both adsorbed and covalent coatings have been used for CIEF. The most commonly employed coat-

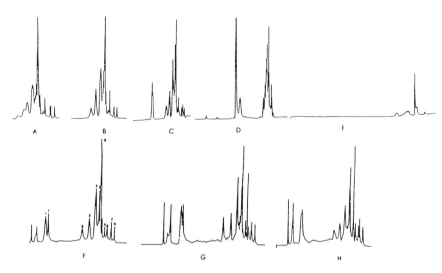

Fig. 24 Capillary isoelectric focusing of IEF standard proteins with and without desalting. The sample mixture consisted of (1) cytochrome C, (2) lentil Lectin, (3) human hemoglobin, (4) equine myoglobin, (5) human carbonic anhydrase, (6) bovine carbonic anhydrase, (7) β-lactoglobulin B, and (8) phycocyanin. Replacement of salt with ampholytes was performed at 30μA constant current and completed when the voltage reached 3 kV. A 3% biolyte (3–10) solution titrated to pH 4.0 and 11.0 served as the anolyte and catholyte, respectively. A–E, 0–100 mM NaCl without desalting; F–H, 100–500 mM NaCl with desalting. (Reproduced from Ref. 208 with permission.)

ing chemistry has been that described by Hjertén [52]. In this procedure, a bifunctional silane such as γ-methacryloporpyltrimethoxysilane is reacted with silanol groups on the internal surface of the capillary. After covalent attachment of this reagent, the acryl group is reacted with acrylamide in the presence of TEMED and ammonium persulfate without any cross-linking agent to form a monolayer coating of linear polyacrylamide covalently attached to the surface.

Additives

The most common type of additive used in CIEF are solubilizing agents. A major problem in IEF is the precipitation of sample components, which causes irreproducibility in migration times, area (thus affecting quantitation), and pattern reproducibility, capillary clogging, unstable current, slow mobilization, and other undesirable effects. Precipitation usually occurs by a combination of factors, including:

1. *Precipitation due to salt removal:* During the focusing process, salts (which have no isoelectric point) exit the capillary, leaving the proteins behind.

2. *Precipitation due to denaturation and electrostatic attraction:* The proteins eventually reach their isoelectric point, concentrating in a small area of the column, where, in the absence of salts, they are free to interact with each other (there is no electrostatic repulsion), forming agglomerates that may eventually precipitate. Some proteins are inherently unstable when they are isoelectric, they denature during focusing, and the denaturation process renders them insoluble. Proteins with these characteristics are difficult to analyze by IEF.

   Strategies to avoid both of the related problems stated above include the reduction of sample concentration, short focusing times, use of chaotropic agents (e.g., urea), nonionic detergents (Brij, Triton X 100, etc.) [201], and increasing in the empholyte concentration. The main principle applied in these strategies is to suppress protein–protein interaction.

3. *Precipitation due to hydrophobic interaction:* Proteins extracted from hydrophobic environments (e.g., membrane proteins) or proteins with hydrophobic patches exposed on the surface of the molecule (e.g., after denaturation) naturally tend to aggregate. These proteins are often extracted using a series of detergents, and if those detergents are nonionic (ionic detergents will either affect the isoelectric point, make the focusing process impossible, or, if unbound to the protein, exit the column upon application of the electric field), they should be added to the ampholyte+sample solution. Small amounts of organic solvents are used for mild cases of aggregation due to hydrophobic interaction.

Hydrophilic Polymers

Hydrophilic polymers are used frequently in CIEF to reduce EOF, either alone or in combination with capillary coatings. Various amounts of hydrophilic polymers are added to sample and electrolyte solutions during hydraulic mobilization. Other uses include:

1. *Fluid stabilization:* High-concentration focused protein zones may have a higher viscosity than the surrounding ampholyte solution, thus giving rise to convective currents. These convection currents can be minimized by increasing the viscosity of the solution through the use of hydrophilic polymers [160].

2. *Increase in viscosity during hydraulic mobilization:* Polymeric additives are used by basically all modes of hydraulic mobilization. The main function of these polymers is to increase the viscosity of the sample matrix. When hydraulic mobilization is used to transport the focused zones to the detection point, the force applied must usually be optimized to a very low value. If higher than optimal pressure is applied, protein zones focused in low-viscosity media are pushed out of the capillary very fast

(to the point that they are not detected) even if only about 1 psi of pressure is applied to a 50 μm capillary (data not published). It is mechanically difficult to accurately control pressure below 0.5 psi. For this reason, Huang et al. [206] used methyl cellulose to increase the viscosity of the sample+ampholyte solution to slow down the flow velocity during pressure-driven mobilization in a neutral, hydrophilic coated capillary. The concentration of methycellulose had to be optimized, and better results were obtained with higher concentrations of polymer. The electric field was maintained during mobilization, and the authors obtained the highest resolution at about 900 V/cm. When the applied force is small enough (e.g., during gravity mobilization), the use of polymers is not necessary [207].

Detection

The strong absorbance of the ampholytes at wavelengths below 240 nm makes detection of proteins in the low-UV region impractical. Therefore, 280 nm is generally used for absorbance detection in CIEF. This results in a loss in detector signal of as much as 50-fold relative to detection at 200 nm, but the high protein concentrations in focused peaks more than compensate for the loss of sensitivity imposed by 280 nm detection. In some instances, proteins possess chromophores that can be detected in the visible range of the spectrum (e.g., hemoglobin and cytochromes). Using modes of detection that require chemical modification (derivatization) of the sample molecules can change the pI and are therefore not widely used.

UV/Vis. Most applications published to date employ online detection of mobilized proteins by absorption in the ultraviolet or visible spectrum at a fixed point along the capillary. In contrast, Hartwick's group [174,175] detected focused protein zones by sliding the capillary in front of a modified detector. In effect, they mobilized the entire capillary instead of its contents. Because most capillaries used in CE are externally coated, the authors employed UV-transparent coated capillaries or removed the external cladding with hot sulfuric acid. To slide the capillary, a string attached to an electrical motor was used. Two related problems had to be solved: noise due to friction and noise due to vibration. The main advantages cited were a uniform rate of mobilization, the capability of performing signal averaging, and shorter analysis time. Although the main purpose of this approach is not to disrupt the equilibrium attained by the system at the completion of focusing, this advantage was compromised by the fact that mobilization was carried out with the voltage turned off, thus allowing the protein zones to diffuse.

Concentration Gradient Detection. Wu and Pawliszyn have reported extensively the use of concentration gradient detectors with various formats [176–185]. The system incorporates a capillary mounted in a holder that aligns the column to a laser beam (He–Ne). A positioning sensor is located at the exit side of the laser beam, and it detects deflections generated by the passage of substances with a refractive index different than that of the background buffer. The main advantages of this detector are its universality and low cost. Because it is a universal detector, ampholytes may produce signals during the mobilization step. It is claimed that the derivative nature of the detector enables recognition of the sharp bands generated by the protein zones against the background of the broader zones produced by the ampholytes. Later the authors [182] performed detection of focused protein zones without mobilization by sliding the detection system along the capillary. Optimization led to fast analysis time (2 min) and detection limits in the 1–5 mg/mL range. Their use of capillary arrays greatly improved throughput [177]. Using the same system, Pawlyszyn et al. [180] analyzed peptides produced by the tryptic digestion of bovine and chicken cytochrome c. The major significance of this work was the detection of peptides not containing aromatic amino acids, which are required for UV absorption detection using 280 nm.

C. Applications

After an initial burst of technique-related developments, most CIEF publications have focused on applications. This does not mean that the method has achieved total optimization but rather that, until recently, CIEF had not received intensive attention. Optimization of parameters affecting reproducibility and issues of protein precipitation during the analysis remain largely unresolved.

Protein Analysis

A number of CIEF applications has been published, including the analysis of human transferrin isoforms [187], recombinant proteins [136,189,190,194], human growth hormone, immunoglobulins [196], human [186,191,193] and bovine hemoglobin variants [188]. Capillary IEF analyses are used to characterize proteins, as well as to determine their purity. Also, it has been suggested that conventional IEF in gels can distinguish conformational states of proteins, although the same has not been reported for CIEF. Capillary IEF provides a rapid method for determining protein isoelectric points, although CZE techniques have also been used for this application [198,199].

Capillary IEF of Hemoglobin Variants and Thalassemias. Analysis of hemoglobin (Hb) variants is of major importance in clinical diagnostics. Hemoglobin is a tetramer of four globin chains: two α-globins, and two β-, δ, or γ-globins. Capillary IEF can easily resolve adult Hb from normal variants such as fetal Hb (Hb F) which is found in blood up to the age of 6 months after birth, at which time it is replaced by Hb A. Capillary IEF can also be used to distinguish abnormal Hb species associated with a variety of blood disorders. Disorders arise from defective genes that code for altered Hb chain sequences, and abnormal levels of individual globin chains produce novel hemoglobin tetramers characteristic of the disorder. A high number of point mutations (about 600) in the Hb chains has been estimated.

Hemoglobin is very soluble in water and is present in erythrocytes at a concentration that can exceed 300 mg/mL; therefore, it is not prone to precipitation under CIEF conditions. Hemoglobin variants usually differ only slightly in composition and are difficult to separate under other modes of CE. Capillary IEF can resolve proteins with pIs that differ by as little as 0.02 pH units. This power of resolution is more than enough to detect subtle shifts in pI in the major Hb variants present in a sample. Examples (Fig. 25) include Hb S, Hb E, Hb G, and Philadelphia C [191].

Abnormalities in the production of globin chains can produce blood disorders collectively known as thalassemias. Thaslaaemias usually arise from the deletion of one or more of the four globin chains. α-Thalassemias affect the pro-

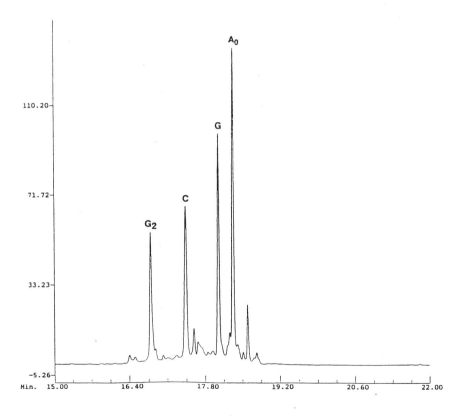

Fig. 25 Capillary isoelectric focusing of hemoglobins from a patient carrying the Hb G Philadelphia and Hb C mutations. (Reproduced from Ref. 193 with permission.)

duction of α-globin chains, and the resulting disorder can range from asymptomatic to lethal anemias. Capillary IEF allows rapid analysis of hemoglobin (Hb) disorders such as Bart's disease (in which the tetramer is composed of four γ-globin chains), Hb H, and Hb Constant Spring (addition of 24–26 amino acids due to a mutation in the termination codon of the α-globin chain). The most severe form of Bart's disease is lethal, and death occurs during fetal stages or soon after birth.

Application of CIEF to analysis of hemoglobinopathies has been extensively described by Zhu et al. [191,193] using chemical mobilization in LPA-coated capillaries. Individual globin chain analysis was also carried out using capillary zone electrophoresis [191] to confirm specific disorders.

In a similar application with a different approach [186], human Hb variants (A1c, A2, F, and S) were analyzed by CIEF with electroosmotic transport (Fig 26.) For different samples examined, good correlation was observed among CIEF, slab gel IEF, and HPLC. HPLC runs were found to be time-consuming as compared with CE. Because a higher number of samples can be analyzed in a

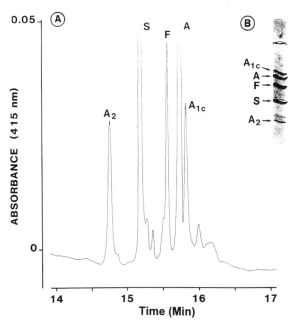

Fig. 26 Dynamic CIEF (A) and gel IEF (B) of an Hb A/F/S sample (hemolysate of a compound heterozygote for Hb S and β-thalassemia, after blood transfusion). Capillary IEF was performed in a 58 cm capillary (39 effective length). (Reproduced from Ref. 186 with permission.)

gel, this method was found to be the one with higher throughput, but it did not provide quantitative information. Important factors that needed optimization in CIEF were carrier ampholyte selection and concentration. Best resolution was obtained using a 1 : 2 v/v mixture of broad range (3.5–10) with narrow-range (6.7–7.7) ampholytes.

Hemoglobin variants [162,205] have been quantified by CIEF with hydraulic mobilization. The main aim of these studies was to validate CIEF as a routine technique for clinical analysis, including quantitation of human hemoglobin variants. Hemoglobin variants present in patient samples were identified by linear regression of pI versus migration time (the calibration plot was created using reported pI values of Hb standards) and using Hb A as a reference peak in each sample. Capillary IEF produced data with statistically significant linear correlation in the quantitation of Hb A2, S, F, and A as compared with results obtained by other methods used by a reference laboratory. Resolution was demonstrated by separation of Hb S and Hb D-Los Angeles (calculated pIs 7.207 and 7.181, respectively). The authors suggested that CIEF could replace multiple conventional assays, simplifying both the operative and regulatory aspects of the analysis.

Capillary IEF was used to follow the process of blood replacement in calves [188]. The study was aimed at potential use of young bovines as blood donors for other bovines used as systems for artificial organ research. Collection of blood from the studied bovines was shown to alter the patterns of hemoglobin production (with a decrease in concentration of Hb A and the appearance of several unidentified peaks), which were restored to normal after about 18 days.

Capillary IEF of Immunoglobulins. Immunoglobulins in the form of monoclonal antibodies have been manufactured commercially for therapeutic and diagnostic uses. Major areas of consideration are thus quality control and bioactivity. Separation of immunoglobulins has proven to be a challenge, even for well-established techniques like HPLC. Difficulties arise from the large size of the antibodies and from their surface properties, which increase their tendency to interact with their surroundings. Monoclonal antibodies have been shown to posses microheterogeneity due to posttranslational modifications, such as glycosylation. Gel IEF is used routinely to analyze different batches of antibodies, but this type of analysis presents several drawbacks which have already been discussed.

Capillary IEF has been used for the analysis of antibodies. One major difficulty is to maintain the antibodies in solution, and the use of additives, short focusing time, low sample concentration, and other "tricks" are a major part of methods development to achieve reproducible high-resolution separations.

The production of recombinant antithrombin III (r-AT III) by cultures of hamster kidney cells was followed using capillary isotachophoresis (CITP), capillary zone electrophoresis (CZE), and CIEF [189]. Recombinant-AT III inhibits serine proteases such as factors IXa, Xa, and XIa, and thrombin. Thrombin is the main target of r-AT III, and it was found in this report that heparin (a polysaccharide) of at least 18 units in length enhances the rate of inactivation. Interference by the media from which the samples were collected posed some difficulties, as some of the media components have similar characteristics to those of the compounds of interest. Detection was also problematic, especially for CITP, because it was performed in the presence of spacers which absorb at UV wavelengths below 250. Sample interferences were minimized by using an ion exchange and an affinity (heparin) purification which also enriched the sample prior to analysis by CZE. Capillary IEF was then used to determine the pIs of the separated components. Three major bands had pIs of 4.7, 4.75, and 4.85, and three minor peaks had pIs of 5.0, 5.1, and 5.3. These data closely resembled data already published from serum AT III based on conventional IEF. Size-based separations were performed to elucidate the molecular size, and the complex formation of r-AT III and thrombin. After 10 s of incubation, only the individual components were detected (thrombin at 32 kD and r-AT III at 59 kD), but after 5 min of incubation, ta third peak (92 kD) was detected with a decrement in the size of the thrombin peak. When heparin was added to the mixture, the 92 kD peak was present even after only 10 s of incubation, which demonstrates the enhancement of the reaction by heparin. These data agreed well with results obtained using slab gels.

Glycoform Analysis. Several applications of CIEF deal with the analysis of glycoproteins, i.e., proteins that contain sugar moieties. A glycoprotein with a single amino acid sequence may exist as a variety of subgroups (glycoforms) which differ in the number, position, or structure of carbohydrate moieties attached to the polypeptide chain. The interest in the analysis of glycoforms is dual: one biological, the other economical. First, the sugar part of the protein may play important roles in cell recognition, protein function, and immunogenicity. Second, production of recombinant glycoproteins as therapeutic agents must then take into account the variation of sugar content, especially because the polypeptides are produced in organisms of different species than those intended to be end beneficiaries.

A second application of glycoform determination is in the analysis of abnormal accumulation of glycosylated proteins. Two examples are the determination of hemoglobin A1c (diabetes mellitus) and elevated glycosylation of transferrin (e.g., alcoholism, pregnancy).

An excellent example [136] of CIEF (and CZE) applied to the analysis of recombinant proteins is the fractionation of human recombinant tissue plasminogen activator (rtPA) glycoforms (Fig. 27). Tissue plasminogen activator is a protein that degrades blood clots, and its recombinant form is produced for the treatment of myocardial infarction. Although rtPA is purified extensively to yield

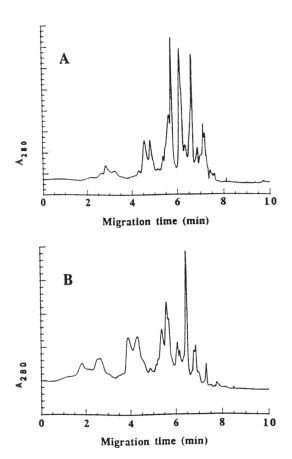

Fig. 27 Microheterogeneity of human recombinant tissue plasminogen activator (rtPA) as analyzed by CIEF. Panel A: type I rtPA, panel B: type II rtPA. Analysis performed in a 14 cm × 25 μm coated capillary. Focusing at 12 kV for 2 min. Mobilization at 8 kV. Ampholyte solution consisted of 2% solids (6–8) + 2% CHAPS + 6 M urea. Online detection at 280 nm. (Reproduced from Ref. 136 with permission.)

high purity of the polypeptide, in some instances up to 20 peaks are observed during CIEF. Treatment of rtPA with neuraminidase (an enzyme that removes sialic acid residues) greatly simplifies the pattern, suggesting that heterogeneity is due to the variation of sialylation.

Analysis of human transferrin isoforms by CIEF has been carried out in glass capillaries with their internal wall modified to minimize electroosmosis [187]. The focusing patterns from normal subjects presented di-, tri-, tetra-, penta-, and hexasialo transferrin glycoforms.

Other Applications

Caslavska et al. [195] studied the behavior of dyes under CIEF conditions. They concluded that aminomethylphenol dyes were suitable as pI markers, exhibiting high solubility in ampholyte mixtures at their pI, long-term stability, no interaction with sample components or ampholytes, and high mobility around their pI. They were also found not to significantly influence the EOF, which the authors employed to transport the protein zones to the detector.

Wu et al. [200] used CIEF to observe changes in pI of transferrin molecules by the binding of iron in vitro using a concentration gradient imaging detection system. First, they focused the transferrin for 2 min in a 4 cm × 100 μm coated capillary by applying 3.5 kV. A 20 sec pulse of iron solubilized in phosphoric acid was applied. With the aid of the concentration gradient detector (scanning the whole capillary) the authors followed the formation and disappearance of focused transferring bands as they incorporated or released the iron ion.

Mazzeo et al [197] demonstrated the use of CIEF in the analysis of peptides. Using two well-characterized proteins (chicken and bovine cytochrome C) and CIEF in the presence of electroosmosis, they first calculated theoretical pIs for the expected tryptic digest peptides, and then compared with results obtained by CIEF. The pIs were determined by generating a standard curve with proteins of known pI values. The authors found that the correlation was only approximate and that although some peptides correlated near perfectly, others showed marked differences from theoretical values. Just as they reported for proteins analyzed by this method in uncoated capillaries, the resolution and peak shape of peptides with acidic pIs were poorer than those for neutral and basic components, due to decreasing EOF as the analysis progressed. To eliminate this problem, a C_8-coated capillary was used. Peak efficiency was increased in the C_8 capillary and longer analysis times were observed (C_8-coated capillaries exhibit only about 40% EOF as compared to uncoated columns). Correlation between theoretical and estimated pI values was still poor.

An ingenious method was developed as an application of CIEF to exploit the concentration effect of the technique, the advantages of affinity recognition, and the detection power of laser induced fluorescence [190]. One very important feature of many biological systems is specific interaction at the molecular level. Antibodies as a group are widely used for molecular recognition (e.g., affinity assays). By labeling an antihuman growth hormone antibody fraction with a fluorescent tag (tetramethylrhodamine-iodoacetamide), the authors were able to detect the presence of growth hormone to a level of 0.1 ng/mL (Fig. 28). By performing the CIEF process in a 75 μm × 15 cm capillary coated with polyacrylamide, the analysis time was ~20 min. Due to the more acidic pI of the hormone, the unattached antibody was easily resolved from the antibody–growth hormone complex due to a shift in pI. Moreover, the system was able to discriminate protein variants, such as sin-

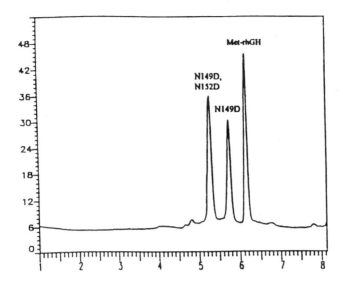

Fig. 28 Simultaneous detection of recombinant human growth hormone (hr-GH) variants. An artificial mixture of met-rhGH, the monodeamidated variant (N149D) and the dideamidated variant (N149D, N152D), each at a concentration of 10 ng/mL, was subjected to affinity probe capillary electrophoresis (APCE) with pharmalyte 3–10 and using TR-Fab' 2 as the affinity probe. Peak identity, from left to right: dideamidated variant, monodeamidated variant, and met-rhGH. (Reproduced from Ref. 190 with permission.)

gle- and double-deamidated growth hormones. By measuring the peak area of the complex after different incubation times, the association time could be estimated. It was found that after only 2 min of incubation, a plateau was already reached.

V. SIEVING SEPARATIONS

The performance of electrophoresis in narrow-bore capillaries obviated most of the functions of gels in electrophoresis [e.g., elimination of convection (through rapid dissipation of joule heat) and reduced diffusion (through a short analysis time)]. However, another important feature of gels is their capability to actively participate in the separation process, mainly by providing a sieving media that differentially affects the migration velocity of sample components according to molecular size. Macromolecules such as nucleic acids and SDS–protein complexes exhibit no significant mobility differences during free-zone electrophoresis and require the presence of an interactive, sieving separation matrix.

Size-based analysis of SDS–protein complexes in polyacrylamide gels (SDS–PAGE) is the most common type of slab gel electrophoresis for the characterization of polypeptides, and SDS–PAGE is one of the most commonly used methods for determination of molecular mass of proteins [229]. The uses of size-based techniques include purity determination, molecular size estimation, and identification of posttranslational modifications [223,234]. Some native proteins studies also benefit from size-based separation.

Due to the importance and broad spectrum of applications of sieving separations, adapting gels to the capillary format has been an effort undertaken by many groups [225,235,237]. Unfortunately, there are several technical difficulties that have limited the use of gel-filled capillaries, such as bubble formation, contamination after repeated runs, and, for polypeptides, poor detection sensitivity in the low UV. A rapidly expanding alternative to gel-filled capillaries is the use of polymer solutions [236]. Advances in both of these methods are described below.

In this chapter, we refer to gel-filled capillaries as those containing matrices that are fixed to the interior of the capillary; that is, they are not replaceable. Replaceable matrices are referred to as "polymer solutions." Gels are typically polymerized in situ, whereas polymer solutions are pumped into the capillary after being premade by the analyst (not all polymer solutions are prepared from monomers). A list of sieving media used for size-based separations in capillary electrophoresis is presented in Table 5.

A. Analysis of Native Proteins

Native proteins consisting of varying numbers of identical subunits or protein conjugates made up of monomers joined by cross-linking agents may be difficult to resolve by free-zone electrophoresis or CIEF, but they can be easily separated by sizing methods. In some instances, it is desirable to maintain the tertiary and quaternary structures of proteins and protein aggregates, which are lost when the polypeptides are denatured. In these cases, sieving of native proteins can be performed using either gel-filled capillaries [230,237] or polymer solutions [231]. Examples of size-based analysis of native proteins by CE include bovine serum albumin (BSA), rat liver proteins [231] (Fig. 29), and human serum albumin [230] (Fig. 30).

Gels other than cross-linked or linear polyacryalmide may be more suitable for the sieving analysis of polypeptides [230]. For instance, low-melting (25.6°C) agarose gels can be used, and by increasing the temperature (40°C) and applying pressure (100 psi), the gel can be extruded from the capillary. A warm solution of agarose is then introduced into the column and allowed to gel. Through this process, the capillary contents are effectively replaced in-between runs.

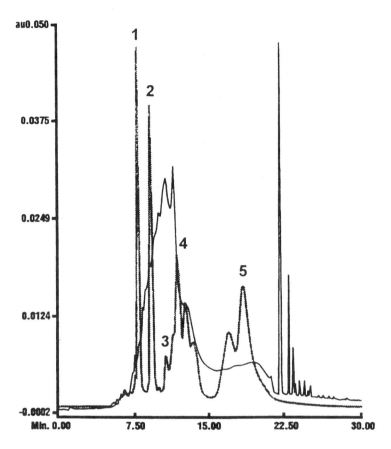

Fig. 29 Separation of native proteins by nongel sieving. A rat liver extract was analyzed (solid trace) and then overlaid with molecular size standards (dotted trace). Separation was performed in a 24 cm × 50 μm coated capillary at 10 kV using a 50 mM phosphate buffer ph 2.5 containing a linear polymer sieving agent. Online detection 200 nm. (Reproduced from Ref. 231 with permission.)

B. Analysis of SDS–Protein Complexes

Sodium dodecylsulfate (SDS) binds to water-soluble proteins approximately in proportion to the polypeptide chain, with one SDS molecule bound per two amino acid residues [209]. Therefore, SDS–protein complexes will possess the same charge-to-mass ratio independent of polypeptide chain length, assuming that the contribution of the charged amino acid side chains is low relative to that of the surfactant sulfate groups. As expected, SDS complexes of proteins with molecular masses greater than 10,000 kDa exhibit identical mobilities in

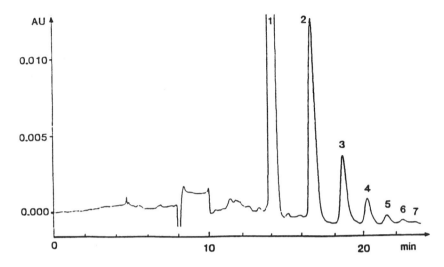

Fig. 30 Capillary electrophoresis of serum albumin. The separation was performed in a methoxylated agarose gel in the presence of SDS. 1, monomer; 2, dimer; 3, trimer; 4, tetramer; 5, pentamer; 6, hexamer; 7, heptamer. (Reproduced from Ref. 230 with permission.)

free solution [210], although proteins which are not fully complexed with SDS may exhibit variable mobilities and my be resolved into multiple species [211].

Size-based analysis by capillary electrophoresis provides similar information and comparable limits of detection than those obtained by SDS–PAGE with Coomassie blue staining [225,228]. Although CE analyzes one sample at a time, total analysis time for multiple samples is shorter for CE than for a 16-lane slab gel [220]. The performance of both electrophoretic techniques for the analysis of polypeptides is superior to size-exclusion chromatography.

An important advantage of using SDS to denatue polypeptides is the solavation power of the detergent. This property allows the study of proteins that easily precipitate under most other conditions (e.g., membrane proteins). Use of SDS permits the study of complex mixtures without the need for extensive sample cleanup.

Protein separations by capillary electrophoresis are often negatively affected by sample–capillary wall interactions, and often require additives or surface modifications to eliminate undesirable interactions. This problem is minimal or nonexistent for SDS–protein complexes, because untreated silica possesses negative charges at pH above 2–3 and the SDS–protein complex is also negatively charged at almost any pH, resulting in electrostatic repulsion, thus eliminating protein adsorption. Nevertheless, in cases where EOF limits resolution or introduces migration-time variations, internally coated capillaries can be used [215]. Capillaries coated with linear polyacrylamide through a C–Si bond were found to be more stable than capillaries coated through siloxane groups [212,213,224]. Uncoated capillaries can be used when the concentration of the polymer ($T = 4\text{--}6\%$) produces sufficient viscosity (>100 cP) to prevent extrusion of the sieving medium from the column by EOF [229]. In some instances, the sieving matrix acts as a surface coating [222].

The SDS–protein complexes have been resolved using gel-filled capillaries or polymer solutions. Next, we describe some of the advantages and disadvantages of each of these methods.

Gel-Filled Capillaries

As stated earlier, we consider as gels those matrices that are fixed to the interior of the capillary and, thus, are not replaced between analyses. The first described use of gel-filled capillaries for analysis of SDS–denatured proteins was in 1983 [237]. Since then, most reports employed either of two types of gels: polyacrylamide cross-linked with bis-acrylamide [235] and linear polyacrylamide. Both gels are polymerized in situ, as their high viscosity precludes pumping them into the narrow-bore column. Because one of the drawbacks of gel-filled capillaries is lifetime, several studies have been aimed at increasing the useful time of the columns. It was found that the lower the degree of cross-linking, the longer the lifetime of the column [221]. Thus, linear polyacrylamide gels (zero cross-linking) were introduced. These gels were also more compatible with high electric fields than cross-linked gels [225]. Unfortunately, even gels with zero cross-linking were able to be used for only 20–40 runs.

The efforts to adapt gels to the capillary format is due to their very high resolving power, which is their only advantage over polymer solutions [225]. Disadvantages of gel-filled columns include a short lifetime, low reproducibility, and poor detection sensitivity due to high UV absorption of the gel matrix. Protein detection in gels is usually accomplished at 280 nm, but the extinction coefficients of proteins are 20–50 times larger at 214 nm. At 214 nm there is also less variability in the intensity of absorbance of proteins [221]. Another drawback is that the composition of the gel-filled capillary cannot be changed, which is a disadvantage in procedures such as the generation of Ferguson plots (see below).

Polymer Solutions

The advantages of using solutions containing polymers as additives to achieve size-based separations include increased reproducibility (because the capillary's content is replaced between runs), increased capillary lifetime, the possibility of using polymers with low absorption in the 200–220 nm range, and ease of storage and handling. Separation parameters are simple to optimize by changing polymer type and concentration, buffer pH, viscosity, and conductivity. As stated above, the main drawback of polymer solutions is that resolution is not as high as that obtained with gel-filled capillaries.

Several types of polymers have been shown to be suitable for separation of a broad molecular-weight range of polypeptides. Care should be exercised when selecting polymers and optimizing analysis conditions [219], as separation parameters such as temperature do not affect all polymers equally [214,227,232]. Resolution also depends on the type, size, and concentration of polymer used. Under optimized conditions, polypeptides differing by as little as 4% in molecular mass can be resolved [222]. Early reports on the use of polymer solutions for the analysis of SDS complexes included dextran and polyethylene glycol (PEG) [221]. Both of these polymers are practically transparent at 214 nm, and thus greatly improve detection over polyacrylamide gels. Migration-time (MT) reproducibility is of prime importance in this technique, as migration times are used to estimate the molecular size of proteins. The use of a replaceable matrix increases MT reproducibility, and RSD values as low as 0.3% were obtained using dextrans (similar values were obtained for PEG matrices). Using polymer solutions, the life of the column was also extended to over 300 analyses.

A list of some polymers used as sieving matrices is displayed in Table 5. One important consideration when selecting a polymer is the viscosity of the final solution. In most cases, low viscosity is desired for easy replacement of the capillary content in-between analyses, but uncoated capillaries will exhibit higher EOF with lower-viscosity polymers [227]. A comparison of resolution achieved using polymer sieving capillary electrophoresis with that obtained using a $T = 12\%$ polyacrylamide slab gel is presented in Fig.31.

Generally, one of the three following mechanisms are considered to explain size-based separations [234]:

Table 5 Sieving Media Used for Size-Based Protein Separations

SIeving media	Ref.
Gels	
Cross-linked acrylamide	225, 235, 237
Linear acrylamide	212, 213, 221
Polymer solutions	
Linear polyacrylamide (LPA)	215, 218, 222, 229
Polyethylene glycol, polyethylene oxide (PEG, PEO)	215, 219, 221, 228, 232–234
Pullulan	224
Poly(vinyl alcohol) (PVA)	227
Dextran	217, 221, 226, 232
Low-melting-point agarose	230

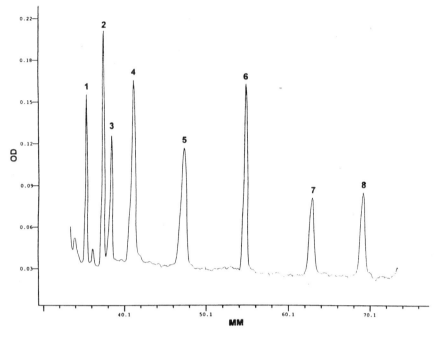
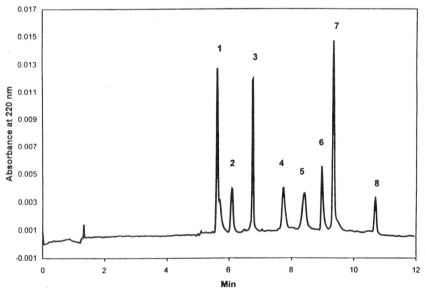

1. The Ogsten model implies true sieving, and when this mechanism is in effect, the pore size of the matrix should be in the same range as the hydrodynamic radius of the sample molecules.
2. The reptation model is used to explain the migration of long, polyionic molecules (e.g., nucleic acids) with a larger molecular radius than the size of the matrix pores.
3. The third model is used when high electric fields are involved, because reptation with stretching may occur.

Guttman [234] suggested that the mechanism of separation for SDS–protein complexes analyzed in solutions containing polyethylene oxide (PEO) is reptation with stretching. Plots generated under various electrophoresis conditions and different polymer characteristics and concentrations were constructed and analyzed, resulting in the elimination of the Ogsten and reptation mechanisms. According to the Ogsten model, a linear relationship is expected for molecular mass and the logarithm of mobility. As shown in Fig. 32, this was not true for any of three different molecular weights of PEO used. The increased curvature of the plots obtained with the high-MW (molecular weight) PEO suggested a possible reptation mechanism. To differentiate these two regimes, the logarithm of mobility of the solute was plotted as a function of the logarithm of the solute molecular mass. A slope value of –1 should be obtained when pure reptation is the mechanism of separation. As indicated in Fig. 33, the values of slopes obtained were much lower (–0.07 to –0.18) than those expected for pure reptation. Three arguments favored a separation mechanism based on reptation while stretching (Fig. 34): (a) lower-than-expected slope values for log mobility versus log molecular mass plots, (b) nonlinear increase of mobility with increasing electric field strength, and (c) variation in extrapolated mobility at zero polymer concentration when the mass of the polymer was increased. According to this model and to the plots obtained, the sieving strength of a polymer solution can be increased by increasing the concentration of a given size polymer and/or by increasing the length of the polymer used (Maintaining the interactions constant).

Karim et al. [226] studied the size and molecular-weight distribution of dextrans and their effect on resolution of SDS–protein complexes. Dextrans of higher MW and narrow distribution were found to produce higher resolution, especially for the larger proteins studied (BSA and phosphorylase B, 66,250 and 97,400 Da, respectively). Unexpectedly, the solutions containing low-MW dex-

Fig. 31 Comparison of resolution for slab electrophoresis (top panel) and capillary sieving electrophoresis employing polymer solutions (bottom panel). Peak identities: 1, lysozyme (14,400 Da); 2, trypsin inhibitor (21,500 Da); 3, carbonic anhydrase (31,000 Da); 4, ovalbumin (45,000 Da); 5, bovine serum albumin (63,000 Da); 6, phosphorylase B (97,000 Da); 7, β-galactosidase (116,000 Da); 8, myosin (200,000 Da).

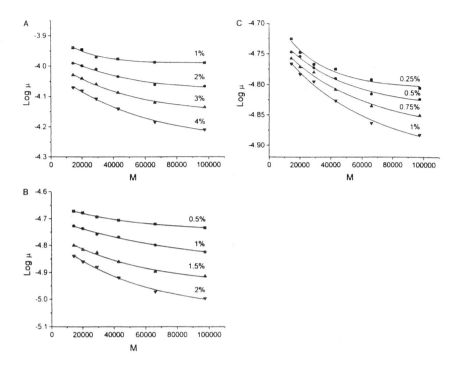

Fig. 32 Relationship between the logarithm of electrophoretic mobility of different sized SDS–protein complexes and the protein molecular weights. Sieving matrices used: (A) 100 kDa polyethylene oxide (PEO), (B) 300 kDa PEO, and (C) 900 kDa. Numbers in the plots correspond to the actual concentration of the sieving polymer solution. Test mixture contained α-lactalbumin (14.2 kDa), soybean trypsin inhibitor (21.5 kDa), carbonic anhydrase (29 kDa), ovalbumin (45 kDa), bovine serum albumin (66 kDa), and phosphorylase b (97.4 kDa). (Reproduced from Ref. 234 with permission.)

trans (1270 Da, and 5220 Da) were also able to resolve these two proteins. This is of interest in regard to the separation mechanism, as such oligosaccharides probably do not form a network structure. In as related study, Simo-Alfonso et al. [227] compared the behavior of sieving solutions containing poly(vinyl alcohol) (PVA) and found that optimal concentration of the polymer showed a strong dependence on the inner diameter of the capillary used. The authors characterized several polymer solutions by using global resolution (R_{sg}, "the product of each individual resolution divided by the average of individual resolutions"). This permitted an easier comparison, especially when polymers exhibited marked differences in resolving power for different sample size ranges. For a 50 μm capillary, R_{sg} was about five times higher than for a 75 μm column, and the

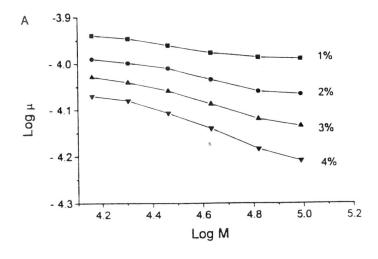

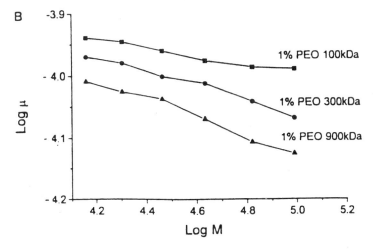

Fig. 33 Double logarithm plots of the electrophoretic mobility and protein molecular mass (*M*) for a protein test mixture. (A) Sieving matrix: 100 kDa PEO, numbers in the plots correspond to the actual sieving matrix concentration. (B) Sieving matrix: 1% PEO (100 kDa), 1% PEO (300 kDa), and 1% (900 kDa). (Reproduced from Ref. 234 with permission.)

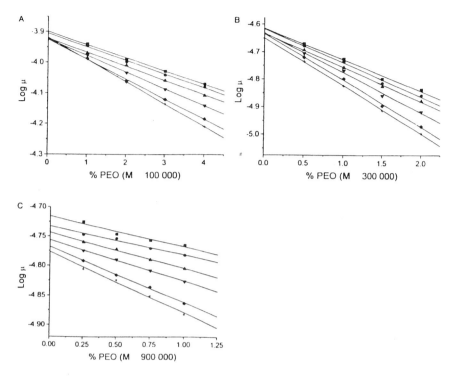

Fig. 34 Ferguson plots of the logarithm of electrophoretic mobility and sieving polymer concentrations of different sized polymer. (A) 100 kDa PEO, (B) 300 kDa PEO, and (C) 900 kDa PEO. The sample used is the same protein mixture described in Fig. 32. (Reproduced from Ref. 234 with permission.)

optimal PVA concentration shifted from about 4% to 6%. The same effect was observed if the number of plates was used instead of R_{sg}. For a 25 µm capillary, the highest R_{sg} obtained was at a concentration of about 1%, which is below the entanglement threshold of PVA (3%). This finding is of significance because it is generally believed that polymer solutions exhibit sieving power only if their concentration is near or above the entanglement threshold. According to De Gennes [239], polymer solutions exhibit two types of regimes: one in which their concentration is low and the individual polymeric chains are isolated from each other, and a second regime, when the polymer concentration is increased and the polymeric chains start to overlap. As the polymers begin to overlap (entanglement threshold) they become densely packed. Polymers in this regime form a dynamic porous structure that can actively participate in the separation process, in a manner similar to cross-linked gels. Barron et al. [240,241] demonstrated that sieving of DNA fragments occurs in extremely dilute hydroxyethyl cellulose (HEC) solutions (≤0.02%), which is below its entanglement threshold of 0.35%. Their findings can be fully explained through the existence of a transient entanglement coupling mechanism. Large DNA filaments will screen a large number of HEC coils (which are in average 10 times smaller than the DNA fragments used) and become entangled, whereas short DNA filaments could not form this coupled system. Still, this theory does not explain Karim and Righetti's findings, as SDS–protein complexes are much shorter than long filaments of DNA. Righetti proposed an additive mechanism of separation: (1) SDS–protein complexes and comicelles are in reality much longer than a polypeptide alone and (2) PVA chains might have a unique entanglement regime due to extensive hydrogen bonding.

Ferguson Analysis

Molecular-weight (MW) estimation of SDS–protein complexes is usually performed by comparing the migration times of sample components with those of proteins of known molecular weights. The standard curve obtained with protein standards may introduce errors in the estimation of MW, especially when the binding of detergent by the protein is anomalous (e.g., membrane proteins, glycoproteins, highly basic proteins). Because detergent binding directly affects protein mobility by changing the mass-to-charge ratio, the MW discrepancy originates from differences in free-solution mobilities of the different polypeptides [228]. One way to avoid such errors is to construct Ferguson plots. These are made by measuring the migration times at different polymer network concentrations and constructing a universal calibration curve [238] by plotting the logarithms of the relative migrations as a function of polymer concentration. According to Ferguson, the logarithm of the protein's mobility varies linearly as a function of the gel concentration employed. The slope of this mobility line yielded a parameter called the retardation coefficient (K_r), which is proportional to the square of the radius of the protein. Universal standard curves are constructed by plotting the logarithm of a known protein MW as a function of the square roots of the retardation coefficients. The slope of the curve represents the retardation coefficient, and the intercept at zero polymer concentration corresponds to the free-solution mobility of a protein. An intercept located at a point other than zero on the concentration scale at zero polymer concentration is an indication of differences in free-solution mobilities of the SDS–protein complexes. Proteins with similar molecular radii show the same slope, independently of where they intersect on the concentration axis. Ferguson analysis for traditional SDS–PAGE is extremely time-consuming, especially because the analysis should be performed using at least six different gel concentrations. Consequently, this method of analysis was practically abandoned until the use of CE with replaceable polymer networks made the Ferguson analysis more feasible. A

Ferguson plot can be generated automatically by CE using different dilutions of the sieving buffer [233,238]. Ferguson plots are also used to elucidate separation mechanism [234].

Application

Collagens. Collagens are a family of structural proteins consisting primarily of two polypeptides (α_1 and α_2) of about 1000 amino acids in length and associated in a triple helix. Deyl and Miksik [242] employed entangled-polymer CE with 4% linear polyacrylamide to separate collagen-SDS complexes. Collagen Type I α_1 and α_2 were resolved, as well as β_{11}, β_{12}, and γ-chain polymers. The separation of glycated collagens produced by incubation with glucose was also achieved.

Cereal Proteins. Werner et al. [243] used a linear polyacrylamide entangled polymer sieving system to separate wheat endosperm proteins extracted from crushed wheat seeds with 1% SDS + 1% 2-mercaptoethanol. This procedure extracted both gliadins and reduced glutenin subunits. Capillary electrophoresis using the SDS–entangled polymer medium supplemented with 5% methanol and 3.8% glycerol yielded separation profiles similar to those obtained with conventional SDS–PAGE and permitted differentiation of several wheat cultivars. It has long been known that SDS–PAGE yields molecular-weight estimates for wheat glutenin subunits which are lower than those obtained by ultracentrifugation or calculated from amino acid sequence data. Similar anomalous molecular-weight values were obtained by Werner [243] using the entangled polymer system described above (but without methanol and glycerol additives). Ferguson-type analyses were performed by determining the mobility of standard proteins and glutenin subunits in a series of sieving media containing variable concentrations of linear polyacrylamide. The slopes of mobility versus polymer concentration were used to construct a plot of retardation coefficients versus molecular weight of protein standards, and the molecular weights of the glutenin subunits were estimated by interpolation. Results agreed with M_r values obtained from sequence data, and suggested that the anomalous mobility of glutenin subunits was due to incomplete SDS binding to the subunits.

Milk Proteins. Quantitative analysis of proteins in bovine whey was carried out by Cifuentes et al [244] using a polyacrylamide-coated capillary and an electrophoresis buffer containing 0.1% SDS with 10% polyethylene glycol 8000 as the sieving polymer. Concentrations of β-lactoglobulins A + B, α-lactalbumin, and bovine serum albumin agreed closely with values obtained by reversed-phase HPLC, and the analysis time was reduced fourfold compared to the gradient chromatographic method.

Kinghorn et al. [245] used a commercial linear polyacrylamide-based polymer sieving system to separate SDS complexes of proteins in liquid bovine whey and reconstituted whey protein concentrate. Pretreatment of the acid whey sample by dilution in 2% SDS of by ultrafiltration was necessary to prevent precipitation of the SDS as potassium dodecylsulfate. The levels of α-lactalbumin and β-lactoglobulin determined by this method agreed satisfactorily with other methods (native PAGE, SDS–PAGE, and size-exclusion, ion-exchange, and affinity HPLC).

VI. CLINICAL APPLICATIONS

Capillary zone electrophoresis and related capillary techniques can achieve the following: (1) fast, high separation power with a large number of theoretical plates, (2) reduction in labor requirements, and (3) full automation. These features are great advantages for use in clinical diagnosis. In addition, these electrophoretic methods require extremely small amounts of samples, a fact that could open up entirely new analytical perspectives in clinical applications (e.g., microanalyses on biopsies). During the last decade, clinical applications of capillary electrophoresis have grown rapidly. Major areas of protein related research include the following:

1. Human serum or plasma
2. Urinary proteins
3. Cerebrospinal fluid
4. Hemoglobin (Hb variants and glycated Hb)
5. Enzymes or isoenzymes
6. Hormones
7. Soluble mediators

There are a number of excellent reviews related to this subject that have been written recently. These review articles covered the area specific to human serum and plasma protein analysis [246] or some routine clinical analyses such as serum proteins, hemoglobins, liproproteins, and isoenzymes [247]. More general reviews covered all clinical applications [248] and a broader range of biomedical applications [249].

A. Serum Proteins

More than 300 proteins have been identified in human serum. Concentrations of these proteins vary from 5 g/dL down to trace amounts. Conventional methods for serum protein analysis employ a solid support matrix, either a microporous membrane of a gel. From the results of analyses using these methods, a "five zone" separation can be defined: (1) albumin, (2) α_1 globulin, (3) α_2 globulin, (4) β-globulin, and (5) γ-globulin. Major proteins which can be detected include albumin, antitrypsin (α_1), haptoglobin (α_2), β-lipoprotein (β), C_3 component (β), transferrin (β), and γ-globulins (γ). The procedures for handling these are quite labor intensive and require operator proficiency. Sample application is the most difficult operation required, as the volumes are in the 3–5 µL range and require precise placement on the support medium. The technician must handle the support matrix through all the steps: buffer preparation, sample application, separation, staining, destaining, and detection procedures. Typically, a skilled, trained technician can process a routine serum protein gel separation from preparation to final results in about 90 min. Because of its complexity and results that are highly skill dependent, a significant variation may results. The sensitivity of the detection by this method is about 1 mg/mL. Thus, the only proteins which could be accurately detected would be the major proteins described above. The nature of capillary electrophoresis, with its higher resolution and precision, as well as walk-away mode automation, would make it ideal for a rapid analysis.

Five-Zone Serum Protein Separation

With regard to replacing conventional gel electrophoresis with five-zone separation, sensitivity would not be a concern because the concentration of the major proteins are all above 1 mg/mL. However, the main concern would come from the nature of the protein, which has a tendency to adhere to the wall of the capillary. When the protein sample is injected into the capillary, the interaction between protein and capillary may change the original nature and composition of the capillary wall. This is because the negatively charged silanol surface of the fused silica may capture some protein molecules. To minimize the interaction, several approaches were performed and the results were obtained successfully. Methods employed included using a high-pH buffer to run the sample, a coated capillary, or an additive. Of these techniques, a high-pH running buffer is probably the easiest method to adapt for clinical applications, as it is the simplest and the results are quite reproducible.

In early 1980s, Holloway et al. analyzed human serum proteins by using capillary isotachophoresis [250]. The experiment was performed on the LKB 2127 Tachophor fitted with a PTFE (polytetrafluoroethylene) capillary (23 cm long and 0.5 mm i.d.). The leading electrolyte system consisted of 5 mmol/L MES (morpholino-ethanesulfonic acid) with 10 mmol/L ammediol as the counterion, and 0.25% hydroxypropyl methycellulose (HPMC) to reduce electroendosmotic effects. The terminator was 10 mmol/L epsilon amino caproic acid (EACA) with 10 mmol/L ammediol as the buffer. The results showed a very high resolution in the gamma region. They compared the isotachopherograms from normal and abnormal sera and found some outstanding peak(s) in abnormal sera. However, the significance and identification of those peak(s) were not defined on this study. In the later 1980s, the capillary electrophoresis (CE) instrument became commercially available. Many investigators began to use capillary zone electrophoresis (CZE) to analyze all kinds of micromolecules and macromolecules including serum proteins. CZE was simple, inexpensive, fast, reproducible, and required minimal sample amounts.

In the early 1990s, two articles were concurrently published, both citing uses of the CE instrument manufactured by Beckman Instruments, P/ACE 2000, to analyze human serum proteins. One of the articles was published by Chen et al. [251]. They used high-pH borate buffer with ionic strength about 80 mM. Serum samples were injected into the capillary tube (75 µm i.d. × 25 cm) by the electrokinetic mode for 3 s at 1 kV, and separations were performed in less than 10 min using a column voltage gradient of 200 V/cm. The separated proteins were detected by 214 nm UV light. The capillary tube was washed and reconditioned with NaOH then water and reequilibrated with the buffer between each run. The results from both normal and abnormal (polyclonal gammopathy) sera were compared with the reference method, "Paragon SPE Gel." The profiles from the CE electropherograms and the densitometric scan of Paragon SPE were very similar (Figs. 35 and 36). The other article was published by Gordon et al. [252]; they performed human serum analysis using the same instrument (P/ACE 2000) and the same buffer at a low concentration (50 mM) and high pH (pH 10). The sample was diluted 40 : 1 with 1 mM boric acid, pH 4.5, containing 20% ethylene glycol. They concluded that ethylene glycol exhibited excellent reproducibility for retention time and peak area by reducing protein–protein interactions. However, the sensitivity was greatly reduced because ethylene glycol absorbs broadly in the UV region (Fig. 37).

Further experimentation showed other operating conditions that could also result in better separations. Reduction of the capillary bore led not only to improved resolution but also to decreased separation times. Chen [253] and Chen and Sternberg [254] demonstrated that a rapid serum protein analysis may be performed within 90 s by using a 20 µm i.d. × 25 cm long capillary and 20 kV of applied voltage (Fig. 38). However, due to excessively rapid migration, some serum proteins may move into the adjacent zone, which may cause a poor correlation with the results from "Paragon SPE Gel."

High-Resolution Serum Protein Separation and Identification
For balancing these two factors, the running conditions were optimized by using high ionic strength running buffer. Serum samples were diluted 1 : 10 in phosphate-buffered saline (PBS) and injected by pressure injection mode for 10 s. The electric field was set at 400 V/cm and electrophoresis was conducted for 6 min. The detection wavelength was 214 nm [247]. At least 10 peaks were separated within 5 min. These 10 peaks were identified by both "spiking" and "immunosub-

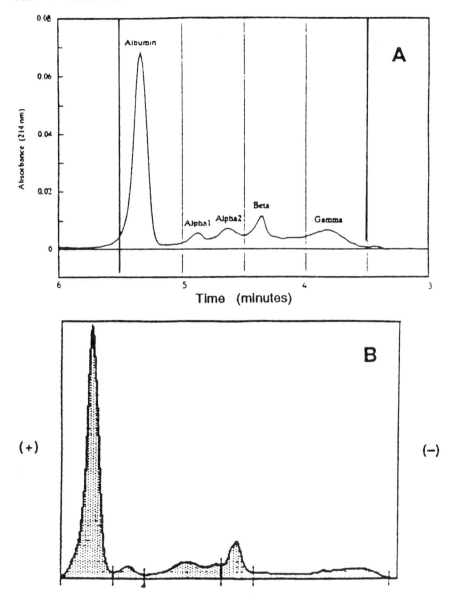

Fig. 35 Capillary electropherogram (A)* and densitometric electrophoresis gel scan (B) from the same normal serum sample. CE running conditions: fused silica capillary, 75 μm × 25 cm; borate buffer (pH 10); 1 kV electrokinetic injection for 1 s at 1 kV; applied voltage, 5kV; detection wavelength, 214 nm.

---

*Time axis has been reversed from the original electropherogram.

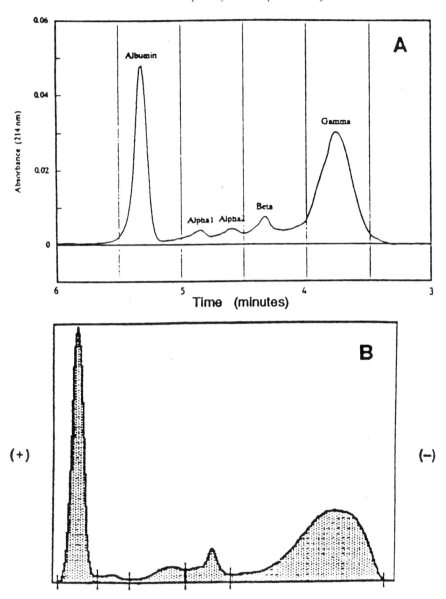

Fig. 36 Capillary electropherogram (A)* and densitometric electrophoresis gel scan (B) from the same abnormal serum sample (polyclonal gammopathic sample). CE running conditions the same as Fig. 1.

---

*Time axis has been reversed from the original electropherogram.

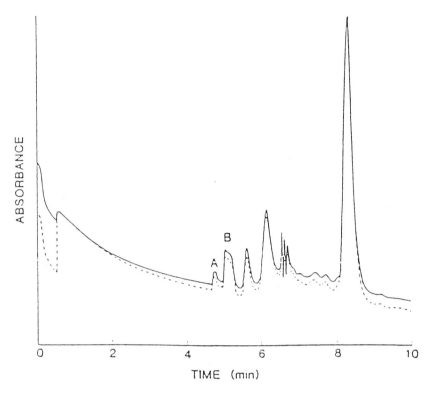

Fig. 37 An overlay of electropherograms from two CZE runs of the same human serum protein sample, showing reproducibility of retention times by addition of ethylene glycol (20%) in the sample diluent.

traction" methods (Figs. 39 and 40). The 10 serum proteins include prealbumin, albumin, α_1 acid glycoprotein, α_1 antitrypsin, haptoglobin, α_2 macroglobulin, β-lipoprotein, transferrin, C_3 complement, and γ-globulin (Fig. 41).

Serum protein analysis is an important part of many screening tests to detect occult disease and certain unsuspected dysproteinemias. The following classification of serum protein patterns may help the clinician to document a definitive diagnosis or interpretation:

  a. Hypoproteinemic
     i. Malnutrition (e.g., Kwashiorkor): decreased albumin
     ii. Protein loss (nonselective): decreased albumin
        iia. Exudative dermatopathies and burns
        iib. Exudative pulmonary disease
        iic. Essential hypoproteinemia

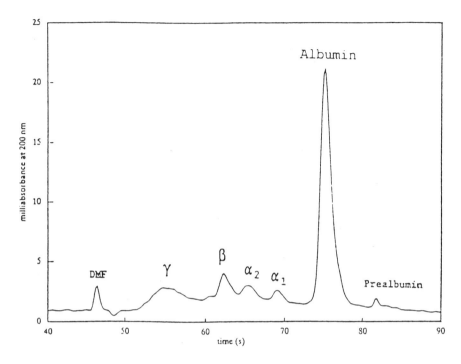

Fig. 38 Electropherogram of capillary electrophoresis with a small-bore capillary (20 μm i.d. × 25 cm) and high applied voltage. This demonstrates that a single analysis required less than 90 s.

      iid. Protein loss gastroenteropathies
      iie. Blood loss and plasmaphoresis
  b. Selective protein loss (nephrotic pattern): decreased albumin and in some cases γ-globulin
  c. Diffuse hepatodegenerative pattern: decreased albumin, α_1, α_2, and β, but increased γ-globulin
  d. Cirrhotic pattern: decreased albumin, α_2, and β–γ bridging
  e. Acute inflammatory and physiological stress type pattern: decreased albumin, increased α_1, and α_2
  f. Chronic inflammatory-type pattern: decreased albumin, increased α_1, α_2 and γ-globulin (polyclonal)
  g. Anemia
      i. Iron deficiency: increased β globulin
      ii. In-vivo hemolysis: decreased α_2 globulin

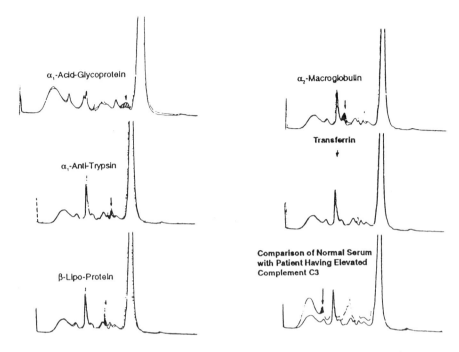

Fig. 39 Two overlaid electropherograms: one of normal serum without spiking and the other spiked with a purified specific protein; (a) with α_1-acid-glycoprotein, (b) α_1-antitrypsin, (c) β-lipoprotein, (d) α_2-microglobulin, (e) transferrin, and (f) C_3 complement. The shaded areas represent the locations of outstanding peaks after spiking with a specific protein.

- h. Polyclonal gammopathy: decreased albumin, increased γ-globulin
- i. Monoclonal gammopathy (MG)
    - i. MG with normal immunoglobulin
    - ii. MG with decreased immunoglobulin of other classes than the monoclone: increased gamma globulin
    - iii. Hypogammaglobulinemia with monoclonal urine protein: decreased γ-globulin
- j. Hyperlipidemic pattern: increased α_2 (β–lipoprotein migrates in α_2 region)
- k. Pregnancy pattern: decreased albumin
- l. Defect dysproteinemia
    - i. Analbuminemia
    - ii. Hypo α_1
    - iii. Atransferrinemia

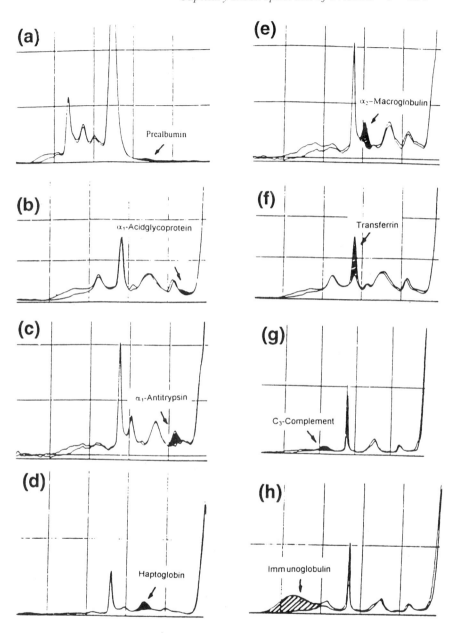

Fig. 40 Two overlaid electropherograms: one of normal serum without immunosubtraction and the other after mixing with a solid support coupled with a specific antibody: (a) antiprealbumin, (b) anti-α_1-acid-glycoprotein, (c) anti-α_1-antitrypsin, (d) antihaptoglobin, (e) anti-α_2-macroglobin, (f) antitransferrin, (g) anti-C_3-complement, and (h) anti-IgG. The shaded areas represent the location of attenuated peaks after mixing with specific protein-coupled solid supports.

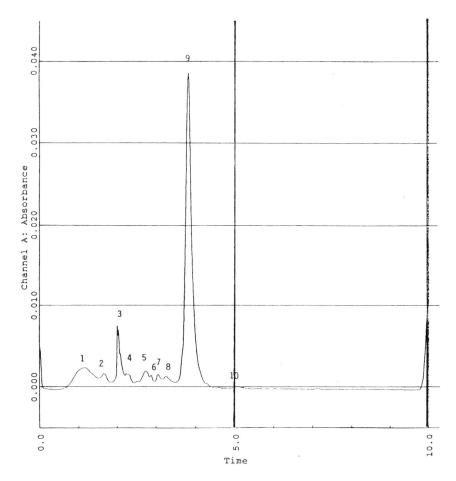

Fig. 41 High-resolution serum protein analysis using a small-bore capillary (25 μm × 27 cm) with high ionic strength and high pH running buffer. The electropherogram showed 10 major serum protein peaks: (1) γ-globulins, (2) complement, (3) transferrin, (4) α_2-macroglobulin, (5) haptoglobin, (6) β_2-lipoprotein, (7) α_1-antitrypsin, (8) α_1 acid glycoprotein, (9) albumin, and (10) prealbumin.

      iv. Analpha-2-β-lipoproteinemia
      v. Afibrinogenemia (plasma)
      vi. Presence of fibrinogen in serum
      vii. Hypogammaglobulinemia
      viii. Hemoconcentration: increase all proteins

ix. Acomplementemia
x. Ahaptoglobinemia

Monoclonal γ-Globulin Identification

Serum protein analysis for detection of "Monoclonal Gammapathies" has become a routine screening test. These diseases comprise a heterogeneous group characterized by the presence in serum or urine of a monoclonal immunoglobulin. This is a product of a single clone of lymphoid cells, with restricted electrophoretic mobility, and appears in serum electropherograms as a narrow band or "spike." Monoclonal gammopathies have been classified as the following:

a. Malignant monoclonal gammopathy
   i. Multiple myeloma
   ii. Waldenstrom's macroglobulinemia
   iii. Solitary plasmacytoma
   iv. Amyloidosis
   v. Heavy chain disease
   vi. Malignant lymphoma
   vii. Chronic lymphocytic leukemia
b. Secondary monoclonal gammopathy
   i. Cancer (nonlymphoreticular)
   ii. Monocytic leukemia
   iii. Hepatobiliary disease
   iv. Rheumatoid disorders
   v. Chronic inflammatory states
   vi. Cold agglutinin syndrome
   vii. Benign hyperglobulinemic purpura of waldenstrom
   viii. Papular cucinosis
   ix. Immunodeficiency
c. Benign monoclonal gammopathy
   i. Transient
   ii. Persistent

Multiple myeloma is the most common disease among the monoclonal gammopathies. The patients may suffer from anemia, bone pain, unexplained weakness, fatigue, elevation of the erythrocyte sedimentation rate, hypercalcemia, Bence Jones proteinuria, renal insufficiency, or recurrent infections. Eighty percent of patients can be diagnosed via serum protein analysis. The diagnosis of multiple myeloma often begins with the recognition of a monoclonal protein in the serum or urine or both. Each monoclonal protein consists of two heavy polypeptide chains of the same class and subclass, and two light polypeptide chains of the same type. Identification of class and type of monoclonal γ-globulin is essential for determining the treatment regimen and prognosis. The two conventional methods for identification are "Immunoelectrophoresis and Immunofixation." Both methods are very cumbersome and time-consuming. Identification of monoclonal myeloma protein (M-protein) by capillary electrophoresis was demonstrated by off-line pretreatment of serum samples with antibody-coupled solid supports [255]. Antibodies against IgG, IgA, IgM, κ, and λ were individually coupled to cyanogen bromide activated Sepharose 4B™. A patient sample was diluted with PBS and then mixed individually with five different solid-phase slurries. After 5 min or incubation and sedimentation was completed, the upper layers of sera were subjected to analysis by CE. All five electropherograms were compared to two control electropherograms; one control was not mixed with the solid support, the other control was mixed with the solid support without being antigen coupled. For identification of the class and type of M-protein, each electropherogram from the pretreated sample was simply overlaid with the control electropherogram. From the missing or attenuated peaks, the class and type of M-protein can be identified.

High-Throughput Capillary Electrophoresis

To complete a single-patient M-protein classification and typing, six runs were required, with each run taking 6 min. This amount of time may be too lengthy to meet the needs of many routine clinical laboratories. For example, some high-volume laboratories may process as many as 500 routine serum separations and may need to process more than 10 M-protein identifications in a single workday. There are two approaches for meeting adequate throughput: (1) reduce the separation time so that high throughput can be achieved (but with a cost in resolution) or (2) run parallel channels at the same time so that more separations can be made without compromising the quality of the separations. Beckman (Brea, CA) has developed a six-channel experimental clinical CE system. Preliminary tests have shown that it can provide the equivalent of one separation per minute with a total run time of 6 min/separation. The optical device and detectors are more complicated on this multiple-channel system. The light from the source lamp must be distributed to all the channels equally. This can be accomplished effectively with fiber optics. An optical system gathers and relays light from the source lamp onto a bundle of fibers. The fibers are then split to provide an optical channel for each capillary. After passing through the capillary, the light must be directed to the detector. Because the nature of each segment of capillaries may not be the same, production of reproducible electropherograms is a challenge that needs to be overcome.

B. Urine Proteins

Bence Jones Proteinuria

The appearance of M-protein in serum is a major immunological feature; however, about 20% of myeloma patients may have normal serum protein patterns, with only Bence Jones (free light chain of immoglobulin) appearing in the urine. Thus, a urinary protein test is essential when the serum examination is negative in patients with clinical symptoms. Bence Jones or light-chain myeloma patients are more likely to be present with severe renal failure. This monoclonal light chain is either a κ or λ type. Usually, patients with λ-type Bence Jones myeloma have more protein in the urine, poor renal function, and shorter overall survival. Therefore, light-chain typing is an important piece of information for the clinician in disease treatment and prognosis. In general, urinary proteins are more difficult to measure by gel or capillary electrophoresis because the concentration of proteins are normally lower than that in serum. In addition, urine contains hundreds of other metabolic waste molecules which may co-migrate with protein and also exhibit high UV absorbance. In conventional gel electrophoresis for urinary protein analysis, sample preconcentration is required. Because of poor detection sensitivity, a 40–80-fold concentration by Minicon concentrator is a standard required process before the sample can be applied to the gel. In contrast, for CE, with its higher sensitivity, concentration may be less important than co-migrating UV-absorbing substances.

A normal urine sample (unconcentrated and without pretreatment) was subjected to CE, and a seismogram-like electropherogram was displayed (Fig. 42). A similar pattern was obtained from analysis of Bence Jones proteinuria urine. However, when both samples were dialyzed with a semipermeable membrane (molecular weight cutoff of 14 kDa), the electropherogram from the normal urine only showed a trace amount of protein, whereas a considerable amount of protein was observed for the urine containing Bence Jones protein (Fig. 43). Membrane dialysis is a very slow process and a less efficient method for removing small molecules from urine. A more efficient and faster method was found by using gel filtration [256]. The gel consists of an open, cross-linked three-dimensional molecular network, cast in bead form for easy column packing. The pores within the beads are of such sizes that some are not accessible by large molecules, but smaller molecules can penetrate all pores. The results showed a good protein recovery and most of the small molecules were removed. The detection limit is about 2 mg/dL for Bence Jones protein and 6 mg/dL for albumin. The identification of Bence Jones protein may follow the same "Immunosubtraction" method as described above. The antibody for coupling solid support can use either "anti-light-chain-specific antibody" or "anti-whole-molecule plus light-chain antibody." The experimental results showed that the latter was more sensitive than the former. In other words, use of anti-light-chain-specific anti-

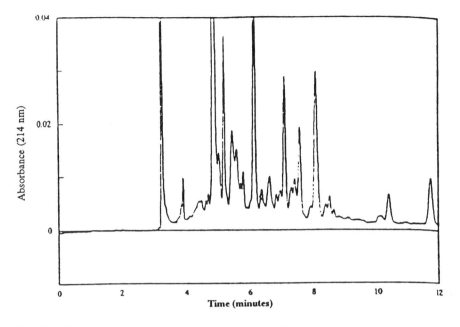

Fig. 42 Electropherogram of urine from a patient with Bence Jones proteinuria. Running conditions same as in Figure 1.

body may cause more false-negative results. Jenkins et al. [257] used Dowex 2, a strong basic anion-exchange resin, to remove anionic species from urine specimens. The urine specimen was mixed with swollen resin for 30 min and the mixture was centrifuged at 9500 × **g** for 2 min (Fig. 44). The results from CE were compared to High-Resolution Agarose Gel Electrophoresis (HRAGE, Helena Titan High-Res.). A good correlation ($r = 0.93$) was found between these two methods. A high-pH and high ionic strength borate buffer was used for most urinary protein analyses because this may minimize protein adhesion to the wall of the capillary. Jenkins et al. [257] evaluated the CE method by examination of 1000 prospective serum samples by both HRAGE and CE. The capillary dimensions were 50 μm i.d. × 72 cm long and the detection wavelength was 200 nm. The borate buffer had an ionic strength of 50 mM, pH 9.7, and contained 1 mM calcium lactate. Each M-protein was evaluated by comparing the two methods. A good correlation was found between the two methods.

Glomerular Proteinuria

The source of proteins found in urine may be due to increased glomerular permeability (glomerular proteinuria), in which the urinary protein is mainly albumin; defective tubular reabsorption (tubular proteinuria), in which the urinary pro-

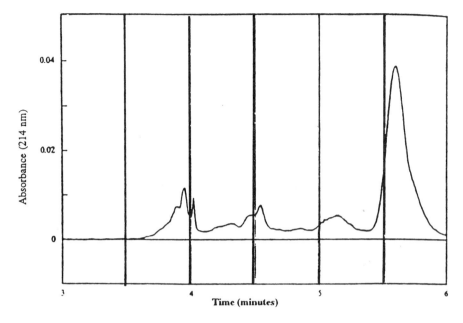

Fig. 43 Electropherogram of the same sample as in Fig. 8 with the sample pretreated by dialysis through a membrane having a molecular-weight cutoff of 14kDa. Running conditions were the same as in Fig. 1.

teins are mainly normal low-molecular-weight plasma proteins; increased concentration in the plasma of an abnormal low-molecular-weight protein such as immunoglobulin light chains (overload proteinuria); and abnormal secretion of protein into the urinary tract (Postrenal proteinuria). The first two are more common and the last two are the least common.

"Glomerular proteinuria" is the most common and serious type of proteinuria. It is the result of an increase in glomerular permeability in numerous conditions characterized by diffuse injury to the kidneys. The diseases causing that pathological change include diabetes, immune complex disease, nephritis associated with systemic lupus erythematosus, and glomerulonephritis. In such progressive diffuse renal diseases, the ability to restrict filtration of the smallest proteins is lost first. Albumin is a major protein with small size, and it appears in urine at an early stage. This condition is called "selective glomerular proteinuria." Progressively severe glomerular lesions produce less selective proteinuria until proteins of all sizes pass the glomerulus. This stage is called "nonselective proteinuria." Recently, small amounts of albumin (microalbumin) in urine was shown to be an early indication of diabetes. This amount of albumin may be under the detection limit of capillary electrophoresis. For detection of microalbumin, either preconcentration of the

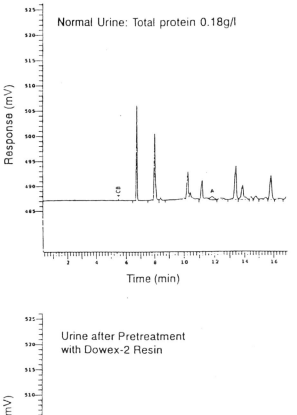

Fig. 44 Electropherograms of a normal human urine before (upper panel) and after pretreatment with Dowex-2 resin (lower panel). The albumin peak is indicated by A at approximately 11.6 min. Bence Jones protein is not found in normal urine specimens. UV detector trace obtained using 18 kV and 54 µA. (Reproduced from Ref. 256 with permission.)

urine sample or detection enhancement by fluorescent labeling and/or by laser-induced fluorescence may be required.

Tubular Proteinuria

The second type of proteinuria, "tubular proteinuria," is characterized by the appearance of low-molecular-weight proteins in the urine because of the defective reabsorption of these compounds in the proximal renal tubules. The low-molecular-weight proteins typically excreted in tubular proteinuria are β_2 microglobulin, lysozyme, retino-binding protein, α_1 microglobulin, and α_1 glycoprotein. The etiology for chronic tubular proteinuria may be hereditary (e.g., Fonconi's syndrome), or a kidney disease such as chronic pyelonephritis, or a systemic disease such as cirrhosis or sarcoidosis; or it may be acquired as a result of drugs such as phenacetin or poisons such as cadmium. The test for tubular proteinuria is used to monitor renal allograft rejection, aminoglycoside and cadmium toxicities, and chronic pyelonephritis; a favorite protein marker is β_2 microglobulin. The conventional method for tubular proteinuria detection is sodium dodecylsulfate (SDS)-PAGE. Urinary proteins are first heated with SDS, to which they bind and thus acquire a uniform negative charge. The polymer dynamic sieving system of CZE has been used for the separation of macromolecules such as peptides, proteins, and nucleic acid molecules [236]. Polymers such as methylcellulose or polyethylene glycol were added into the separation medium. These polymers generate a molecular sieving effect which facilitates the separation. This approach has several advantages over gel-filled capillaries such as low UV absorption and no degradation of the sieving medium in the capillaries. This system has the potential for clinical applications, and proteinuria determination may be one of them.

C. Cerebrospinal Fluid

Cerebrospinal fluid (CSF) is secreted by the choroid plexuses, around the cerebral vessels, and along the walls of the ventricles of the brain. CSF fills the ventricles and cisternae and bathes the spinal cord. In the meantime, CSF is reabsorbed into the blood through the arachnoid villi to maintain equilibrium. Because CSF is mainly an ultrafiltrate of plasma, low-molecular-weight plasma proteins such as prealbumin, albumin, and transferrin normally predominate. In pathological conditions, such as bacterial- or viral-induced meningitis, the total CSF proteins increased significantly. This is due to the malfunction of the blood-brain barrier, increasing its permeability. Another cause of increased CSF protein is "intrathecal synthesis of immunoglobulins in demyelinating diseases of the central nervous system (CNS)," especially multiple sclerosis. In multiple sclerosis, patchy deterioration of myelin sheaths of axons occurs in the CNS and causes lesion sites. B-Lymphocytes infiltrate these lesion sites and synthesize

IgG (and occasionally other immunoglobulins). Because axons of the CNS are in intimate contact with the CSF, the immunoglobulins produced in the lesion appear in the CSF. The synthesis of immunoglobulins by B-lymphocytes is restricted to only a few clones (oligoclonal banding) and the reason for this is still unknown. The conventional method for CSF oligoclonal banding detection is "high-resolution electrophoresis" which is conducted on this agarose gels using a high-voltage separation with concurrent cooling. CSF samples are concentrated 40-fold before being applied to the gel. This technique and instrumentation are commercially available from either Beckman Instruments (Brea, CA) or Helena Laboratories (Beaumont, TX). IgG oligobanding has been demonstrated by CZE without sample pretreatment. The same running conditions for serum and urinary protein were applied to CSF analysis without sample preconcentration (Liu, unpublished data). The results showed poor resolution in the γ-globulin separation because of the low concentration.

D. Hemoglobin

Hemoglobin Variants

The two modes of capillary electrophoresis that have been widely used for hemoglobin variants analysis are capillary zone electrophoresis (CZE) and capillary isoelectric focusing (CIEF).Because CIEF has been discussed extensively in another section of this chapter, the following paragraphs will concentration only on some recent progress in hemoglobin variant analysis by CZE with uncoated capillary tubes. There are two major advantages for using CE to analyze hemoglobin. First, hemoglobin absorbs not only in the UV but also in visible wavelengths (415 nm). Because of this unique characteristic, hemoglobin is more easily detected without interference from other substances. This also means greater choice in running buffers without concern for high background absorption. The second advantage is that the concentration of hemoglobin is always high enough (12–16 g/dL in whole blood) so no sensitivity problem will be encountered.

More than 300 different kinds of mutant hemoglobins have been found in humans. The replaced amino acid may bear a different charge to the original amino acid, so it may slightly affect the charge density of the whole molecule. A familiar example is sickle-cell hemoglobin, whose structural difference from normal hemoglobins was elucidated decades ago. In sickle-cell hemoglobin, the two forms of α-chains are identical, whereas the glutamic acid residue at position 6 in the β-chain of normal hemoglobin is replaced by a valine residue. Table 6 lists some of the many mutations; the names of these abnormal forms are derived from the location of their discovery.

Table 6 Amino Acid Replacements in Human Hemoglobins

Abnormal Hb	Position	Normal residue	Replacement
α-chain			
I	16	Lys	Glu
G$_{Honolulu}$	30	Glu	Gln
Norfolk	57	Gly	Asp
M$_{Boston}$	58	His	Tyr
G$_{Philadelphia}$	68	Asn	Lys
O$_{Indonesia}$	116	Glu	Lys
β-chain			
C	6	Glu	Lys
S	6	Glu	Val
G$_{San Jose}$	7	Glu	Gly
E	26	Glu	Lys
M$_{Saskatoon}$	63	His	Tyr
Zurich	63	His	Arg
M$_{Milwaukee}$	67	Val	Glu
D$_{Punjab}$	121	Glu	Gln

Some hemoglobin mutations are lethal; that is, the patient may never live to maturity because the amino acid replacement results in a functionally defective molecule. On the other hand, some limit the physiological function of the hemoglobin less seriously, and some are apparently harmless. However, under stress conditions such as massive blood loss or hypoxia, these heterozygous genotype combinations may cause life-threatening conditions. Hemoglobin mutation identification tests may help warn the clinicians to take precautions in the care of these patients.

As most hemoglobin variants only involve a single amino acid replacement, the charge : mass ratio differs slightly. For separating hemoglobin variants, higher resolution power is needed. Using conventional agarose gel electrophoresis, two different kinds of gel have to be employed in order to complete the analysis. The reason for this is that in one of the gel systems, "Alkaline Gel Electrophoresis" (pH - 8.6), Hb S co-migrates with Hb D or Hb G, and HB A$_2$, C, O-Arab, and E cannot be differentiated on the basis of electrophoretic mobility. In addition, Hb F may be poorly resolved from Hb A. The complementary gel "Acid Gel Electrophoresis" (pH = 6.2), is able to separate Hb S from D or G very well and can separate A$_2$ from O or C. However, Hb A$_2$ co-migrates with A, E, D, G, and Lepore which can be resolved on an alkaline gel. High-resolution capillary zone electrophoresis has been shown to separate several Hb variants (Hb A, F, S, and C) by using a high-pH and high-ionic-strength buffer [251] (Fig. 45). A high-pH buffer may prevent the interaction between hemoglobin and the capillary wall more effectively and a high-ionic-strength running buffer may slow down the electroosmotic flow (EOF). Moreover, the relatively lower-ionic-strength sample plug was injected into a column filled with a buffer of higher ionic strength. Consequently, sample ions migrate rapidly into the run buffer under the applied voltage, resulting in stacking in front of the water boundary. As the sample concentration increases, the local conductivity of the stacking region becomes higher, resulting in a further drop in the electric field strength. Thus, the leading edge of the sample region slows down and further enhances the stacking effect. This sample stacking effect concentrates analytes in a narrow band inside the capillary. The use of a high-ionic-strength buffer under a high electric field induces joule heat, which may contribute to band-broadening and deterioration of resolution. Fortunately, capillary electrophoresis exhibits more efficient heat dissipation than a convention slab gel due to a higher relative contact area. This may minimize the band-broadening. Zhu et al. [191] separated hemoglobin variants by capillary isoelectric focusing (CIEF) and separated globin chains by CZE. The capillary was coated internally with covalently attached linear polyacrylamide for minimizing EOF. Isoelectric focusing was carried out in a 17 cm × 25 µm i.d. coated capillary. The capillary was purged with water and 10 mM phosphoric acid between separations. Hemoglobin samples were mixed with pH 3–10 ampholytes to a final ampholyte concentration of 2% and a total hemoglobin concentration of about 1 mg/mL. The mixtures were injected by pressure (60 s at 100 psi). Focusing was carried out at 7 kV constant voltage for 5 min using 40 mM sodium hydroxide as the catholyte and 20 mM phosphoric acid as the anolyte. Cathodic mobilization was performed by replacing the catholyte with a proprietary zwitterionic solution (Bio-Rad Labs.). The results showed (Fig. 46) an excellent separation among Hb A, F, S, and C within 15 min. However, the throughput is insufficient to compete with current HPLC methods.

Glycated Hemoglobin

Measurement of glycated Hemoglobin provides a retrospective index of the serum or plasma glucose values over a long period of time. Glycated hemoglobin is a nonenzyme-mediated, posttranslational modification, occurring slowly in erythrocytes [191]. Human adult hemoglobin (Hb) usually consists of Hb A (97% of the total), Hb A_2 (2.5%), and F (0.5%). Hb A consists of four polypeptide chains: two α-chains and two β-chains. Glycated Hb is formed by the condensation of glucose with the N-terminal valine of each β-chain of Hb A to form an unstable Schiff base (aldimine) which then undergoes an

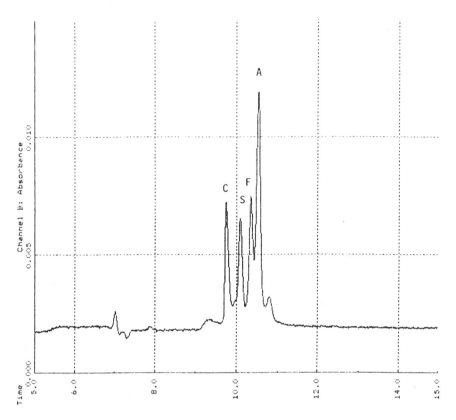

Fig. 45 Electropherogram of a mixture of hemoglobin A, F, S, and C (from IsoLab Hb AFSC standard). Capillary, fused silica (25 cm × µm i.d.); injection, electrokinetic for 1 s at 3 kV; running conditions, 5 kV; Beckman proprietary buffer at pH 8.5; detection at 415 nm.

Amadori rearrangement to form a stable and irreversible ketoamine, Hb A_{1c}. Phosphosugars, such as fructose-1,6-diphosphate and glucose-6-phosphate, may also attach to some other amino acids to form Hb A_{1a1} and Hb A_{1a2}, respectively. The exact structure of Hb A_{1c} remains uncertain. However, Hb A_{1c} is the major fraction of glycated Hb which constitutes approximately 80% of Hb A_1. The Hb A_{1c} level in blood depends on both the life span of the red blood cell (average 120 days) and the blood glucose concentration. The amount of Hb A_{1c} therefore represents the integrated value for glucose over the preceding 6–8 weeks and provides an additional criterion for assessing glucose control. Recently, glycated Hb monitoring has become a routine diag-

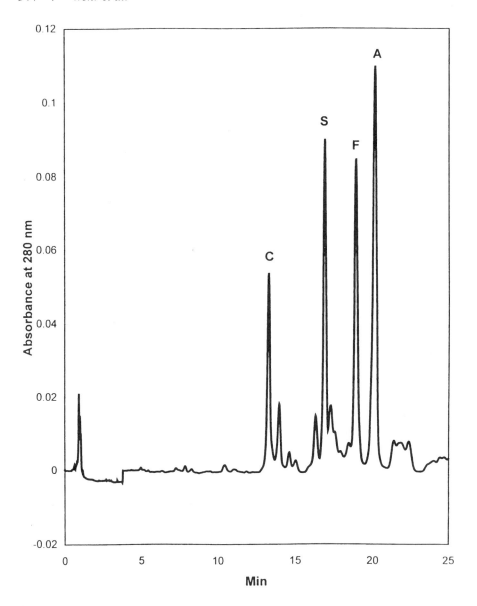

Fig. 46 Separation of hemoglobins A, F, S, and C using capillary isoelectric focusing with wide-range (pH 3–10) ampholytes. Conditions: capillary, 22 cm × 75 μm coated with AAEE; focusing for 210 s at 15 kV with 20 mM phosphoric acid + polymer additive as anolyte and proprietary buffer (pH 8.8) containing polymer additive as catholyte; gravity mobilization at 15 kV by reducing the catholyte volume from 1.5 to 0.5 mL; detection at 280 nm.

nostic test administrated every 3–4 months. In certain clinical situations, such as diabetic pregnancy or a major change in therapy, more frequent monitoring should provide useful information.

Conventional methods for the determination of glycated hemoglobin include ion-exchange chromatography, high-performance liquid chromatography (HPLC), gel electrophoresis, isoelectric focusing, and affinity chromatography. These methods are either temperature or pH dependent and also measure only total glycated Hb; they are not specific to Hb A_{1c}.

An alternative to the conventional methods of separation is capillary electrophoresis. Chevigne et al. [259] demonstrated that Hb A1c, Hb A1a, Hb A1b, Hb A1d, methemoglobin, Hb S, and Hb C can be separated within 7 min with an optimized Hb running buffer. The results showed a good correlation with Diatrac, agarose gel electrophoresis. The composition of the Chevigne's running buffer is under patent consideration and has yet been published.

E. Enzymes or Isoenzymes

Enzyme assays for clinical diagnosis are one of the most important fields of contemporary clinical chemistry. All of the hundreds of different enzymes present in the human body are synthesized intracellularly, and most of them carry out their functions within the cells in which they are formed. When tissue damage occurs, a specific enzyme may leak from the specific cell population into the bloodstream. Detection of tissue specific enzyme levels in the blood is essential for prognosis and diagnosis. Many enzymes may have different types of posttranslational modification which may give rise to multiple forms or isoenzymes. These isoenzymes may differ in their physico-chemical properties such as electrophoretic mobility, resistance to inactivation, or solubility. However, the catalytic properties of these isoenzymes are virtually the same. The most commonly exploited difference between isoenzymes is the difference in the net molecular charge that results from the altered amino acid composition of the molecules. Because the separation mechanism of capillary zone electrophoresis (CZE) is based on the charge : mass ratio, CZE may be perfectly suitable for isoenzymes analysis. However, the major problem in enzyme analysis by CZE is that the enzyme concentration in serum is always very low and under the detection limit of CZE. In addition, a large amount of serum proteins may also co-migrate with the enzymes, potentially making detection even more difficult. One alternative method based on the properties of enzymes may overcome these problems—enzymatic amplification of the end product. If the enzyme reacts with an excess amount of substrate, the enzymatic reaction will continue, forming large amounts of end product. The amount of end product is directly proportional to the turnover number of the enzyme and to the incubation time. So, the signal can be amplified with a longer incubation time. Again, this hypothetical idea may face two major hurdles: (1) capillary electrophoresis is a closed system, so adding substrate into the capillary after enzyme or isoenzyme separation would be difficult, and (2) the enzymes or isoenzymes (or enzyme–substrate complex) will exhibit different mobility (migration time) from the end product.

Lactate dehydrogenase (LD) is composed of four polypeptide chains of two types called H (heart) or M (muscle) subunits. There are five combinations of these subunits that give rise to the five electrophoretically separable fractions LD1, LD2, LD3, LD4, and LD5. The LD1 and LD2 types are associated with cardiac muscle, kidney, and red blood cells (RBCs). LD4 and LD5 are associated with skeletal muscle and liver, and LD3 is associated with endocrine glands, spleen, lung, lymph tissue, platelets, and nongravid uterine muscle. Serum LD isoenzyme levels are closely related to myocardial infarction; total LD activity rises 8–12 h after onset of chest pain, reaches a maximum 24–48 h after the episode, and remains elevated for some 7 or more days. The LD1/LD2 ratio has also been advocated as an additional test for myocardial infarction. It has been suggested that both tests are of equivalent diagnostic power during the first 24 h after an infarction. Recently, some other myocardial infarction markers, such as troponin, CK-MB, and its isoforms may substitute for the LD isoenzymes test. Nevertheless, LD isoenzymes analysis by CZE has established a model for enzyme assay by capillary electrophoresis. LD is a hydrogen-transfer enzyme that catalyzes the oxidation of L-lactate to pyruvate with the mediation of NAD+ as a hydrogen acceptor and reducing it to NADH. The absorbance spectrum of the product NADH (340 nm) is uniquely different from that of either the assay reagents or the enzyme. Hsien et al. [260] first demonstrated LD isoenzymes separation by CZE. They added an excess amount of substrates (lactate and NAD+) to the running buffer (Beckman Dri-Stat LD reagent with a fivefold dilution with distilled water, pH 8.7). Equal units of purified five human LD isoenzymes (purchased from Sigma and Aalto) were mixed and injected (1 s 1 kV, electrokinetic injection) into the capillary (75 μm i.d. × 37 cm long). After 2 min separation (running voltage was 7.5V), high voltage was turned off for 2–10 min, then resumed. Hsien et al. referred to the process as "parking." During this parking period, the enzymatic reaction keeps going and the end product, NADH, accumulates in each separated isoenzyme zone. When the potential was resumed, each NADH zone moved through the detection window (Fig. 47). The results showed six distinct peaks on the top of a gradually elevated baseline. Because LD5 has the fewest negative charges, its migration counter to electroosmotic flow (EOF) is lowest and thus its apparent migration velocity is faster than the other isoenzymes. In contrast, LD1, with the most negative charge, exhibited the highest rate of migration counter to EOF and thus had an apparent migration velocity slower than the other isoenzymes. LD2, LD3, and LD4 migrated in that order between LD1 and LD5. These five isoen-

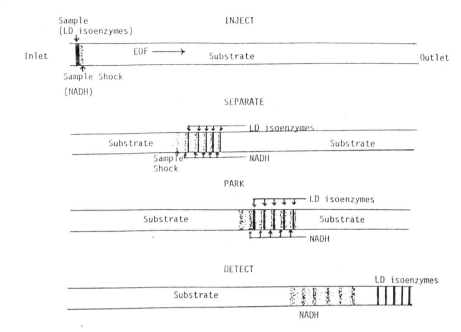

Fig. 47 Schematic diagram for demonstration of "parking" model for LD isoenzymes analysis.

zymes were separated into five distinct zones after the first 2 min of separation, so the first five NADH peaks reflected the amount of LD5 to LD1 reading from left to right. One extra peak to the right of the LD1-induced NADH was formed when the enzyme-containing sample was injected into the capillary. At this time, all of the isoenzymes were exposed to the substrate at the boundary between the sample volume and the buffered substrate. An immediate reaction between the isoenzymes and the substrate occurred. This is termed "sample shock." Because the net flow of NADH to the window is slower than for any of the isoenzymes, it appeared last. The gradually elevated baseline represented the NADH formed during continuous contact of migrating isoenzymes with the substrate in the running buffer (Fig. 48).

The deficiency of glucose-6-phosphate dehydrogenase (G-6-PDH) is an inherited trait which may cause hemolytic anemia and is widely distributed in approximately 200 million people throughout the world. G-6-PDH deficiency is particularly prevalent among people of southeast Asia, in blacks, in the people of the Mediterranean littoral, and in people of India. Bao and Regnier

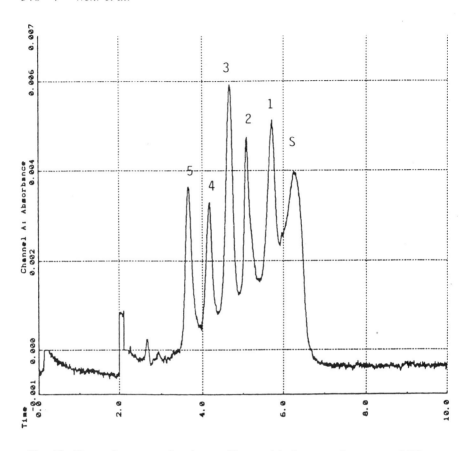

Fig. 48 Electropherogram of a mixture of lactate dehydrogenase isoenzymes, LD1, LD2, LD3, LD4, and LD5. The six peaks represent five isoenzymes and a "sample shock" peak. Each peak was identified from the migration time of a single isoenzyme electropherogram.

[70] demonstrated glucose-6-PDH assay using a coated capillary. The basic theory is almost identical as described in the previous paragraph. The capillary was filled with enzyme-saturating concentrations of substrate and coenzyme in the running buffer. Then G-6-PDH was injected and the enzyme activity was assayed by electrophoresis of the product, NADPH, to the detector where it was detected at 340 nm. The amount of product was amplified by switching the potential to zero before the enzyme eluted from the capillary. After 5-min zero-potential incubation, the potential was reapplied for product migration to the

detector. This approach was used to generate an NADH peak accumulated from a sample estimated to contain 4.6×10^{-17} mol of G-6-PDH. This "zero-potential assay" idea is the same as the "parking" idea described above. Later, this mode of enzyme assay (in which electrophoretic mobility issued to separate protein analytes, mediate the mixing of assay reagents, and transport products to a detector) was referred as electrophoretically mediated microassay (EMMA).

Wu and Regnier [71] used a polyacrylamide gel-filled capillary to measure alkaline phosphatase (ALP) and β-galactosidase because the product diffusion during "zero-potential" during the assay seriously diminished the resolution using an open tubular capillary. Polyacrylamide gel is highly viscous and permits enzymes to be incubated up to 2 h in the capillary without significant band spreading. The lowest detection limit achieved was 5.2×10^{-20} mol of ALP.

Xue and Yeung [261] detected a trace amount of lactate dehydrogenase within single human erythrocytes by laser-induced fluorescence. The isoenzymes signals were amplified by detecting their end product, NADH, with prolonged reaction time. Washed red blood cells (RBCs) were suspended in the same wash solution just before injection. Then the RBCs were easily lysed in the running buffer because of the low ionic strength. Different isoenzymes were separated for 1 min under 30 kV. After the different isoenzymes migrated into different zones according to their electrophoretic mobilities, the high voltage was turned off for 2 min. During this incubation period, NADH accumulated in the different isoenzyme zones due to their enzymatic activities. Finally, high voltage was reapplied to move the components past the detection window. The running buffer contained 5mM lactate, 5 mM NAD$^+$, and 30 mM phosphate at a pH of 7.3. Because only a small amount of enzymes was injected, even for 36 runs the effect on the net charge of the capillary wall was very minimal. Because human LDH isoenzymes activities are related to metastatic cancers and to cell age, their separation and individual determination may have broad clinical implications.

Meller et al. [262] examined the utility of time-resolved laser-induced fluorescence (TRLIF) detection to circumvent biological sample matrix interference. They used dual collection windows to generate two electropherograms from one injection, A 200-point fluorescence decay waveform was transferred to the data-collection program every second. Two independently adjustable integration windows were used to sum the values from the desired portion of the decay waveform. These two sums were plotted versus elapsed time to generate two electropherograms. The delay value refers to the time delay in nanoseconds from the peak minimum of the fluorescence decay. They chose leucine aminopeptidase (LAP) as a model to demonstrate their idea because the enzyme has clinical significance for detecting the function of liver, bile duct, and pancreas. In addition, this enzyme cleaves the nonfluorescent leucine amide derivative, L-leucine-4-MBNA, to produce 4-MBNA (4-methoxy-β-naphthylamine), which has a 26 ns lifetime. The fluorescence of this product is easily distinguished from fluorescent species with shorter lifetimes such as the interfering substances in matrix. Thus, TRLIF allows low levels of enzyme detection with short incubation times. LAP enzyme levels as low as 6×10^{-13} M can be detected using this method. This method involved a very sophisticated procedure and high-cost instrumentation, so the utility value in the clinical environment is still questionable, at least at the current stage.

F. Peptide Hormones

Most hormone assays have been performed by immunoassay due to ease of processing and the commercial availability of automated instruments. However, some hormones, or their releasing factors, and some neurohypophyseal peptides are analyzed by HPLC mainly due to lack of commercially available immunoassay. CE holds some advantages, such as low cost, small sample volume, and high resolution which may allow it to replace HPLC. Idei et al. [263] used an uncoated capillary to analyze synthetic somatostatin and its analog peptides. They suggested that CZE can be applied very advantageously in peptide analysis. Its performance in terms of selectivity, resolution, theoretical plate number, duration, and cost of the analysis is comparable or sometimes superior to that of HPLC. Sutcliffe and Corran [264] used an uncoated capillary and MECC to analyze purified oxytocin, vasopressin, and its analogues. Tagliaro et al. [264] used free-solution CE to separate calcitonin from human, salmon, and rat. The amino acid composition between human and rat are relatively conserved in phylogenesis (e.g., only two modifications exist between these two species). They found that when CE was carried out at pH 2.5, the salmon calcitonin (sCT), ell calcitonin (eCT), and human calcitonin (hCT) were separated quite well, but chicken calcitonin (cCT) and sCT co-migrated. Arcellini et al. [266] separated synthetic human growth hormone, the dimer form, and a variant by uncoated CZE. Overall, use of CZE to analyze hormones is still in its infancy, with most studies so far performed using purified hormones.

G. Soluble Mediators from Immune Cells

Growth factors are mediators which may either stimulate or inhibit a specific immune cell population. Growth factors may have potential to become an indicator for cancer diagnosis. Yowell et al. [267] analyzed a recombinant GMCSF by both CZE and capillary isoelectric focusing. However, not one in-vivo growth factor analysis by CE has yet to be reported.

H. Lipoproteins

Lipoproteins play important functions in the transport of water-insoluble lipids and in cholesterol and triglyceride metabolism. Lipoprotein metabolism is of particular interest to clinicians concerned with the diagnosis and treatment of atherosclerosis. Lipoproteins are separated electrophoretically because of their differing surface electrical charges. Conventional electrophoretic methods for lipoprotein separations employ agarose, cellulose acetate, polyacrylamide, and starch gels. Tadey and Purdy [268] separated two major apolipoproteins: high-density lipoproteins(HDL) and low-density lipoproteins (LDL) from human serum. Separations were carried out in bare silica and polyacrylamide-coated capillaries. The dimensions of the capillary were 50 μm i.d. × 75 cm and the detection wavelength was 220 nm. The running buffer was 30 mM borax at pH 8.3. The sample was centrifuged at 142,200 g and 5°C and the lipoproteins were isolated by their different densities. HDL could be separated in an uncoated capillary filled with borax buffer containing 0.2% SDS. Using a coated capillary, a mixture of HDL and LDL apolipoproteins was resolved in less than 12 min. Because of the sample preparation that is required, this method may not be ready for routine clinical usage. Alternatively, Schmitz et al. [269,270] used an isotachophoretic CE method to partition the plasma lipoproteins into 14 subfractions. Serum or plasma was prestained with a lipid stain and injected into the capillary with a series of spacers. This procedure determined not only HDL and LDL but also intermediate density lipoprotein (IDL) and very-low-density lipoprotein (VLDL).

Apo A-I and A-II are two major protein constituents of human high-density lipoprotein. HDL is involved in the transport of cholesterol from peripheral tissues to the plasma and liver. In addition, it is of particular interest because HDL is inversely correlated with the development of premature cardiovascular diseases, particularly in predicting the risk of myocardial infarction. Conventional methods for HDL assay include immunoassay and polyacrylamide gel electrophoresis (PAGE). However, these methods are either nonquantitative or inaccurate due to cross-reactions of antigen-antibody. Lehmann et al. [271] determined serum apolipoproteins A-I and A-II using a BioFocus 3000 CE (Bio-Rad). This CZE method used a fused silica capillary (50 cm × 50 μm i.d.), a voltage of 20 kV, and a proprietary LLV buffer (Bio-Rad Laboratories). Apo A-I could be detected from serum directly without an ultracentrifugation step. Sample preparation was simplified by diluting the serum with LLV buffer before injection. However, Apo A-II co-migrated with albumin, so only Apo A-II could be quantitatively measured from 1 mg/dL to 100 mg/dL.

REFERENCES

1. W. G. Kuhr, *Anal. Chem.*, 62: 403R (1990).
2. W. G. Kurh and C. A. Monnig, *Anal. Chem.*, 64: 389R (1992).
3. C. A. Monnig and R. T. Kennedy, *Anal. Chem.*, 66: 280R (1994).
4. M. V. Novotny, K. A. Cobb, and J. Liu, *Electrophoresis*, 11: 735 (1990).
5. Z. Deyl and R. Struzinsky, *J. Chromatogr.*, 569: 63 (1991).
6. C. Schöneich, S. K. Kwok, G. S. Wilson, S. R. Rabel, J. F. Stobaugh, T. D. Williams, and D. G. Vander Velde, *Anal. Chem.*, 65: 67R (1993).
7. C. Schöneich, A. F. R. Hühmer, S. R. Rabel, J. F. Stobaugh, S. D. S. Jois, C. K. Larive, T. J. Siahaan, T. C. Squier, D. J. Bigelow, And T. D. Williams, *Anal. Chem.*, 67: 155R (1995).
8. H. H. Lauer and D. McManigill, *Anal. Chem.*, 58: 166 (1986).
9. R. M. McCormick, *Anal. Chem.*, 60:2322 (1988).
10. J. S. Green and J. W. Jorgenson, *J. Chromatogr.*, 63 (1989).
11. F. A. Chen, L. Kelly, R. Palmieri, R. Biehler, and H. Schwartz, *J. Liq. Chromatogr.*, 15: 1143 (1992).
12. M. M. Bushey and J. W. Jorgenson, *J. Chromatogr.*, 480: 301 (1989).
13. V. Rohlick and Z. Deyl, *J. Chromatogr.*, 494: 87 (1989).
14. M. J. Gordon, K.-J. Lee, A. A. Arias, and R. N. Zare, *Anal. Chem.*, 63: 69 (1991).
15. W. G. H. M. Muijselaar, C. H. M. M. de Bruihn, and F. M. Everaerts, *J. Chromatogr.*, 605: 115 (1992).
16. A. Emmer, M. Jansson, and J. Roeraade, *J. Chromatogr.*, 547: 544 (1991).
17. A. Emmer and J. Roeraade, *J. Liq. Chromatogr.*, 17: 3831 (1994).
18. G. Mandrup, *J. Chromatogr.*, 604: 267 (1992).
19. J. P. Landers, R. P. Oda, B. J. Madden, and T. C. Spelsberg, *Anal. Biochem.*, 205: 115 (1992).
20. M. Taverna, A. Baillet, D. Biou, M. Schlüter, R. Werner, and D. Ferrier, *Electrophoresis*, 13: 359 (1994).
21. L. Song, Q. Ou, and W. Yu, *J. Liq. chromatogr.*, 17: 1953 (1994).
22. N. Cohen and E. Grushka, *J. Chromatogr. A*, 661: 305 (1994)
23. D. Corradini, A. Rhomberg, and C. Corradini, *J. Chromatogr, A*, 661: 305 (1994).
24. D. Corradini, G. Cannarsa, E. Fabbri, and C. Corradini, *J. Chromatogr. A*, 709: 127 (1995)
25. N. Guzmen, J. Moschera, K. Iqbal, and A. W. Malick, *J. Chromatogr.*, 608: 197 (1992)

26. X.-H. Fang, T. Zhu, and V.-H. Sun, *J. High Resolut. Chromatogr.*, *17*: 749 (1994).
27. M. A. Strege, and A. L. Lagu, *J. Liquid Chromatogr.*, *16*: 51 (1993).
28. M. A. Strege and A. L. Lagu, *J. Chromatogr.*, *630*: 337 (1993).
29. U. J. Yao and S. F. Y. Li, *J. Chromatogr. A*, *663*: 97 (1994).
30. R. P. Oda, B. J. Madden, T. C. Spelsberg, and J. P. Landers, *J. Chromatogr. A*, *680*: 85 (1994).
31. N. Cohen, and E. Grushka, *J. Cap. Electrophoresis*, *1*: 112 (1994).
32. G. N. Okafo, H. C. Birrell, M. Greenaway, M. Haran, and P. Camilleri, *Anal. Biochem.*, *219*: 201 (1994).
33. T. Wehr, *LG–GC*, *11*: 14 (1993).
34. K. A. Turner, *LC–GC*, *9*: 350 (1991).
35. J. R. Mazzeo and I. S. Krull, *Biochromatography*, *10*: 638 (1991).
36. J .E. Wiktorowicz and J. C. Colburn, *Electrophoresis*, *11*: 769 (1990).
37. J. W. Jorgenson and K. D. Lukacs, *Science*, : 226 (1983).
38. A. Emmer, M. Jasson, and J. Roerrade, *J. High Resolut. Chromatogr.*, *14*: 778 (1991).
39. A. Cifuentes, J. M. Santos, M. de Frutos, and J. C. Diez-Mesa, *J. Chromatogr. A*, *652*: 161 (1993).
40. D. Schmalzing, C. A. Piggee, F. Foret, E. Carrilho, and B. L. Karger, *J. Chromatogr. A*, *652*: 149 (1993).
41. P. Sun, A. Landman, G. E. Barker, and R. A Hartwick, *J. Chromatogr. A*, *685*: 303 (1994).
42. M. H. A. Busch, J. C. Kraak, and H. Poppe, *J. Chromatogr. A*, *695*: 287 (1995).
43. S. Hjertén and K. Kubo, *Electrophoresis*, *14*: 390 (1993).
44. Y. Mechref and Z. El Rassi, *Electrophoresis*, *16*: 617 (1995).
45. K. Cobb, V. Dolnik, and M. Novotny, *Anal. Chem.*, *62*: 2478 (1990).
46. D. Belder and G. Schomburg, *J. High Resolut. Chromatogr.*, *15*: 686 (1992).
47. M. Gilges, M. H. Kleesmiss, and G. Schomburg, *Anal. Chem.*, *66*: 2038 (1994).
48. J. K. Towns and F. E. Regnier, *Anal. Chem*, *63*: 1126 (1991).
49. J. L. Liao, J. Abramson, and S. Hjertén, *J. Cap. Electrophoresis*, *2*: 4 (1995).
50. M. Huang, J. Plocek, and M. V. Novotny, *Electrophoresis*, *16*: 396 (1995).
51. M. Chiari, C. Micheletti, M. Nesi, M. Fazio, and P. G. Righetti, *Electrophoresis*, *15*: 177 (1994).
52. S. Hjertén, *J. Chromatogr.*, *347*: 191 (1985).
53. S. Hjertén, L. Valtcheva, and Y.-M. Li, *J. Cap. Electrophoresis.*, *1*: 83 (1994).
54. F. Foret, E. Szoki, and B. L. Karger, *J. Chromatogr.*, *608*: 3 (1992).

55. T. Hirokawa, A. Ohmori, and Y. Kiso, *J. Chromatogr. A*, *634*: 101 (1993).
56. R. L. Chien and D. S. Burgi, *J. Chromatogr.*, 559: 141 (1991).
57. P. Jandik and W. R. Jones, *J. Chromatogr.*, *546*: 431 1991).
58. N. A. Guzman, M. A. Trebilock, and J. P. Advis, *J. Liq. Chromatogr.*, *14*: 997 (1991).
59. S. Hjertén, K. Elenbring, F. Kilár, J. L. Liao, A. J. C. Chen, C. J. Siebert, and M. Zhu, *J. Chromatogr.*, *403*: 47 (1987).
60. F. Foret, E. Szoko, and B .L. Karger, *Electrophoresis*, *14*: 417 (1993).
61. F. M. Everaerts, J. L. Beckers, and Th. P .E. M. Verheggen, *Chromatogr.*, *6*: 7–23 (1976).
62. S. Hjertén, J-L. Liao, and R. Zhang, *J. Chromatogr. A*, *676*: 409 (1994).
63. J-L. Liao, R. Zhang, and S. Hjertén, *J. Chromatogr. A*, *676*: 421 (1994).
64. F. Foret, E. Szoko, and B. L. Karger, *Electrophoresis*, *14*: 417 (1993).
65. D. T. Witte, S. Nagard, and M. Larsson, *J. Chromatogr. A*, *687*: 155 (1994).
66. R.Chien and D. S. Burgi, *Anal. Chem.*, *64*: 489A (1992).
67. N. A. Guzman, M. A. Trebilcock, and J. P. Davis, *J. Liq. Chromatogr.*, *14*: 997 (1991).
68. J. H. Beattie, R. Self, and M. P. Richards, *Electrophoresis*, *16*: 322 (1995).
69. L. J. Cole and R. T. Kennedy, *Electrophoresis*, *16*: 549 (1995).
70. J. Bao and F. E. Regnier, *J. Chromatogr.*, *608*: 217 (1992).
71. D. Wu, and F .E. Regnier, *Anal. Chem.*, *65*: 2029 (1993).
72. B. J. Harmon, D. H. Patterson, and F. E. Regnier, *Anal Chem.*, *65*: 2655 (1993).
73. D. Wu, F .E. Regnier, and M. C. Linhares, *J. Chromatogr. B*, *657*: 357 (1994).
74. P. Sun and R. A. Hartwick, *J. Chromatogr.*, *695*: 279 (1995).
75. Y. H. Chu, L. Z. Avila, H. A. Biebuyck, and G. M. Whitesides, *J. Med. Chem.*, *65*: 2915 (1992).
76. L. Z. Avila, Y. H. Chu, E. C. Blossey, and G. M. Whitesides, *J. Med. Chem.*, *36*: 126 (1993).
77. F. A. Gomez, L. Z. Avila, Y. H. Chu, and G. M. Whitesides, *Anal. Chem.*, *66*: 1785 (1994).
78. J. C. Kraak, S. Busch, and H. Poppe, *J. Chromatogr.*, *608*: 257 (1992).
79. S. Honda, A. Taga, K. Suzuki, S. Suzuki, and K. Kakehi, *J. Chromatogr.*, *597*: 377 (1992).
80. Y. H. Chu, J. L. Watson, A. Stassinopoulos, and C. T. Walsh, *Biochemistry*, *33*: 10616 (1994).
81. R. G. Nielsen, E. C. Rickard, P. F. Santa, D. A. Sharknas, and G. S. Sittampalam, *J. Chromatogr*, *539*: 177 (1991).

82. O. W. Reif, R. Lausch, T. Scheper, and R. Freitag, *Anal. Chem.*, *66*: 4027 (1994).
83. K. Nadeau, S. G. Nadler, M. Saulnier, M. A. Tepper, and C. T. Walsh, *Biochemistry*, *33*: 2561 (1994).
84. J. Liu, K. J. Volk, M. S. Lee, E. H. Kerns, and I. E. Rosenberg, *J. Chromatogr. A*, *680*: 395 (1994).
85. P. Sun. A. Hoops, and R. A. Hartwick, *J. Chromatogr. B*, *661*: 335 (1994).
86. V. T. Chadwick, A. C. Cater, and J. J. Wheeler, *J. Liq. Chromatogr.*, *16*: 1903 (1993).
87. J. C. Olivier, M. Taverna, C. Vauthier, P. Couvreur, and D. Baylocq-Ferrier, *Electrophoresis*, *15*: 234 (1994).
88. T. J. Pritchett, R. A. Evangelista, and F.-T. A. Chen, *J. Cap. Electrophoresis*, *2*: 145 (1995).
89. M. Novotny, H. Soini, and M. Stefansson, *Anal. Chem.*, *66*: 646A (1994).
90. S. G. Allenmark and S. Andersson, *J. Chromatogr. A*, *666*: 167 (1994).
91. H. Nishi and S. Terabe, *J. Chromatogr. A*, *694*: 245 (1995).
92. D. K. Lloyd, S. Li, and P. Ryan, *J. Chromatogr. A*, *694*: 285 (1995).
93. S. Birnbaum and S. Nilsson, *Anal. Chem.*, *64*: 2872 (1992).
94. G. E. Barker, P. Russo, and R. A. Hartwick, *Anal. Chem.*, *64*: 3024 (1992).
95. P. Sun, N. Wu, G. Barker, and R. A. Hartwick, *J. Chromatogr.*, *648*: 475 (1993).
96. P. Sun, G. Barker, and R. A. Hartwick, *J. Chromatogr. A*, *652*: 247 (1993).
97. R. Vespalec, V. Sustácek, and P. Bocek, *J. Chromatogr.*, *638*: 255 (1993).
98. T. Arai, N. Nimura, and T. Kinoshita, *Biomed. Chromatogr.*, *9*: 68 (1995).
99. T. Arai, M. Ichinose, H. Kuroda, N. Nimura, and T. Kinoshita, *Anal. Biochem.*, *217*: 7 (1994).
100. S. Li and D. K. Lloyd, *Anal. Chem.*, *65*: 3684 (1993).
101. Y. Ishihama, Y. Oda, N. Asakawa, Y. Yoshida, and T. Sato, *J. Chromatogr. A*, *666*: 193 (1994).
102. Y. Tanaka, N. Matsubara, and S. Terabe, *Electrophoresis*, *15*: 848 (1994).
103. S. Busch, J. C. Kraak, and H. Poppe, *J. Chromatogr,* , *635*: 119 (1993).
104. D. Wistuba, H. Diebold, and V. Schurig, *J. Microcol. Sep.*, *7*: 17 (1995).
105. L. Valtcheva, J. Mohammad, G. Pettersson, and S. Hjertén, *J. Chromatogr.*, *638*: 263 (1993).
106. Y. Tanaka and S. Terabe, *J. Chromatogr. A*, *694*: 277 (1995).
107. R. S. Rush, A. Cohen, and B. L. Karger, *Anal. Chem.*, *63*: 1346 (1991).
108. M. A. Strege and A. L. Lagu, *J. Chromatogr. A*, *652*: 179 (1993).
109. M. A. Strege and A. L. Lagu, *Am. Lab.*, *26*: 48C (1994).
110. F. Kilár and S. Hjertén, *J. Chromatogr.*, *638*: 269 (1993).
111. V. J. Hilser, G. D. Worosila, and E. Freire, *Anal. Biochem.*, *208*: 125 (1993).

112. V. J. Hilser and E. Freire, *Anal. Biochem.*, *224*: 465 (1995).
113. G. R. Paterson, J. P. Hill, and D. E. Otter, *J. Chromatogr. A*, *700*: 105 (1995).
114. I. Recio, E. Molina, M. Ramos, and M. de Frutos, *Electrophoresis*, *16*: 654 (1995).
115. R. Rodriguez-Diaz, M. Zhu. V. Levi, R. Jimenez, and T. Wehr, *7th Symposium on Capillary Electrophoresis*, Würzburg, Germany, 1995.
116. M. Kanning, M. Castella, and C. Olieman, *LC–GC*, *6*: 701 (1993).
117. R. Jimenez-Flores and A. Ulibarri, in press.
118. K. R. Kristiansen, J. Otte, M. Zakora, and K. B. Qvist, *Milchwissenscaft*, *49*: 683 (1994).
119. F-T. A. Chen and A. Tusak, *J. Chromatogr. A*, *685*: 331 (1994).
120. N. M. Kinghorn, C. S. Norris, G. R. Paterson, and D. E. Otter, *J. Chromatogr. A*, *700*: 111 (1995).
121. T. M. I. E. Christensen, K. R. Kristiansen, and J. S. Madsen, *J. Dairy Res.*, *56*(5): 823 (1989).
122. M. P. Richards and J. H. Beattie, *J. Cap. Electrophoresis*, *1*: 196 (1994).
123. R. S. Rush, A. S. Cohen, and B. L. Karger, *Anal. Chem.*, *63*: 1346 (1991).
124. T. T. Lee and E. S. Yeung, *Anal. Chem.*, *64*: 3045 (1992).
125. H. Kajiwara, *J. Chromatogr.*, *559*: 345 (1991).
126. J. H. Beattie., M. P. Richards, and R. Self, *J. Chromatogr.*, *632*: 127 (1993).
127. G.-Q. Liu, W. Want, and X.-Q. Shan, *J. Chromatogr. B*, *653*: 41 (1994).
128. M. P. Richards and P. J. Aagaard, *J. Cap. Electrophoresis*, *1*: 90 (1994).
129. M. P. Richards and J. H. Beattie, *J. Chromatogr. B*, *669*: 27 (1995).
130. M. P. Richards, *J. Chromatogr. B*, *657*: 345 (1994).
131. J. H. Beattie and M. P. Richards, *J. Chromatogr.*, *664*: 129 (1994).
132. Z. Zhao, A. Malik. M. L. Lee, and G. D. Watt, *Anal. Biochem.*, *218*: 47 (1994).
133. F. S. Markland, S. Morris, J. R. Deschamps, and K. B. Ward, *J. Liq. Chromatogr.*, *16*: 2189 (1993).
134. J. Y. Cai and Z. El Rassi, *J. Liq. Chromatogr.*, *16*: 2007 (1993).
135. F. Kilár and S. Hjertén, *J. Chromatogr.*, *480*: 351 (1989).
136. K. Yim, *J. Chromatogr.*, *559*: 401 (1991).
137. A. D. Tran, S. Park, and P. J. Lisi, *J. Chromatogr.*, *542*: 459 (1991).
138. E. Watson and F. Yao, *Anal. Biochem.*, *210*: 389 (1993).
139. D. E. Morbeck, B. J. Madden, and D. McCormick, *J. Chromatogr. A*, *680*: 217 (1994).
140. P. M. Rudd, H. C. Joao, E. Coghill, P. Fiten, M. R. Saunders, G. Opdenakker, and R. A. Dwek, *Biochemistry*, *33*: 17 (1994).
141. K. Yim, J. Abrams, and A. Hsu, *J. Chromatogr. A*, *716*: 401 (1995).
142. J. A. Bietz, in *Gluten Proteins*, Association of Cereal Research, Detmold, Germany, 1993, pp. 404–413.

143. J. A. Bietz and E. Schmalzried, *Lebensm.-Wiss.-Technol.*, *28*: 174 (1995).
144. W. E. Werner, J .E. Wiktorowicz, and D. D. Kasarda, *Cereal Chem.*, *71*: 397 (1994).
145. G. L. Lookhart and S. R. Bean, *Cereal Chem.*, *72*: 42 (1995).
146. G. Lookhart and S. Bean, *Cereal Chem.*, *72*: 527 (1995).
147. G. L. Lookhart and S. R. Bean, *Cereal Chem.*, *72*: 312 (1995).
148. G. L. Lookhart and S. R. Bean, *Cereal Chem.*, in press.
149. G. L. Lookhart, S. R. Bean, R. Graybosch, O. K. Chung, B. Morena-Sevilla, and S. Baenziger, *Cereal Chem.*, in press.
150. I. Shomer, G. Lookhart, R. Salomon, R. Vasiliver, and S. Bean, *J. Cereal Sci.*, in press.
151. T. M. Wong, C. M. Carey, and S. H. C. Lin, *J. Chromatogr. A*, *680*: 413 (1994).
152. F. Kilár, in *Handbook of Capillary Electrophoresis*, CRC Press, Boca Raton, FL, 1992, pp. 95–109.
153. S. F. Y. Li, *J. Chromatogr.*, *52*: 341 (1992).
154. S. Hjertén, in *Capillary Electrophoresis*, P.D. Grossman and J.C. Colburn, Eds., Academic Press, San Diego, 1992, pp. 191–214.
155. R, Weinberger, *Practical Capillary Electrophoresis*, Academic Press, Boston, 1993, pp. 81–97.
156. Y. Xu, *Anal. Chem.*, *65*: 12 (1993).
157. H. Waetzig and C. Dette, *Pharmazie*, *46*(2–3): 80–96 (1994).
158. G. J. M. Bruin and A. Paulus, *J. Anal. Methods Instrum.*, *2*: 3 (1995).
159. S. Hjertén and M. de Zhu, *J. Chromatogr.*, *346*: 265 (1985).
160. W. Thormann, A. Tsai, J. P. Michaud, R. A. Mosher, and M. Bier, *J. Chromatogr.*, *389*: 75 (1987).
161. W. Thormann, M. A. Firestone, M. L. Dietz, T. Cecconie, and R. A. Mosher, *J. Chromatogr.*, *461*: 95 (1989).
162. S. M. Chen and J. E. Wiktorowicz, *Anal. Biochem.*, *206*: 84 (1992).
163. T. Nelson, *J. Chromatogr.*, *623*: 357 (1992).
164. X-W. Yao and F. E. Regnier, *J. Chromatogr.*, *632*: 185 (1993).
165. X-W. Yao, D. Wu, and F. E. Regnier, *J. Chromatogr.*, *636*: 21 (1993).
166. R. Mazzeo and I. R. Krull, *Anal. Chem.*, *63*: 2852 (1991).
167. C. A. Bolger, M. de Zhu, R. Rodriguez, and T. Wehr, *J. Liq. Chromatogr.*, *14*: 895 (1991).
168. M. A. Firestone and W. Thormann, *J. Chromatogr.*, *436*: 309 (1988).
169. S. Hjertén, J. L. Liao, and K. Yao, *J. Chromatogr.*, *387*: 127 (1987).
170. W. Thormann, J. Caslavska, S. Molteni, and J. Chmelik, *J. Chromatogr.*, *589*: 321 (1992).
171. J. R. Mazzeo and I. S. Krull, *J. Chromatogr.*, *606*: 291 (1992).
172. J. Chmelik and W. Thormann, *J. Chromatogr.*, *632*: 229 (1993).
173. J. Wu and J. Pawliszyn, *Anal. Chem.*, *67*: 2010 (1995).

174. N. Wu, P. Sun, J. H. Aiken, T. Wang, C. W. Huie, and R. Hartwick, *J. Liq. Chromatogr.*, *16*: 2293 (1993).
175. T. Wang and R. A. Hartwick, *Anal. Chem.*, *64*: 1745 (1994).
176. J. Wu and J. Pawliszyn, *J. Chromatogr.*, *608*: 121 (1992).
177. J. Wu and J. Pawliszyn, *Electrophoresis*, *14*: 469 (1993).
178. J. Wu and J. Pawliszyn, *Electrophoresis*, *16*: 670 (1995).
179. J. Wu and J. Pawliszyn, *J. Chromatogr. B*, *657*: 327 (1994).
180. L. Vonguyen, J. Wu, and J. Pawliszyn, *J. Chromatogr. B*, *657*: 333 (1994).
181. J. Wu and J. Pawliszyn, *Anal. Chem.*, *66*: 867 (1994).
182. J. Wu and J. Pawliszyn, *Anal. Chem.*, *64*: 224 (1994).
183. J. Wu and J. Pawliszyn, *Anal. Chem.*, *64*: 219 (1992).
184. J. Wu and J. Pawliszyn, *Anal. Chem.*, *64*: 2934 (1992).
185. J. Wu and J. Pawliszyn, *J. Liq. Chromatogr.*, *16*: 3675 (1993).
186. S. Molteni, H. Frischknecht, and W. Thormann, *Electrophoresis*, *15*: 22 (1994).
187, F. Kilar and S. Hjertén, *Electrophoresis*, *10*: 23 (1989).
188. S. B. Harper, W. J. Hurst, and C. M. Lang, *J. Chromatogr. B*, *657*: 339 (1994).
189. O-S. Reif and R. Freitag, *J. Chromatogr. A*, *680*: 383 (1994).
190. K. Shimura and B. L. Karger, *Anal. Chem.*, *66*: 9 (1994).
191. M. Zhu, R. Rodriguez, T. Wehr, and C. Siebert, *J. Chromatogr.*, *608*: 225 (1992).
192. J. Pedersen, M. Pedersen, H. Soeberg, and K. Biedemann, *J. Chromatogr.*, *645*: 353 (1993).
193. M. Zhu, T. Wehr, V. Levi, R. Rodriguez, K. Shiffer, and Z. A. Cao, *J. Chromatogr. A*, *652*: 119 (1993).
194. G. G. Yowell, S. D. Fazio, and R. V. Vivilecchia, *J. Chromatogr. A*, *652*: 215 (1993).
195. J. Caslavska, S. Molteni, J. Chmelik, K. Slais, F. Matulik, and W. Thormann, *J. Chromatogr. A*, *680*: 549 (1994).
196. C. Silverman, M. Komar, K. Shields, G. Diegnan, and J. Adamovics, *J. Liq. Chromatogr.*, *15*: 207 (1992).
197. J. R. Mazzeo, J. A. Martineau, and I. S. Krull, *Anal. Biochem.*, *208*: 323 (1993).
198. K. Kleparnik, K. Slais, and P. Bocek, *Electrophoresis*, *14*: 475 (1993).
199. Y. J. Yao, K. S. Khoo, M. C. M. Chung, and S F. Y. Li, *J. Chromatogr. A*, *680*: 431 (1994).
200. J. Wu and J. Pawliszyn, *J. Chromatogr. A*, *652*: 295 (1993).
201. M. Zhu, R. Rodriguez, and T. Wehr, *J. Chromatogr.*, *559*: 479 (1991).
202. S. Molteni and W. Thormann, *J. Chromatogr.*, *638*: 187 (1993).
203. K. Slais and Z Friedl, *J. Chromatogr. A*, *661*: 249 (1994).
204. P. G. Reghetti and C. Gelfi, *J. Cap. Electrophoresis*, *1*: 27 (1994).

205. J. M. Hempe and R. D. Craver, *Clin. Chem.*, *40*: 2288 (1994).
206. T. L. Huang, P. C. H. Shieh, and N. Cooke, *Chromatographia*, *39*: 543 (1994).
207. R. Rodriguez and C. Siebert, *Symposium on Capillary Electrophoresis*, San Diego, 1994.
208. J. L. Liao, and R. Zhang, *J. Chromatogr. A*, *684*: 143 (1994).
209. C. Tanford, *The Hydrophobic Effect. Formation of Micelles and Biological Membranes*, 2nd ed., Wiley, New York, 1980, pp. 159–164.
210. M. R. Karim, S. Shinagawa, and T. Takagi, *Electrophoresis*, *15*: 1141 (1994).
211. K. Sasa, and K. Takeda, *J. Colloid Interf. Sci.*, *157*: 516 (1993).
212. M. Nakatani, A. Shibukawa, and T. Nakagawa, *Biol. Pharm. Bull.*, *16*: 1185 (1993).
213. M. Nakatani, A. Shibukawa, and T. Nakagawa, *Anal. Sci.*, *10*: 1 (1994).
214. K. Tsuji, *J. Chromatogr. A*, *661*: 257 (1994).
215. P. C. H. Shieh, D. Hoang, A. Guttman, and N. Cooke, *J. Chromatogr. A*, *676*: 219 (1994).
216. K. Hebenbrock, K. Schugerl, and R. Freitag, *Electrophoresis*, *14*: 753 (1993).
217. R. Lausch, T. Scheper, O-W. Reif, J. Schosser, J. Fleischer, and R. Freitag, *J. Chromatogr. A*, *654*: 190 (1993).
218. A. Widhalm, C. Schwer, D. Blaas, and E. Kenndler, *J. Chromatogr.*, *549*: 446 (1991).
219. A. Guttman, P. Shieh, D. Hoang, J. Horvath, and N. Cooke, *Electrophoresis*, *15*: 221 (1994).
220. A. Guttman, J. A. Nolan, and N. Cooke, *J. Chromatogr*, *632*: 171 (1993).
221. K. Ganzler, K. S. Greve, A. S. Cohen, B. L. Karger, A. Guttman, and N. C. Cooke, *Anal. Chem.*, *64*: 2665 (1992).
222. W. E. Werner, D. M. Demorest, J. Stevens, and J. E. Wiktorowicz, *Anal. Biochem.*, *212*: 253 (1993).
223. K. Tsuji, *J. Chromatogr. A*, *652*: 139 (1993).
224. M. Nakatani, A. Shibukawa, and T. Nakagawa, *J. Chromatogr. A*, *672*: 213 (1994).
225. K. Tsuji, *J. Chromatogr. B*, *662*: 291 (1994).
226. M. R. Karin, J-C. Janson, and T. Takagi, *Electrophoresis*, *15*: 1531 (1994).
227. E. Simo-Alfonso, M. Conti, C. Gelfi, and P. G. Righetti, *J. Chromatogr. A*, *689*: 85 (1995).
228. K. Benedek and S. Thiede, *J. Chromatogr. A*, *676*: 209 (1994).
229. D. Wu and F. Regnier, *J. Chromatogr.*, *608*: 349 (1992).
230. S. Hjertén, T. Srichaiyo, and A. Palm, *Biomed. Chromatogr*, *8*: 73 (1994).
231. M. Zhu, V. Levi, and T. Wehr, *Am. Biotech. Lab.*, *11*: 26 (1993).
232. A. Guttman, J. Horvath, and N. Cooke, *Anal. Chem.*, *65*: 199 (1993).

233. A. Guttman, P. Shieh, J. Kindahl, and N. Cooke, *J. Chromatogr. A, 676*: 227 (1994).
234. A. Guttman, *Electrophoresis, 16*: 611 (1995).
235. A. S. Cohen and B. L. Karger, *J. Chromatogr, 397*: 409 (1987).
236. M. Zhu, D. L. Hansen, S. Burd, and F. Gannon, *J. Chromatogr., 480*: 311 (1989).
237. S. Hjertén, *Electrophoresis '83*, Hirai, H., Ed., Walter de Gruyter & Co., New York, 1994, pp. 71–79.
238. W. E. Werner, D. M. Demorest, and J.E. Wiktorowicz, *Electrophoresis, 14*: 759 (1993).
239. P. G. De Gennes, *Scaling Concepts in Polymer Chemistry*, Cornell University Press, Ithaca, NY, 1979.
240. A. E. Barron, H. W. Blanch, and D. S. Soane, *Electrophoresis, 15*: 597 (1994).
241. A. E. Barron, D. S. Soane, and H. W. Blanch, *J. Chromatogr. A, 652*: 3 (1993).
242. Z. Deyl and I. Miksik, *J. Chromatogr. A, 698*: 369 (1995).
243. W. E. Werner, *Cereal Chem., 72*: 248 (1995).
244. A. Cifuentes, M. de Frutos, and J. C. Diez-Masa, *J. Dairy Sci., 76*: 1870 (1993).
245. M. N. Kinghorn, C. S. Norris, G. R. Paterson, and D. E. Otter, *J. Chromatogr. A, 700*: 111 (1995).
246. O. W. Reif, R. Lausch, and R. Freitag, in *Advances of Chromatography*, Marcel Dekker, Inc., New York, 1994, Vol 34, pp. 1–55.
247. G. L. Klein, and C. R. Jollif, in *Handbook of Capillary Electrophoresis*, J. P. Landers, Ed., CRC Press, Boca Raton, FL, 1993.
248. J. P. Landers, *Clin. Chem., 41*: 495 (1995).
249. Z. Deyl, F. Tagliaro, and I. Miksid, *J. Chromatogr., 656*: 3 (1994).
250. C. J. Holloway, W. Heil, and E. Henkel, in *Electrophoresis '81*, R. C. Allen and P. Arnaud, Eds., Walter de Gruyter & Co., New York, 1981, pp. 753–766.
251. F.-T. A. Chen, C.-M. Liu, Y.-Z. Hsieh, and J. C. Sternberg, *Clin. CHem., 37*: 14 (1991).
252. M. J. Gordon, K.-J. Lee, A. A. Arias, and R. N. Zare, *Anal. Chem., 63*: 69 (1991).
253. F.-T. A. Chen, *J. Chromatogr., 559*: 445 (1991).
254. F.-T. A. Chen and J.C. Sternberg, *Electrophoresis, 15*: 13 (1994).
255. C.-M. Liu, H. P. Wang, F.-T. A. Chen, J. C. Sternberg, and G. L. Klein, U.S. Patent No. 5,228,960, July 20, 1993.
256. C.-M. Liu and H.-P. Wang, U.S. Patent No. 5,491,834, March, 1996.
257. M. A. Jenkins, T. D. O'Leary, and M. D. Guerin, *J. Chromatogr. B, 662*: 108 (1994).

258. M. A. Jenkins, E. Kulinskaya, H. D. Martin, and M. D. Guerin, *J. Chromatogr. B*, *672*: 241 (1995).
259. R. Chevigne, J. Janssens, and P. Louis, *The Sixth Annual Frederick Conference on Capillary Electrophoresis*, 1995.
260. Y. Z. Hsieh, F. T. A. Chen, J. C. Sternberg, G. Klein, and C.-M. Liu, U.S. Patent Nos. 5,145,567, 1992 and 5,264,095, 1992.
261. Q. Xue and E. S. Yeung, *Anal. Chem.*, *66*: 1175 (1994).
262. K. J. Meller, I. Leesong, J. Bao, F. E. Regnier, and F. E. Lytle, *Anal. Chem.*, *65*: 3267 (1993).
263. M. Idei, I. Mezö, Z. Vadàsz, A. Horvàth, Teplàn, and G. Kèri, *J. Chromatogr.*, *648*: 251 (1993).
264. N. Sutcliffe, and P. H. Corran, *J. Chromatogr.*, *636*: 95 (1993).
265. F. Tagliaro, M. Moffa, M. M. Gentile, G. Clavenna, R. Valentini, S. Ghielmi, and M. Marigo, *J. Chromatogr. B*, *656*: 107 (1994).
266. C. Arcelloni, I. Fermo, G. Banfi, A. E. Pontiroli, and R. Paroni, *Anal. Biochem.*, *212*: 160 (1993).
267. G. G. Yowell, S. D. Fazio, and R. V. Vivilecchia, *J. Chromatogr. A*, *652*: 215 (1993).
268. T. Tadey and W. C. Purdy, *J. Chromatogr. A*, *652*: 131 (1993).
269. G. Schmitz and E. Williamson, in *Current Opinion in Lipodology*, Current Science, Philadelphia, 1991, Vol. 2, pp. 177–189.
270. G. Schmitz, U. Borgman, and G. Aassmann, *J. Chromatogr.*, *320*: 253 (1985).
271. R. Lehmann, H. Liebich, G. Grübler, and W. Voelter, *Electrophoresis*, *16*: 998 (1995).

7
Chiral Micelle Polymers for Chiral Separations in Capillary Electrophoresis

Crystal C. Williams, Shahab A. Shamsi, and Isiah M. Warner
Louisiana State University, Baton Rouge, Louisiana

I.	INTRODUCTION	364
II.	FUNDAMENTALS OF CAPILLARY ELECTROPHORESIS	367
	A. Chiral Separation Using CZE	371
III.	MEKC CONCEPTS	378
	A. Surfactants and Micelles	378
	B. MEKC Theory with Chiral Surfactants	379
	C. Monomeric Chiral Surfactants for MEKC Separation of Enantiomers	383
IV.	POLYMERIC MICELLES	398
	A. Concept of Surfactant/Micelle Equilibria	398
	B. Factors Affecting Polymerization Rate and Polymer Structure	400
	C. Examples of a Cationic Polymerized Micelle	402
	D. Example of an Anionic Polymerized Micelle	402
	E. Applications of Polymerized Micelles	404
V.	CONCLUSION	419
	REFERENCES	419

I. INTRODUCTION

Chiral compounds, defined as molecules with nonsuperimposable mirror images, are common to many synthetic and most biological systems. It is well established that such chiral molecules are often associated with precise control of biological processes, protein structure, and function, as well as enzyme activity [1].

In general, organic molecules that have the ability to rotate plane-polarized light are termed optically active. Optical activity is a well-established property of chiral molecules. The extent of rotation and the direction of rotation of polarized light by optically active molecules is typically measured by use of a polarimeter. Optically active molecules that rotate light to the left are said to be levorotatory (L). If the light is rotated to the right, the molecule is dextrorotatory (D) [2]. In contrast, the R and S notation describe the stereochemical configuration around the chiral carbons. Using the Cahn–Ingold–Prelog sequence rules, if the priority of the substituents is in a clockwise direction, then the configuration is considered to be R, meaning right or rectus in Latin. If the priority of the substituents are in a counterclockwise rotation, the chiral center has the S configuration, left or sinister [2].

The separation of chiral molecules has long been regarded as a tedious and difficult task. Louis Pasteur was the first to physically separate crystals of racemic tartarate as early as 1848. Today, we call these nonsuperimposable mirror images, optical isomers or enantiomers. Enantiomers are identical in all their normal physical properties such as melting points, boiling points, solubility, and spectroscopic properties [e.g., nuclear magnetic resonance (NMR)]. However, as noted above, enantiomers have opposite signs of optical rotation [2].

Enantiomeric differences arise from an asymmetric element, which may be a center, an axis, or a plane of asymmetry present in the molecule [1]. Racemates (an equal mixture of D- and L- molecules) are common in synthetic products such as pharmaceuticals, herbicides, pesticides, and some natural products [1]. It has been acknowledged that often only one enantiomer is active while the others may be less active, inactive, or have delitorius side effects (toxic). In the case of pharmaceuticals, the pharmacokinetic characteristics of individual enantiomers of a chiral drug may be quite different [3], possibly causing physiological problems. This phenomenon was noted with the racemic form of the drug Thalidomide, which was prescribed for pregnant women during the 1950s. It was later observed that the R form of the drug was medically beneficial, whereas the S form was responsible for the birth defects observed in newborn infants. In other cases (e.g., the perception of pheromones by certain insects), the correct ratio of enantiomers are necessary for maximum response [1].

As shown in the examples above, chemical and pharmacokinetic properties of enantiomers may differ. Therefore, administration of the wrong enantiomer may cause less than desired responses. Therefore, methods for measuring and separating enantiomers have been developed. During the early years of research on chiral separations, the resolution of different chiral enantiomers was often achieved by the introduction of an additional chiral center into the molecule of interest, thus giving diastereomers [1]. The difference between enantiomers and diastereomers is that enantiomers have opposite configuration at all chiral centers, whereas diastereomers have opposite configurations at some (one or more) chiral centers, but the same configuration at the other chiral centers [2]. Conversion of enantiomers to diastereomers will produce molecules with different physical and chemical properties. Therefore, the use of diastereomeric separation to achieve chiral separation is not the preferred approach.

In order to avoid the use of diastereomers, other methods have been investigated to separate enantiomers without altering the molecule. The application of both capillary gas chromatography (GC) and high-performance liquid chromatography (HPLC) for the separation of enantiomers has a fairly long history [4]. In the 1950s–1960s, developments in chiral separations were dominated by GC [1,4]. In the 1980s, the search for optimum systems in enantiomeric separation included high-performance liquid chromatography (HPLC) and supercritical fluid chromatography (SFC) [1].

In general, chromatographic chiral separations require manipulation of either the mobile phase or stationary phase. Currently, a variety of HPLC and GC chiral stationary phases are commercially available. When separating racemates, the selectivity of solute–stationary interaction is crucial because of differences in the free energy, $-\Delta(\Delta G°)$, of interaction. Because the interaction between D- and L- solutes and the solvent–stationary-phase environment are small, the key to enantiomeric separation is the choice of a chiral selector with sufficient selectivities (α values). It has been established by the use of thermodynamic relationship $-\Delta(\Delta G°) = RT \ln \alpha$, that the larger the difference in $-\Delta(\Delta G°)$, the better is the separation. Although advances are being made in chiral analysis by capillary GC and SFC, the temperature variable in the free-energy equation is a major problem for capillary GC and SFC methods; however, these approaches have a high number of theoretical plates (kinetically favored) compared to HPLC [1]. Currently, most chiral separations are performed by HPLC [4] despite the fact that method development in HPLC can be time-consuming. This is because the mode of separation is often not understood. In addition, HPLC methods are often plagued with poor efficiency, thus causing a decrease in sensitivity.

Recently, capillary electrophoresis (CE) has shown great promise for enantiomeric separations [1,5]. The practical benefits of using CE are the use of minimal sample, small chiral selector consumption, and inexpensive column replacement. Furthermore, CE has a large theoretical plate number as well as a choice of many chiral selectors without thermal stability limitations which is inherent to GC [1].

It should be noted that mechanisms of chiral separations are still not completely understood. However, the "three-point rule" is a widely accepted axiom of current chiral recognition strategies [6]. Chiral recognition requires a minimum of three simultaneous interaction between the chiral phase and at least one of the enantiomers to be separated. It is further stipulated that at least one of these interactions must be stereochemically dependent. This concept is depicted schematically in Fig. 1. In this figure, we show the interactions of an enantiomer analyte (A) with another chiral species. It should be noted that this latter specie (chiral discriminator) could be bound to the stationary phase or it could be in the mobile phase. Either case allows three points of interaction with the given enantiomer. One can easily rationalize that the antipode (enantiomer B) of this enantiomer analyte would not be able to achieve the same three interactions. Thus, this difference in interactions would allow separation. It should be noted that the three interactions need not all be attractive interactions. The "three-point rule" was successfully employed in 1971 in the design of a chiral stationary phase for the separation of the antipodes of L-DOPA (L-dihydroxyphenylalanine) [7].

Direct chiral separations in CE are achieved either through the use of an immobilized chiral phase [5] or through the addition of chiral selectors as mobile-phase additives [5]. In this review, we explore the use of chiral selectors as mobile-phase additives in CZE (capillary zone electrophoresis) and MEKC (mi-

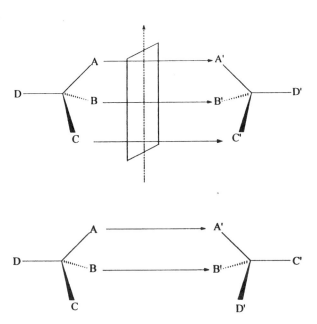

Fig. 1 Schematic representation of the three-point rule.

cellar electrokinetic capillary chromatography). We include a brief discussion of (1) host–guest chiral selectors and (2) chiral metal complexes, with particular emphasis on (3) chiral surfactants and chiral micelle polymers.

II. FUNDAMENTALS OF CAPILLARY ELECTROPHORESIS

One of the main advantages of CE is the use of simple instrumentation. Figure 2 represents a schematic diagram of a basic CE which contains a high-voltage power supply, two buffer reservoirs, a capillary, and detector. However, the basic instrumentation for CE can be enhanced with the use of autosamplers, multiple injection devices, sample and capillary temperature control, programmable power supply, multiple detectors, fraction collection, and computer interfacing [8]. Numerous instrumental variations that have different operative and separative characteristics can be accomplished by use of the basic principles of CE [6]. These techniques are (A) capillary zone electrophoresis (CZE), (B) micellar electrokinetic capillary chromatography (MEKC or MECC), (C) capillary gel electrophoresis (CGE), (D) capillary electrochromatography (CEC), (E) capillary isoelectric focusing (CIEF or IEF), and (F) capillary isotachophoresis (CITP). This chapter provides a brief overview of the fundamentals of CZE and MEKC as applied to chiral separations of charged and uncharged chiral solutes, respectively. Emphasis will be placed on the utility of novel chiral polymerized micelles for chiral separations in MEKC.

Electrophoretic separation may be conducted in continuous or discontinuous electrolyte systems. In a continuous electrolyte system, the background electrolyte forms a continuum along the migration path. The continuum does not change with time and provides an electrically conducting medium for flow of electric current and the formation of an electric field across the migration path. Separation is achieved either through a kinetic or steady-state process. The back-

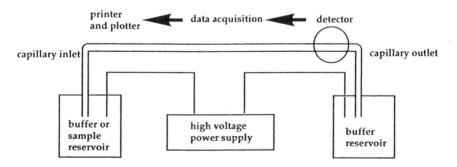

Fig. 2 Schematic representation of a simple capillary electrophoresis system.

ground electrolyte is usually a buffer which can selectively influence the effective mobility of the analyte [8].

In a continuous kinetic process, the composition of the background electrolyte is constant along the migration path; therefore, the electric potential and effective mobilities of the resolved analytes are constant. Consequently, different analytes migrate with constant but different velocities when constant current is passed through the system. The CZE and MEKC methods are examples of continuous kinetic processes. In the steady-state process, the composition of the background electrolyte is not constant. Therefore, the electric field and effective mobilities may change along the migration path [8]. In a discontinuous electrolyte system, the sample migrates between two different electrolytes as a distinct individual zone. There is a front zone formed by the leading electrolyte, whereas the end electrolyte forms a rear zone [8].

Capillary Zone Electrophoresis

Capillary zone electrophoresis (CZE), also known as free-solution capillary electrophoresis, is the simplest form and most commonly used technique in CE. Separation in CZE is based on differences in charge-to-mass ratio and electrophoretic mobilities of ionic species at a given pH [8]. When the fused silica capillary column is conditioned with 1 N NaOH, free silanol groups are ionized to the anionic form (SiO^-). Upon flushing with an aqueous buffer solution, the inner wall of the capillary carries an overall negative charge. At the interface between the capillary wall and the solution, negative charges are balanced by the positive ions in solution. Some of these positive ions will be adsorbed to the wall, forming an immobilized compact layer. The remaining counterions (positive, negative, as well as neutrals) will be distributed into a diffuse layer. The arrangement of positive ions in a static and a diffuse layer give rise to an "electric double layer" (Fig. 3). When an electric field is applied across the capillary, positively charged hydrated ions in the diffuse layer migrate toward the cathode, or negative electrode. Solvent molecules and neutrals as well as negative ions are pulled along with the cations giving

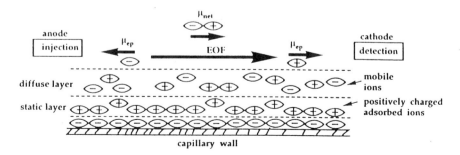

Fig. 3. Schematic representation of the "electrical double layer."

rise to a pumping mechanism called "electroosmotic flow" (EOF) (Fig. 4). In contrast, positive ions which are adsorbed on the compact layer of the capillary will not move toward the cathode. The corresponding potential across the layers (i.e., between compact and diffuse layers) is called the zeta potential, denoted by ζ, and given by the Helmholtz equation:

$$\zeta = \frac{4\pi\eta\mu_{eo}}{\varepsilon} \qquad (1)$$

where η is the viscosity, ε is the dielectric constant of the solution, and μ_{eo} is the mobility of the EOF. The double layer is typically a very thin layer (up to several hundred nanometers) relative to the radius of the capillary. Therefore, the EOF is considered to flow from the walls of the capillary as shown in Fig. 4 relative to the pumped process mentioned earlier [8]. The equations for EOF are similar to those developed for electrophoretic migration because both phenomena are complementary. The electroosmotic velocity (v_{eo}) or electrophoretic velocity (v_{ep}) is given by

$$v = \mu_{ep(eo)} E = \mu_{ep(eo)} \frac{V}{L} \qquad (2)$$

where μ_{ep} and μ_{eo} are the electrophoretic and electroosmotic mobility, respectively; E is the field strength (V/L); V is the voltage applied across the capillary, and L is the length of the capillary [8]. It is important to note that it is possible to change the charge-to-mass ratio of ions by changing the pH of the buffer and thus influence the ionization and electrophoretic mobility.

The electroosmotic velocity v_{eo} can also be expressed as

$$v_{eo} = \left(\frac{\varepsilon}{4\pi\eta}\right) E\zeta \qquad (3)$$

Equation (3) shows that v_{eo} can be adjusted by changing the pH (zeta potential and flow increase), viscosity (changes velocity), the ionic strength (affects the zeta potential), the voltage (flow is proportional to voltage), and the dielectric

Fig. 4 Flow profiles of pumped flow versus electroosmotic flow. Reprinted from S.F.Y. Li, *Capillary Electrophoresis: Principles, Practice, and Applications*, 1992, pg. 15, with kind permission from Elsevier Science.

constant of the buffer. This equation was derived by solving for the electrophoretic (electroosmotic) mobility term in the form of v in Eq. (2) and substituting back into Eq. (1), and rearranging to solve for v_{eo} [8].

One of the advantages of CZE and MEKC is that there is no need for a pressure-driven flow (Fig. 4). Thus, the flow profile is essentially flat. As a result of this, the EOF does not contribute significantly to band-broadening in the manner that the parabolic flow profile does in liquid chromatography. Both anions and cations can be separated in the same run. Cations are attracted toward the cathode and their speed is augmented by the EOF. As the magnitude of the EOF toward the cathode is very large, anions are swept toward the cathode with the bulk flow of the electrophoretic medium or EOF. Cations with the highest charge-to-mass ratio migrate first, followed by cations with the smallest ratios. All unresolved neutral components migrate at the rate of EOF because their charge-to-mass ratio is zero. Finally, anions with smaller charge-to-mass ratio migrate earlier than anions with large charge-to-mass ratios (Fig. 5) [8].

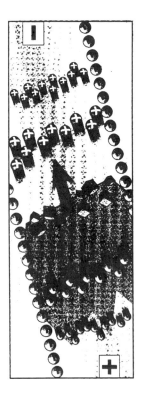

Fig. 5 Diagramatic representation of capillary zone electrophoresis.

Because no analyte should have retention on a stationary phase or capillary in conventional CZE, a more descriptive term is migration. The time required for a solute to migrate to the point of detection is called the migration time (t_m). The t_m is measured from the electropherogram and is related to the apparent mobility of the analyte by:

$$\mu_a = \frac{L_d L_t}{t_m V} \qquad (4)$$

where μ_a is apparent mobility, which is the actual mobility measured in the presence of EOF, L_d and L_t are effective capillary length (cm) from the injection to the center of the detector window and total capillary length (cm), respectively, t_m is the migration time (min), and V is the applied voltage (V). The effective electrophoretic mobility (μ_e) can be obtained from the μ_a and by independently measuring the EOF (e.g., $\mu_e = \mu_a - \mu_{eo}$). Note that μ_e values will be positive for cations and negative for anions. The EOF is usually measured using a neutral species such as dimethyl sulfoxide or mesitylene oxide which moves at a velocity of the EOF.

A. Chiral Separation Using CZE

Now that we have a brief overview of the theory and the parameters that affect CZE, we can examine some examples where chiral mobile-phase additives such as host–guest and metal complexes may be used to enhance enantiomeric separation.

Host–Guest Complexations

Host–guest complexes are complexes in which an analyte (typically the guest molecule) is spatially enclosed by a ligand (host molecule). In CE, the two groups that are most commonly used to form inclusion complexes with enantiomers are cyclodextrins and their derivatives [9–18] and chiral crown ethers [9].

Over the past 5 years, there has been an explosive increase in the use of cyclodextrins (CDs) for enantiomeric separation. The CDs were introduced to CE in 1988 by Snopek et al. [10].

Cyclodextrins (CD) are cyclic, oligossaccharides consisting of six, seven, or eight glucopyranose units bonded through nonreducing α-(1,4)-linkages (Fig. 6). These CD units are named α-, β-, or γ-CD, depending on the number of glucopyranose units present. The structure of CD can be represented by a truncated cone with hydrophilic edges and a hydrophobic cavity. Host–guest complexations of CDs are size and geometry related, depending on the guest molecule. Derivatives of CDs are used to increase the solubilities in buffer solutions and to improve the selectivities of some drugs. The CDs are typically not charged. Thus, these molecules move with the EOF. Consequently, the migration time of the

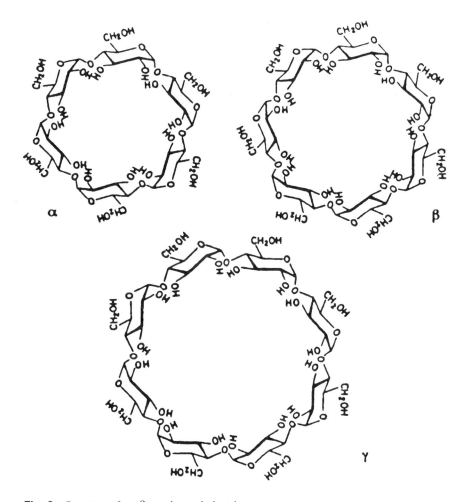

Fig. 6 Structure of α-, β-, and γ-cyclodextrins.

solute is typically controlled by electrophoretic mobilities and the interaction of enantiomers with the host [9].

It has also been demonstrated that cationic and anionic analytes can be easily separated using 10 mM γ-CD (Fig. 7a and 7b) with various buffers [9]. It has also been shown that the separation factor of enantiomers increases with increasing CD concentration [11–13]. Kuhn and Karger found a linear decrease of resolution and separation factor with increasing temperature [12,13]. Figure 8 shows the resolution achieved by Heuermann and Blaschke [14] for a mixture of six chemically different racemic drugs with a basic nitrogen. The examples given

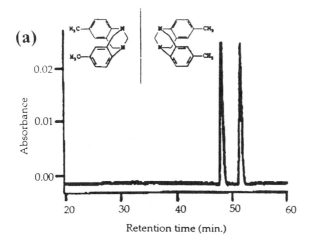

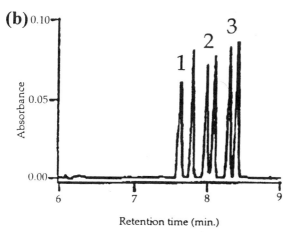

Fig. 7 (a) Electropherogram of the enantiomers of Tröger's base by CZE with β-cyclodextrin. Conditions were 10 mM β-CD in 50 mM phosphate buffer (pH 2.5); fused silica capillary, 75 μm i.d., 50 cm to the detector; applied field strength 260 V/cm; detection at 214 nm, and pressure sample injection for 1 s. (b) Electropherogram of DNS-D,L-leucine (1), DNS-D,L-methionine (2), and DNS-D,L-threonine (3). Experimental conditions: same as Fig. 7a except used 10 mM γ-CD in 50 mM sodium tetraborate (pH 9.0). Reprinted from Ref. 9 with permission from Friedr. Vieweg and Sohn.

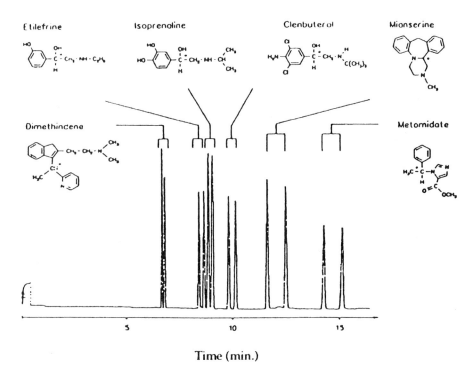

Fig. 8 Electropherogram of a mixture of six chemically different basic racemic drugs (figures and names shown) using HP-β-CD. Run buffer: HP-β-CD, 30 mg/mL in 50 mM phosphate buffer (pH 3.3). Reprint from *J. Chromatogr.*, 648:267, G. Blaschke, 1994 with permission from Elsevier Science.

show that CDs as well as derivatives of CDs can separate small chiral molecules, amino acids, and mixtures of chiral drugs. Numerous other examples of separations of chiral compounds using CDs are available [15–17]. The optimum separation depends on the CD type, concentration, pH of the buffer, and temperature.

Crown Ethers

Chiral crown ethers are examples of other host–guest molecules for chiral separation. These molecules were first discovered in 1967 by Pederson [19]. Kuhn et al. [12,18] reported the use of the macrocyclic polyether ring system of 18-crown-6 tetracarboxylic acid (18C6) consisting of six oxygen joined by ethylene bridges (Fig. 9) for CE separations of enantiomers. The four substituents in 18C6 are arranged perpendicularly in a plane formed by oxygens. The cavity is able to form inclusion complexes with potassium, ammonium, and primary alkylamine cations [9]. It is believed that a tripod arrangement is formed with

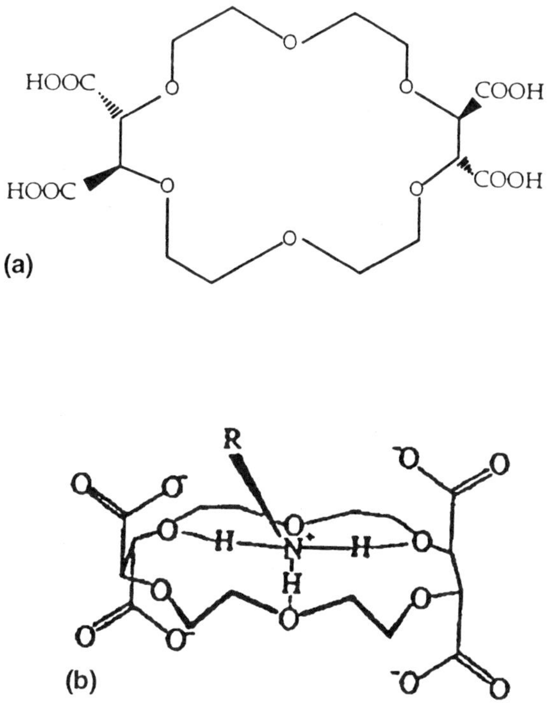

Fig. 9 (a) Structure of 18-crown-6 tetracarboxylic acid (18C6). (b) Tripod arrangement of 18C6. Reprinted from *J. Chromatogr.*, 666:367, R. Kuhn, 1994 with permission from Elsevier Science.

the three hydrogens on the quartenary amine analyte. The primary interactions occur by hydrogen bonding of the analyte to the dipoles of the oxygen atoms in the crown ether [20] (+NH ••• O) (Fig. 9) [21]. Figure 10 shows the separation of six primary amines using this approach. The observed peak tailing is attributed to the strong interaction of both enantiomers (3-amino-3-phenyl propionic acid) with the crown ether ring. It is important to note that a buffer composed of $18C_6H_4$ contains only protons, which have a higher mobility than the cationic analytes. To minimize electrophoretic dispersion it is important for the ionic buffer species to match the analyte mobility as closely as possible. Running buffers containing sodium, potassium, and ammonium ions should be avoided in order to prevent competition between the cations and the sample (primary amines) for complexation with crown ether [21]. As with CDs, the concentration of the crown ether and the pH of buffer were found to play a major role in the separation [9].

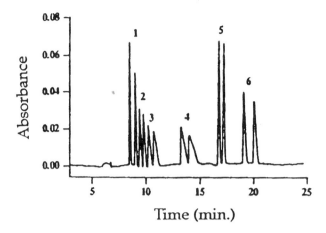

Fig. 10 Chiral separation of six primary amines with 18C6 as the chiral selector. Peaks: (1) azatryptophan, (2) norephedrin, (3) phenylglycinol, (4) 3-amino-3-plhenyl-propionic acid, (5) tryptophan, and (6) homophenylalanine. Experimental conditions: 10 mM 18C6, 10 mM Tris, adjusted to pH 2.5 with citric acid. Reprinted from *J. Chromatogr.*, *666*:367, R. Kuhn, 1994, with permission from Elsevier Science.

Ligand Exchange Complexation

The third category of chiral mobile-phase additives used in CZE is ligand-exchange complexation. In ligand-exchange electrophoresis (LEE), a multicomponent complex consist of a central ion (e.g., Cu^{2+}, Ni^{2+}) and at least two chiral bifunctional ligands. Two modes are possible. First, one chelator (e.g., an optically pure amino acid) is added to the buffer electrolyte with the central ion to form a hemicomplex. The different stabilities of the bidentate analyte enantiomers with the hemicomplex determines the extent of electrophoretic separation. Second, the chelator concentration can be chosen such that all coordination positions of the central ion are saturated. The analyte enantiomers replace one chelator by forming a ternary complex:

$$[CL]_n \cdot [M] + [CA] \longrightarrow [CL]_{n-1} \cdot [M] \cdot [CA] + [CL]$$

where CL is the chiral ligand, M is the central ion, and CA is the chiral analyte [9]. Zare and co-workers achieved separation of a mixture of four D- and L-dansyl amino acids (Fig. 11) using this approach [8]. Separation of the chiral compounds was based on diastereomeric interactions with the Cu(II)-L-histidine complex [8,9]. Studies were also conducted using a Cu(II)–aspartame complex [9], different pHs, temperature, and stoichiometric ratios of Cu(II) [12,13].

The use of host–guest complexes and ligand-exchange complexation mobile-phase additives have proved to be very successful when applied to CZE.

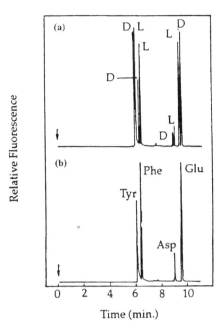

Fig. 11 Electropherogram of D,L-dansyl amino acids with (a) Cu(II) L-histidine electrolyte and (b) 1 : 1 Cu(II) D- and L-histidine electrolyte both at pH 7.0. Reprinted from S.F.Y. Li, *Capillary Electrophoresis: Principles, Practice, and Applications*, 1992, pg. 281, with permission from Elsevier Science.

The latter have also been used in conjunction with chiral surfactants and micelles.

Macrocyclic Antibiotics

Macrocyclic antibiotics have been recently introduced as new types of chiral selectors for CZE. Armstrong and co-workers [22] introduced rifamycin B, a specific class of macrocyclic antibiotics with favorable binding properties. The presence of various functionalities such as hydroxy and carboxy methyl groups, amide bonds, as well as carboxylic acids in rifamycin B enhance the chiral recognition ability of this chiral selector. The migration time, apparent and effective electrophoretic mobility, as well as enantiomeric resolution of a variety of β-amino alcohols were studied. A systematic study was performed to optimize the above separation parameters with respect to pH, rifamycin B concentration, and various organic modifiers (acetonitrile, methanol, ethanol, and 1- and 2-propanol). It was proposed that various multiple interactions (namely charge–charge, hydrogen bonding, and hydrophobic inclusion dominate in hydroorganic solvents) enhanced chiral recognition. These data suggest that only 2-

propanol, out of several organic solvents tested, could provide enhanced resolution. The improvement in resolution (using 10% propanol) occurred at the expense of migration which was as long as 63 min for terbutalene. In contrast, when the ionic strength of rifamycin B was increased, a longer migration time deteriorated the enantiomeric resolution. Rifamycin B has a substantial ultraviolet (UV) absorbance at 254 nm; therefore, direct photometric detection was found to be difficult. Because β-amino alcohols usually have lower molar absorptivity than rifamycin B at 254 nm, indirect photometric detection (IPD) is possible. It appears that the electrostatic attractive interaction of negatively charged rifamycin B with positively charged β-amino alcohols facilitated the IPD method. This contrasts with a charge-displacement process that usually occurs between the chromophore electrolyte and analyte counterions of similar charges in conventional IPD. Interestingly, the authors also noted that they were unable to resolve the racemates of anionic carboxylates. However, it was not clear whether an unresolved carboxylate peak was observed by either direct or IPD. Further research on the separation and detection aspect is needed to demonstrate the versatility of the rifamycin and other antibiotic selectors for IPD.

III. MEKC CONCEPTS

In 1984, Tsuda [23] and Terabe and co-workers [24,25] introduced a novel technique in CE known as micellar electrokinetic capillary chromatography (MEKC or MECC). The MEKC approach provides another dimension to the selectivity of CZE separations, enabling neutral and charged (anions and cations) molecules to be separated in a single run. The primary separation mechanism of MEKC is based on solute partitioning between a micellar phase and the solution phase. It should be noted that neutral and/or ionic compounds may also be used to form micelles. Separation of the analyte is dependent on a combination of charge-to-mass ratios, hydrophobicity, and charge interactions at the surface of the micelle [8].

A. Surfactants and Micelles

Surfactants (also known as amphiphiles or detergents) form molecular aggregates in solution above a narrow concentration range [26]. This aggregation is due to the presence of a polar head group (ionic and nonionic) attached to a hydrocarbon tail. Thus, surfactants are molecules with both a hydrophilic and a hydrophobic part. At low concentrations and at temperatures above the critical micelle (Krafft temperature point), the surfactant is dispersed in the aqueous medium. As the surfactant concentration surpasses a minimum value, the molecules aggregate to form micelles. The average number of molecules per micelle is termed the aggregation number. Each micelle is typically composed of 40–140 molecules [27]. The concentration range above which aggregation occurs is called the critical micelle concentration (CMC) [26,27].

Surfactants are classified on the basis of the charge of the polar head group [26]. For example, anionic ($R - X^- - M^+$), cationic ($R - N^+(CH_3)_3X^-$), zwitterionic [$R - (CH_3)_2N^+CH_2X^-$], or nonionic ($R(OCH_2CH_2))_nOH$, where R is a long aliphatic chain, M^+ is typically a metal ion, X^- is typically a halogen, CO_3^{2-} or SO_4^-, and n is an integer [27]. The aggregation process depends on the surfactant species and the conditions of the system in which the surfactants are dissolved [26].

The aggregation of surfactants to form micelles may be explained as a balance between hydrocarbon chain attraction and ionic repulsion. For nonionic surfactants, the hydrocarbon chain attraction is opposed by the requirements of the hydrophilic groups for hydration and space. Micellar structure is determined by an equilibrium between the repulsive forces among hydrophilic groups and short-range attractive forces among hydrophobic groups. In addition, it is well established that the chemical structure of a surfactant determines the size and shape of the micelle. For example, McBain suggested that lamella and spherical micelles may coexist [28]. Hartley suggested that micelles are spherical with charged groups situated at the micellar surface [29]. Debye and Anacker proposed that micelles are rod shaped rather than spherical or dislike [30]. Finally, there is also the Menger model which considers many features absent in the Hartley model. Menger's nuclear magnetic resonance (NMR) studies showed that micelles are more disorganized than believed in the past. Menger's model has a rough surface, water-filled pockets, chain looping, nonradial distribution of chains, and contact of terminal methyl groups with water [31]. However, the spherical form is generally accepted as a representation of the actual structure (Fig. 12) [26].

As the surfactant concentration is increased, the shape of the ionic micelle changes in sequence to spherical–cylindrical–hexagonal–lamella (Fig. 13). For nonionic micelles, the shape changes from spherical directly to lamella with increasing concentration. The net charge of an ionic micelle is less than the degree of micellar aggregation, indicating that large fractions of counterions remain associated with the micelle. These counterions form the so-called Stern layer at the micellar surface. It is crucial to understand that in solution, micelles exist in dynamic equilibrium with the monomers from which they are formed [26]. Thus, surfactants exist there in solution as monomers up to species (micelles) well above the average aggregation number. Hence, micelles are polydispersed species due to dynamic equilibria. Therefore, such polydispersity would be detrimental to chromatographic separations.

B. MEKC Theory with Chiral Surfactants

In recent years, increasing interest has been shown in the MEKC approach to chiral separation by use of chiral surfactants. This technique involves the introduction of a chiral surfactant into the running buffer of CZE at a concentration

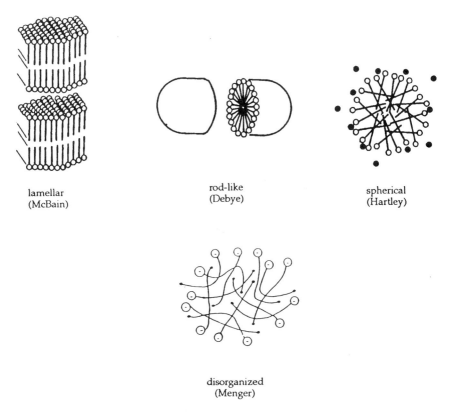

Fig. 12 Different proposed structures and shapes of the micelle. Reprinted from Refs. 26 and 31 with permission from Plenum Press and VCH Verlagsgesellschaft mbH.

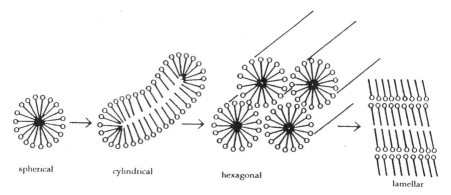

Fig. 13 Changes in micelle shape and structure with respect to change in surfactant concentration. Reprinted from Ref. 26 with permission from Plenum Press.

above the CMC. The chiral micellar run buffer serves as a phase for chromatographic separation. This chiral micellar phase usually contains a chiral polar head group and a hydrocarbon tail that acts as a pseudo-stationary phase or a separation carrier [32]. Because the surface of the chiral micelle is usually charged, this gives the chiral micelle an electrophoretic mobility in CZE, at a velocity different from the surrounding aqueous phase [32].

If there are no capillary wall interactions, then negatively charged chiral micelles migrate electrophoretically toward the anode and positively charged chiral micelles migrate electrophoretically toward the cathode. Thus, the electrophoretic migration of a negatively charged micelle is normally opposed by the EOF. Because the electroosmotic velocity is toward the cathode (e.g., detector end in normal polarity CE) and is higher than the electrophoretic velocity, the net migration of negatively charged chiral micelles will also be toward the cathode [32]. However, positively charged chiral micelles interact with the capillary surface and may reverse the direction of EOF toward the anode (e.g., injection end in normal polarity CE). Therefore, for faster separations, the polarity should be switched when cationic surfactants are used. With reverse polarity, both EOF and the net velocity of cationic chiral surfactants will be in the opposite direction. A neutral chiral solute which is not separated by CZE migrates at the velocity of the EOF where it does not interact (solubilize) with the pseudo-stationary phase. When a chiral solute is incorporated into the chiral pseudo-stationary phase, the solute interacts with the micelle on the surface through polar–polar interactions as well as hydrophobic interactions with the core of the micelle [27,33,34]. However, in most cases, the hydrophilic polar group of the surfactant is responsible for chiral discrimination. Consequently, separation and retention time is based on differential solubilization (interaction) as well as chiral recognition in the micellar phase.

As chiral MEKC is considered similar to micellar liquid chromatography (MLC) equations used in MLC or MEKC [e.g., $t_R = (1 + k')t_0$] can be used as models in chiral MEKC with few modifications. Neutral solutes will have a retention time between the retention time of a solute that has no interaction with the micelles that moves with the electroosmotic velocity and the retention time of a solute [i.e., solubilized (interacts) with the micelle] [26]. To solve for the migration time of the neutral solute, x, we must determine the capacity factor, k', which is defined by

$$k' = \frac{\eta_{mc}}{\eta_{aq}} \tag{5}$$

where η_{mc} and η_{aq} are the total number of solute molecules incorporated into the micelle and the total number dissolved in the surrounding aqueous phase, respectively. The neutral solute migration time, t_{Rx}, is related to the capacity factor by

$$t_{Rx} = \frac{(1+k')t_0}{1} + \left(\frac{t_0}{t_{mc}}\right)k' \tag{6}$$

where t_0 is the migration time of the bulk solution and t_{mc} is the migration time of the micelle. The migration times t_0 and t_{mc} can be measured by using methanol or formamide as aqueous-phase tracers and Sudan III or IV as the micelle tracers, respectively. The selectivity, α, for two solutes (enantiomers) x and y can be written as

$$\alpha_{xy} = \frac{k'_y}{k'_x} \tag{7}$$

The resolution, R_{xy}, equation of chiral MEKC is then written as

$$R_s = \frac{N^{1/2}}{4} \frac{\alpha_{xy}-1}{\alpha_{xy}} \frac{k'_y}{1+k'_y} \left(\frac{1-\left(\frac{t_0}{t_{mc}}\right)}{1+\left(\frac{t_0}{t_{mc}}\right)k'_x} \right) \tag{8}$$

where N is the theoretical plate number and α is the selectivity factor (separation factor) defined as k'_y/k'_x (≥ 1) and k'_x and k'_y are the capacity factors for solute x and solute y, respectively.

As mentioned earlier, another factor that affects resolution is the effect of the EOF in CZE. The net migration of the micelle is usually in the same direction as the EOF but at a slower velocity, as the EOF is much stronger than the electrophoretic mobility of the micelle. However, Otsuka and Terabe [35] have shown that pH affects the v_{eo} and electrophoretic velocity of the micelle (v_{mc}). Figure 14a illustrates that in the pH range 5.5–9.0, a slight increase of v_{eo} is observed for the 0.2 M sodium dodecylsulfate solution. At pH values below 7.0, the electrokinetic velocity is constant. However, Figure 14b shows that in the pH range 3.0–7.0, the v_{eo} decreases with a decrease in pH below 5.5, whereas the electrophoretic velocity of sodium dodecylsulfate is essentially constant throughout the range 3.0–7.0. Therefore, the v_{mc} of the anionic micelle decreases with a decrease in pH, and its direction changes toward the cathode at pH 5.0, suggesting that the direction is the same as the v_{eo} above pH 5.0 and opposite for a pH below 5.0 [8]. If the conditions are such that one is working at a low pH (acidic conditions), we may assume that the ratio t_0/t_{mc} is negative. Consequently, it is possible to obtain an extremely high resolution when the EOF can be controlled at the expense of a longer separation time. Therefore, the ratio t_0/t_{mc} is expressed as

$$\frac{t_0}{t_{mc}} = \frac{v_{mc}}{v_{eo}} = \frac{1+\mu_{ep}(mc)}{\mu_{eo}} \tag{9}$$

where v_{mc} is the migration velocity of the micelle, which is equal to $v_{eo} + v_{ep}(mc)$; $v_{ep}(mc)$ is the electrophoretic velocity of the micelle, and $\mu_{ep}(mc)$, is the electrophoretic mobility of the micelle. The mobility, $\mu_{ep}(mc)$, is almost constant

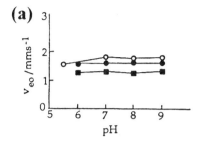

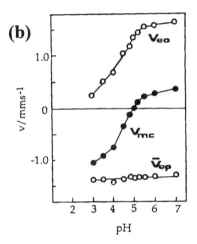

Fig. 14 (a) Effect of pH on the electrokinetic velocity. Experimental conditions: applied voltage was 15 kV, capillary was 50 μm i.d., 190 mm o.d., 650 mm long. Velocities and solutions: v_{eo} in 0.1 M SDS in 0.05 M phosphate–0.1 M borate buffer (•), v_{eo} in 0.2 M SDS in the same buffer as above (°), and v_{ep} in 0.1 M SDS in the same buffer (block). (b) Dependence of electrokinetic velocities on pH: v_{ep} is the electrophoretic velocity in the micelle all other symbols are the same as in the text. Conditions are the same as in (a). Reprinted from S.F.Y. Li, *Capillary Electrophoresis: Principles, Practice, and Applications*, 1992, pg. 242, with permission from Elsevier Science.

for a given micelle, whereas μ_{eo} is dependent on conditions such as pH change, the addition of an organic solvent, or the use of a coated capillary [32].

C. Monomeric Chiral Surfactants for MEKC Separation of Enantiomers

Over the last 6 years, a number of chiral MEKC methods have been developed to provide the separation of various enantiomers. Thus, a brief review of this technology is appropriate. It should be noted that a variety of monomeric chiral surfactants have been employed: (1) N-dodecanoyl-L-valinate, (2) bile salts, (3) digitonin, (4) saponins, and (5) glucopyranoside-based phosphate and sulfate surfactants.

Sodium-N-Dodecanoyl-L-Valinate

One of the first chiral surfactants used in MEKC was sodium N-dodecanoyl-L-valinate (SDVal) (Fig. 15) [36–40]. It was found that chiral amide-terminated monolayers anchored to a silica gel surface gave a hydrophobic interfacial phase which diminished the liquid–solid interfacial area in aqueous media, allowing hydrogen-bond association to occur, which leads to enantiomeric resolution [36]. With this model, Dobashi and co-workers [36] incorporated chiral hydrogen-bonding amide functionality into the hydrophobic core in order to provide a more defined interaction of the enantioselective micelle with a solute molecule. Thus, N-dodecanoyl-L-amino acids (L-valine and L-alanine) sodium salts were effectively employed for enantiomer separation of amino acid derivatives. SD-Val micelles are distinctively different from structured chiral cavity chiral selectors such as CDs, where the solute binds hydrophobically. In the case of SDVal, there is no inclusion complex with the solute and micelle. It is believed that enantioselectivity occurs when a solute binds by hydrogen bonds to the amide functional group in the micellar inner core [36].

Enantiomeric resolution of a mixture containing amino acid derivatives such as N-(3,5-dinitrobenzoyl) o-isopropyl esters and their derivatives has been achieved with SDVal (Fig. 16). The D-enantiomer eluted faster than the corresponding L-enantiomer. This suggests that the L-enantiomer binds to the chiral micelle more strongly than the D-enantiomer. The elution order of the amino acid derivatives is in accordance to the extent of the increase in the hydrophobicity of the amino acid side chain, with the least hydrophobic side chains (methyl) eluting first. In contrast, those with more hydrophobic side-chain characteristic (phenyl) eluted last [36]. Similarly, the use of relatively less hydrophobic surfactants such as sodium N-dodecanoyl-L-alaninate (SD-Ala) resulted in smaller capacity factors for all amino acid enantiomers. Mixed micellar media involving common micelle-forming detergents, such as anionic sodium dodecylsulfate (SDS) in conjunction with SDVal, also reduced the enantiomeric selectivity.

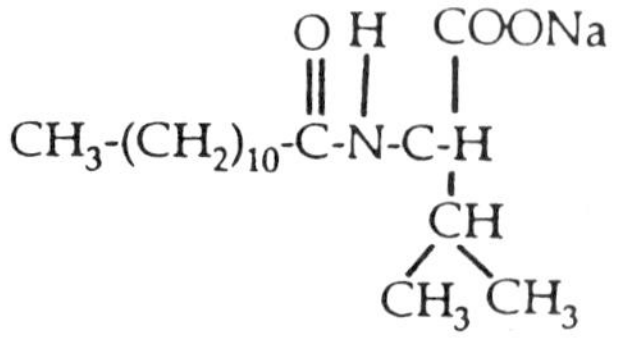

Fig. 15 Structure of sodium-N-dodecanoyl-L-valinate.

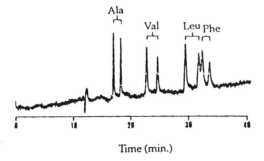

Fig. 16 Electropherogram of a mixture containing four pairs of amino acids as N-(3,5-dinitrobenzoyl) o-isopropyl ester derivatives. Experimental conditions: fused silica capillary 50 cm long, 50 μm i.d., 0.025 M sodium N-dodecanoyl-L-valinate (SDVal) in 0.025 M borate–0.05 M phosphate buffer (pH 7.0); applied voltage 10 kV, 26 μA current; detection, UV at 230 nm. Reprinted from *J. Chromatogr., 480:*413, A. Dobashi, 1989, with permission from Elsevier Science.

Sodium N-dodecanoyl-L-valinate (SDVal) was also used to separate phenylthiohydantoin (PTH) derivatives of DL-amino acids by Terabe et al. [37,38]. When SDVal was used alone, tailing peaks were observed, resulting in poor efficiency. An extended migration-time window was needed to improve resolution. This was done by modifying the micelle buffer solution with a small amount of methanol. The addition of methanol to micellar solutions reduced the solute retention as determined by measurement of the capacity factor. This is due in part to the introduction of phase polarity in the hydrophobic interfacial phase [36]. Also, the electroosmotic velocity is reduced and this, in turn, extends the migration-time window. In Fig. 17, enantiomeric resolution of four PTH-DL-amino acids with 25 mM SDVal, pH 7.0, and 10% methanol were resolved with poor peak shapes; however, better peak shapes were observed than the run without the addition of 10% methanol [38]. Urea was also added to the micellar solution to achieve better resolution. Urea improves resolution as a result of adsorption of urea to the inside wall of the fused silica capillary, which prevents the adsorption of solutes. As shown in Fig. 18, the addition of urea improved peak shape, but efficiency was still not optimal and PTH-DL-Nva/PTH-DL-Met comigrate [38]. Nevertheless, under different conditions such as 50 mM SDVal–30 mM SDS–0.5 mM urea, pH 9.0, and 10% methanol (v/v), a mixture of six different PTH-DL-amino acids (Fig. 19) were resolved with perfect peak shape [37].

The SDVal solutions can provide enantiomeric separation for some compounds; however, peak tailing is usually observed. Therefore, the addition of methanol and urea gave improved peak shapes and resolution by increasing the capacity factor and extending the migration-time window [37,38].

Fig. 17 Electropherogram of a mixture of four PTH-D,L-amino acid derivatives. Peaks correspond to (1) Nva, (2) Met, (3) Trp, (4) Nle; micellar solution, 25 mM SDVal, pH (7.0) containing 10% methanol; capillary length was 500 mm, 0.05 mm i.d.; applied voltage 15 kV, 10 µA current; detection wavelength 260 nm. Reprinted from Ref. 38 with permission from VCH Verlagsgesellschaft mbH.

Bile Salts

Bile salts are a class of anionic surfactants containing a hydroxy-substituted steroid backbone (Fig. 20) [45]. They possess hydrophilic and hydrophobic faces which results in the formation of aggregates. The bile salt micelles have a helical structure [41], with the hydrophilic region concentrated toward the interior of the micelle, and the hydrophobic tail facing the aqueous solution, forming a reversed micelle. Due to the polarity of the bile salt, a general reduction in the solute k' occurs. As the degree of hydroxyl substitution increases, the polarity increases, producing a decrease in k'; thus, sodium cholate (SC) with three hydroxyl substituents gives lower k' values than the dihydroxy substituted sodium deoxycholate (SDC) [42]. Bile salts have been used for a variety of different classes of compounds, namely dansylated (Dns)-amino acids [43], optical isomeric drugs such as diltiazem and trimetoquinol [44–46], binaphthyl analogs, as well as some polycyclic aromatic hydrocarbons (PAHs) [42].

It has been suggested that bile salts prefer a rigid, planar structure for chiral recognition [42]. Figure 21a shows enantiomers of 1,1′–bi-2-naphthol (BNOH) and 1,1′–binaphthyl diyl hydrogen phosphate (BNPO$_4$) separated under alkaline (pH 9.0) mobile phase at various concentrations of SDC. At low concentrations (5 mM) of SDC, the enantiomers are not resolved and the efficiency is poor. In

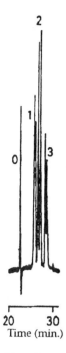

Fig. 18 Electropherogram of PTH derivatives of three D,L-amino acids: Peaks (0) acetonitrile, (1) Nva, (2) Trp, (3) Nle; micellar solution, 25 mM SDVal, pH 7.0 containing 10% methanol and 5 M urea; current 7.3 µA. Other conditions same as Fig. 17. Reprinted from Ref. 38 with permission from VCH Verlagsgesellschaft mbH.

electropherograms (Figs. 21b–21d), high-efficiency separations of BNOH and BNPO$_4$ were obtained at higher surfactant concentrations of 20 mM. An increase in surfactant concentration to 50 mM deteriorated the resolution. This suggests that (1) the bile salt micelles are nearly monodisperse or (2) the kinetics of the micelle–monomer exchange is rapid on a chromatographic time scale. However, the addition of methanol improves the resolution further at the expense of a longer elution time [42].

The utility of bile salts as natural chiral surfactants was first reported by Terabe and co-workers in 1989 to separate optically active dansylated D,L-amino acids (Dns-DL-AA) (Fig. 22). Initial efforts to achieve chiral recognition of Dns-D,L-AA using SDC and sodium taurolithocholate were unsuccessful due to the formation of gelatin inside the fused silica capillary. However, the addition of sodium taurocholate (STC) or sodium taurodeoxycholate (STDC) with 50 mM phosphate buffer (pH 3.0) resulted in baseline separation of two out of six enantiomeric pairs of Dns-AA (Fig. 22). The electrostatic attraction between the neg-

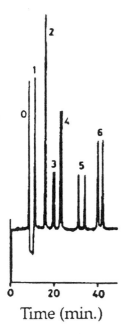

Fig. 19 Chiral separation of six PTH-D,L-amino acids: (1) Ser, (2) Aba, (3) Nva, (4) Val, (5) Trp, (6) Nle, (0) acetonitrile; micellar solution, 50 mM SDVal, 30 mM SDS, 0.5 M urea (pH 9.0) containing 10% (v/v) methanol; capillary dimensions, 650 mm × 0.05 mm i.d.; 500 mm long; total applied voltage 20 kV; current 17 µA; detection wavelength 260 nm. Reprinted from *J. Chromatogr., 559:*209, K. Otsuka, 1991, with permission from Elsevier Science.

atively charged group of STDC and the positively charged amino group of Dns-AAs was thought to influence chiral recognition. A major disadvantage of this approach was the long analysis time (>60 min). In order to shorten the analysis time without a concomitant decrease in resolution, sodium dodecylsulfate (SDS) (an achiral surfactant) was added (Fig. 23). Unfortunately, the resolution was impaired. However, the separation time was decreased by almost a factor of 2 [43]. Once again, bile salts provided reasonable enantiomeric separation.

It should also be noted that several optical isomeric drugs have been separated by use of bile salts [45,46]. Enantiomers of diltiazem hydrochloride, trimetoquinol hydrochloride, carboline derivatives, 2,2'-dihydroxy-1,1'-dinaphthyl derivatives, and certain analogs were separated successfully under neutral or alkaline conditions. Enantiomers of diltiazem hydrochloride and trimetoquinol hydrochloride were resolved with only STDC solutions and in both drugs the (+)-isomers, (SS)-diltiazem hydrochloride and (R)-trimetoquinol hydrochloride, eluted faster than the corresponding (−)-isomer. Studies on the influence of buffer pH and the effects of applied voltage were also conducted [45].

Bile Salts	Symbol	R_1	R_2	R_3	R_4
Sodium cholate	SC	OH	OH	OH	ONa
Sodium taurocholate	STC	OH	OH	OH	$NHCH_2CH_2SO_3Na$
Sodium deoxycholate	SDC	OH	H	OH	ONa
Sodium taurodeoxycholate	STDC	OH	H	OH	$NHCH_2CH_2SO_3Na$

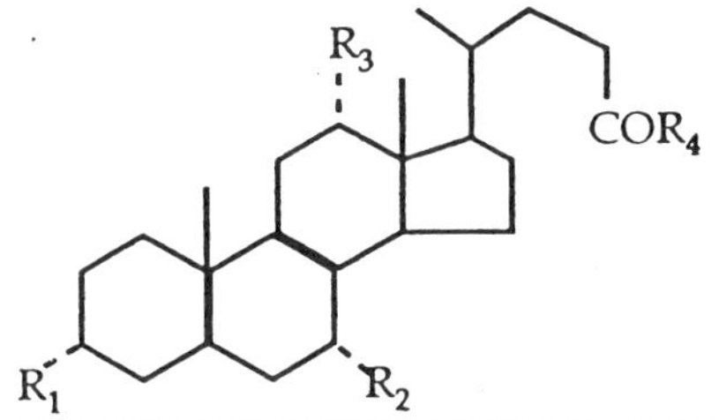

Fig. 20 Structure of bile salts.

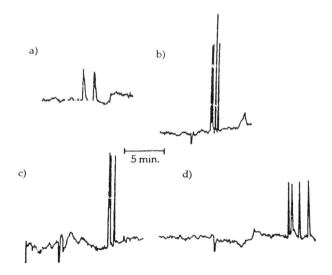

Fig. 21 Separation of BNOH and $BNPO_4$ enantiomers at pH 9.0 using a mobile chiral phase containing: (a) 0.005 M SDC, (b) 0.01 M SDC, (c) 0.05 M SDC, and (d) 0.05 M SDC with 12% methanol. Reprinted from Ref. 42 with permission from the author.

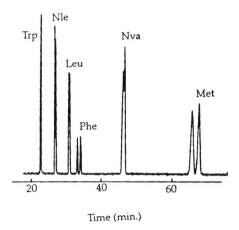

Fig. 22 Separation of Dns-D,L-amino acids (labeled in electropherogram) with 50 mM taurodexycholate in 50 mM phosphate buffer (pH 3.0); in a 500 mm × 0.05 mm i.d. capillary; current 50 µA. Reprinted from *J. Chromatogr., 480:*403, S. Terabe, 1989, with permission from Elsevier Science.

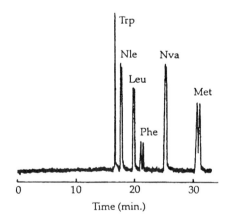

Fig. 23 Separation of Dns-D,L-amino acids. Conditions the same as Fig. 22 except the electroosmotic velocity is slightly faster. Reprinted from *J. Chromatogr., 480:*403, S. Terabe, 1989, with permission from Elsevier Science.

Polymers for Chiral Separations in CE / 391

The resolution at pH 9.0 of four carboline derivatives [(S), (R), (SS), and (RR) forms] (Fig. 24a) and (RS)-2,2'-dihydroxy-1-1'dinaphthyl (Fig. 24b) were similar to resolution at pH 7.0. However, the α values for the latter were obtained in SDS or STDC solutions of pH 7.0, which were superior to those at pH 9.0. This is believed to be a result of ionic repulsion between the phenolic

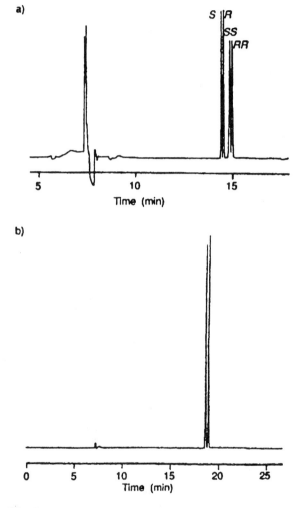

Fig. 24 Separation of 2,2'-dihydroxy-1,1'-dinaphthyl. Experimental conditions: 0.005 M deoxycholate in a 0.02 M phosphate–borate buffer (pH 9.0); applied voltage 20 kV. (a) and (b) as defined in text. From H. Nishi, *J. Microcol. Sep.*, *1*:234, © 1989. Reprinted by permission of John Wiley and Sons, Inc.

hydroxyl group of the solute and the anionic micelle which affects the chiral separation at pH 9.0 [45]. It was found that for chiral separation of diltiazem hydrochloride, trimetoquinol hydrochloride, and (RS)-1-napthylethylamine (using a pH 7.0, 5 mM STDC) solution, the k' values gradually increased as applied voltage was decreased, although the α values did not change. However, resolution, R_s, was enhanced by the decrease in applied voltage (Fig. 25) [45].

Digitonin

Otsuka and Terabe explored enantiomeric resolution of phenylthiohydantoin (PTH) amino acids using a nonionic compound, digitonin (Fig. 26) [47]. Digitonin is a glycoside of digitogenin, which is used for the determination of cholesterol.

Because digitonin is electrically neutral, an ionic micelle is added to the digitonin solution to form charged mixed micelles [47,48]. No resolution of any PTH-DL-amino acids was achieved when a 25 mM digitonin–50 mM SDS solution of pH 7.0 was used. In this case, the v_{eo} of digitonin–SDS mixed micelle caused the migration-time range to be too narrow for adequate resolution [47]. In order to suppress the EOF and to extend the migration-time range, an acidic micellar solution (pH 3.0) was used. Under acidic conditions, the PTH derivatives of six amino acids [tryptophan (Trp), norleucine (Nle), norvaline (Nva), valine (Val), α-aminobutyric acid (Aba), and alanine (Ala)] were separated and mostly resolved (Fig. 27) with a

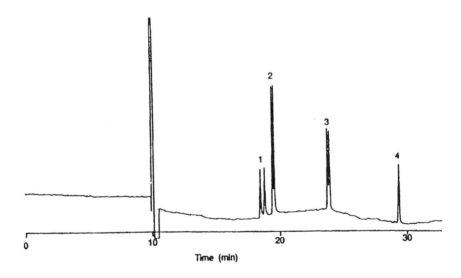

Fig. 25 Chiral separation of (1) trimetoquinol hydrochloride, (2) 1-naphthylethylamine, (3) diltiazem hydrochloride, and (4) 2,2,2-trifluoro-1-(9-anthryl)ethanol. Running buffer: 0.005 M taurodeoxycholate in a 0.002 M phosphate–boron buffer (pH 7.0); applied voltage 15 kV. From H. Nishi, *J. Microcol. Sep.*, 1:234, © 1989. Reprinted by permission of John Wiley and Sons, Inc.

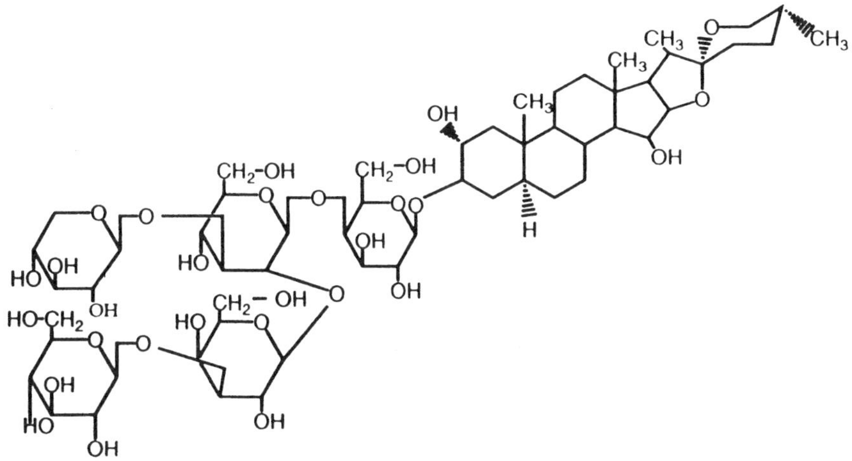

Fig. 26 Structure of digitonin.

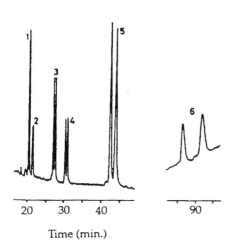

Fig. 27 Separation of six PTH derivatives of D,L-amino acids. Corresponding amino acids: (1) Trp, (2) Nle, (3) Nva, (4) Val, (5) Aba, and (6) Ala. Micellar solution: 25 mM digitonin, 50 mM SDS (pH 3.0); capillary; 630 × 0.05 mm i.d.; length of capillary for separation: 490 mm, applied voltage 20 kV; current 34 μA; detection wavelength 260 nm. Reprinted from *J. Chromatogr., 515:*221, K. Otsuka, 1991, with permission from Elsevier Science.

longer analysis time [47]. Separation of PTH-DL-amino acids using digitonin in combination with STDC additives have also been reported [48].

Saponins

Both glycyrrchic acid (a tricarboxylic acid) and β-escin belong to a class of natural chiral surfactants which are called saponins. Structurally, these chiral compounds are composed of triterpene and glucose moieties, respectively, providing the hydrophobic and hydrophilic region within the molecule (Fig. 28). As these triterpene glucoside surfactants are not effective as single micelles, they have been used as binary or ternary micellar systems to induce chiral recognition [49]. A ternary micelle comprised of glycyrrchic acid (GRA), octyl β-D-glucoside, and SDS was found to produce a stable pseudophase and generated moderate electrophoretic mobility of the GRA micelle. Some Dns-DL-AAs were resolved using the above micellar system under neutral pH conditions; however, the retention time was as long as 80 min for the separation of Dns-DL-leucine. Because GRA has appreciable absorbance in the UV region due to the presence of a carbonyl group in conjugation with a carbon–carbon double bond, a fluorescence detector was used. However, for β-escin, SDS was still needed. The PTH-DL-AAs were separated with higher peak capacity and a shorter separation time of 35 min (Fig. 29).

Glucopyranoside

One of the latest and perhaps the most versatile class of monomeric chiral surfactants is the dodecyl β-D-glucopyranoside [50]. These surfactants contain either a phosphate or a sulfate functional group in the 6-position of the sugar residue, which are easily ionizable with a low pK_a in the range 1–2 (Fig. 30). Furthermore, the glucopyranoside-band phosphate or sulfate surfactants have a very low CMC (0.5 and 1.0 mM), respectively. These CMCs are about an order of magnitude lower than SDS. Other advantages reported for these surfactants include higher water solubility and no appreciable absorbance even at wavelengths as low as 200 nm. Figure 31 shows electropherograms for the separation of Dns-DL-AAs. The chiral surfactant used in this separation was dodecyl β-D-glucopyranoside monophosphate, which is found to provide greater separation than the monosulfate anionic surfactants. It is believed that linking the phosphate group to the hydroxyl group in the 4-position of the sugar moiety results in a more rigid bicyclic structure in contrast to that of a more flexible sulfate group. Various classes of chiral compounds that differed in stereochemical structure, acidity/basicity, and hydrophobicity were successfully resolved.

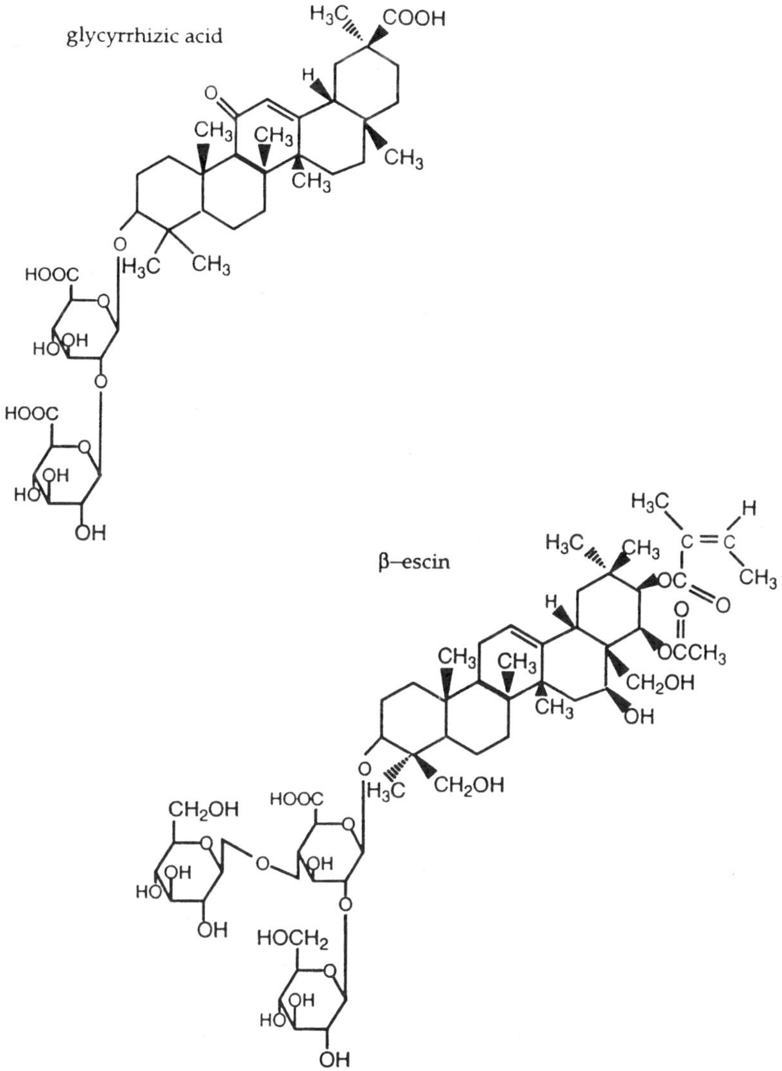

Fig. 28 Structure of saponins.

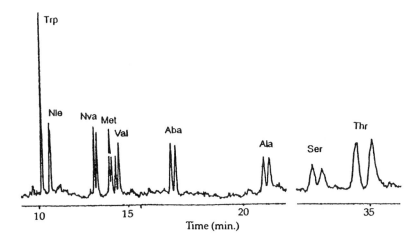

Fig. 29 Separation of PTH-DL-amino acids by MEKC with a buffer containing 25 mM β-escin and 50 mM SDS in a 50 mM phosphate (pH 3.0). Capillary 50 μm i.d. × 65 cm (effective length 50 cm); separation voltage +20 kV; detection wavelength at 220 nm. Reprinted from Ref. 49 with permission from the author.

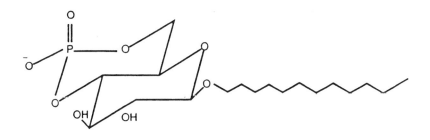

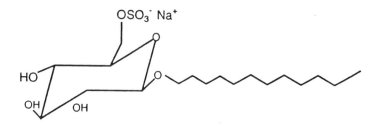

Fig. 30 Structure of glucopyranoside surfactants.

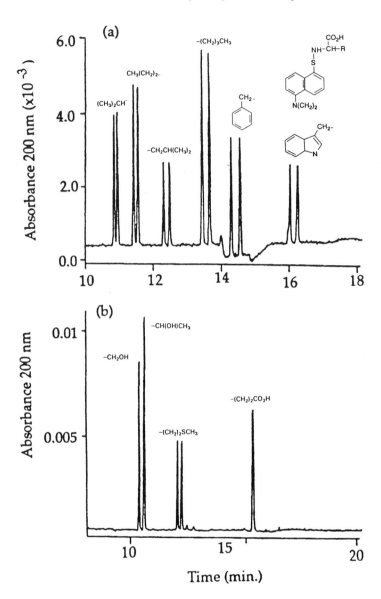

Fig. 31 Enantiomeric separation of Dns-DL-amino acids by MEKC with a buffer containing n-dodecyl β-D-glucopyranoside 4,6-hydrogen monophosphate, 25 mM phosphate buffer (pH 8). Capillary 50 μm i.d. × 57 cm (effective length 50 cm); separation voltage +20 kV; detection wavelength at 200 nm; temperature 25°C. Reprinted with permission from P. Camilleri, *Anal. Chem.*, 66:4121 (1994), © 1994 American Chemical Society.

IV. POLYMERIC MICELLES

A. Concept of Surfactant/Micelle Equilibria

Although we have seen several examples of successful enantiomeric separations using CZE and MEKC, there are problems with using conventional micelles in chromatography. Many of these problems will be identified later and may be improved by use of polymerized micelles in MEKC. Wang and Warner have suggested that using polymerized micelles in MEKC instead of conventional micelles would eliminate the dynamic equilibrium that normally exists, minimize the effects of organic modifiers, eliminate concentration dependence, reduce peak-broadening, and decrease joule heating [5,51].

Micelles are involved in a highly dynamic equilibrium, being constantly formed and dissolved, as demonstrated in kinetic studies [52]. It is essential to recognize this dynamic equilibrium of the system in experimental studies. Anisson and Wall have proposed a kinetic model for micelle formation, which is now generally accepted and has been adopted as the basis for the relaxation process of micelles [53,54]. In their model, the following equilibria are observed:

$$S_1 + S_1 \underset{K_{-2}}{\overset{K_2}{\rightleftharpoons}} S_2$$

$$S_2 + S_1 \underset{K_{-3}}{\overset{K_3}{\rightleftharpoons}} S_3 \qquad (10)$$

$$S_{n-1} + S_1 \underset{K_{-n}}{\overset{K_n}{\rightleftharpoons}} S_n$$

where S_1 refers to the surfactant monomer, S_2 to the dimer, and S_n to the n-mer. From stepwise formation process, it can be deduced that a micellar solution will contain aggregates with values of n from 2 to far above the average n value. Thus, normal micelles are polydispersed aggregates. Such polydispersity will normally contribute to poor resolution in chromatographic separations.

Thermodynamics of Micellization

If we assume a simple association equilibrium between surfactant monomers and micelles with aggregation number n, we can derive [55]

$$\Delta G^\circ = RT \ln(\text{CMC}) \qquad (11)$$

where R is the gas constant and T is the temperature on the Kelvin scale. The numerical value ΔG° depends on the units of concentration used. Similarly, the free energy of micellization for the ionic surfactant can be derived as

$$\Delta G° = RT\left(2 - \frac{p}{n}\right)\ln(CMC) \qquad (12)$$

where p is the effective charge on the micelle [55].

Micelle–Substrate Interaction

Earlier we have defined the thermodynamics of micellization. As discussed previously, a process of primary importance to this review is the interaction of ligands with micelles. For the purpose of our discussion here, we will define the partition coefficient for association of an arbitrary ligand with a micelle. Thus, the partition coefficient, K, is defined as

$$K = \frac{C_m}{C_w} \qquad (13)$$

where C_m represents the concentration of the substrate which has partitioned into the micelle and C_w represents the concentration of the substrate dissolved in the aqueous phase. Therefore, under standard state conditions, the free energy of association is given by

$$\Delta G° = -RT \ln K \qquad (14)$$

Of course, partitioning into and interacting with the micelle must also depend on the micellization process, as interaction with micelles cannot occur if no micelles are present. Thus, the entire process of ligand interaction with the micelle is less complicated if the micellization process is eliminated. We will revisit this concept in our later discussions on chiral interactions and separations with chiral micelles [5,51].

The observations and hypotheses cited above suggest that control of dynamic equilibria may yield improved discrimination for molecules with small differences in ΔG of interaction (e.g., racemic mixtures). This hypothesis serves as the focus of this review on the use of chiral micelle polymers for improved chiral separations.

As mentioned earlier, as the concentration of surfactant changes, the micelle shape also changes (Fig. 13). In 1972, Kammer and Elias [58] suggested fixing the micellar structure by polymerizing micelles from amphiphiles with polymerizable groups in the hydrophobic part [56].

Polymerization of a micelle is most often achieved by one of two methods. Using the first method, surfactants containing a vinyl group are put into solution above the CMC-forming micelles. The latter, containing the vinyl group surfactants, is polymerized. In the second method, the micelle is formed by surfactant monomers which contain polymerizable groups other than vinyl groups [59–69].

B. Factors Affecting Polymerization Rate and Polymer Structure

In principle, three main effects influence the polymerization kinetics and polymeric structure of polymerized micelles. These are (1) the concentration effect, (2) the medium effect, and (3) the topochemical effect [62]. The concentration effect is a measure of polymerizable groups inside the micelles. This concentration could have values 100 times larger than the values of the analytical concentration of the surfactant in the solutions. Furthermore, concentration influences the reaction rate [62]. For example, polymerization barely proceeds below the CMC because of low micellar concentration.

The rate of dissociation of a single amphiphile depends on its solubility and critical micelle concentration (CMC). The dynamics of micellization falls into two general categories. One is the millisecond region where whole micelles dissolve and reform. The second is the microsecond region and below where an amphiphile molecule exchanges between the micelle and solution [59]. The question of whether the lifetime of a micelle is sufficiently long compared to the lifetime of the active polymer chain in the micelle had to be answered in order to determine if polymerization could occur before micellar dissociation. Two rate processes were found to be important for the formation and dissociation of micelles [57,58]. The relaxation times were between 10^{-2} and 10^{-6} s. The kinetic processes (fast and slow) are similar to the average residence time of the surfactant in the micelle and the average lifetime of a micelle [56].

Chain radicals of 1000 monomeric units in an isotropic medium have an average lifetime range from 0.1 to 10 s [65]. The lifetime of polymer radicals with an association number between 10 and 100 should range between 10^{-1} and 10^{-4} s. Fortunately, these times are compatible with the relaxation times of micelles (10^{-2} and 10^{-6} s). Therefore, for some micelle-forming amphiphiles with polymerizable groups, polymerization of monomeric micelles with aggregation number, N, should be possible where the degree of polymerization is equal to N [56].

Topochemical polymerization involves the linking of monomer (micelle) species in a fixed geometric orientation or arrangement such that the polymer geometry closely resembles that of the original monomer geometry [59]. The medium effect is important because the polymerizable groups experience different environments in the micellar and singly dispersed state. Physiochemical properties tell us that different environments affect the rate of polymerization and the structure and properties of the polymeric micelle [62].

Two main types of polymer micelles are possible. They are termed H-type for head and T-type for tail polymers. If the polymerizable double bond is located on the ionic head of the surfactant, then that polymer is characterized as type H. However, T-type polymers are produced if the double bond is at the end of the aliphatic hydrocarbon chain (Fig. 32) [62]. T-type polymers have a hydrophobic backbone, whereas H-type polymers have hydrophilic backbones, and their behavior in aqueous solutions are quite different from the behavior of monomeric micelles. It is preferable to have some separation between the charge head and the double bond because this facilitates polymerization. As the separation increases, the packing constraints are relaxed. In type T polymers, the constraints should be minimal. However, in H-type polymerized micelles, the constraints are not totally removed due to the closeness of the ionic head group and vinyl group [62].

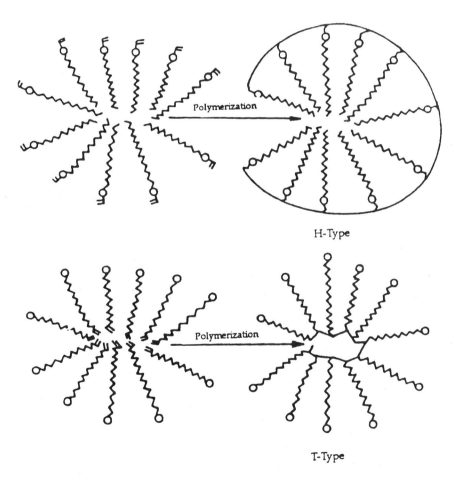

Fig. 32 Schematic representation of type H and type T micellar polymerization in an aqueous media. Reprint from Ref. 62 with permission from the author.

C. Examples of a Cationic Polymerized Micelle

One of the first examples of polymerization in the micellar state were the vinylpyridinium salts. These include, 1,2-dimethyl-5-vinylpyridinium methyl sulfate, 1-methyl-2-vinylpyridinium methyl sulfate, and 1-methyl-4-vinylpyridinium methyl sulfate (Fig. 33a). These cationic surfactants aggregate at high concentrations and exhibit some unusual kinetic properties upon spontaneous polymerization in aqueous solution [66]. At high concentrations of the monomer (greater than 2.5 M) the overall reaction is found to be first order with respect to the monomer. The initial rate of monomer conversion $(dq/d_t)_0$, dropped sharply when the initial monomer concentration $[M_0]$ was decreased despite the fact that $(dq/d_t)_0$ should be independent of $[M_0]$ for first-order reactions. Furthermore, the average degree of polymerization, expressed in terms of intrinsic viscosity, $[\eta]$, was reduced when the initial monomer solution was diluted [62]. According to Kabanov [66,67], these events result from chemical attraction between the molecules (intermolecular forces). He hypothesized that organized aggregates of the monomers which formed had lifetimes comparable to the time of propagation of the polymerization process.

There does not appear to be a consensus on the spontaneous polymerization of quarternized vinylpyridinium salts. Therefore, Salamone et al. [68] studied 1,2-dimethyl-5-vinylpyridinium methyl-sulfate in order to derive specific conclusions regarding these polymers. It was found that the polymerization behavior of this monomer at concentrations above 1.0 M was due to kinetic effects rather than to aggregation/organization of monomers in micellar structures. It was also found that even at concentrations below the CMC (such as 0.75 M), the polymeric product was formed after prolonged polymerization. Salamone believes this problem is resolvable if polymerization is performed in solvents in which micelles are not formed. In the latter case, if comparable conditions were used and spontaneous polymerization occurs, then obviously it is not the aggregation and organization of the monomers that influence polymerization but a kinetic effect.

For aromatic cationic surfactants such as 4-vinyl-pyridinium protonated salts, it was demonstrated that polymer formation is dependent on the monomer surfactant concentration. Different products, depending on the monomer concentration, have been reported in polymerization of the protonated vinylpyridinium salts in aqueous solutions [69–74]. At high concentrations, a 1,2-addition polymer is obtained when the 4-vinylpyridinium protonated salts are spontaneously polymerized. In contrast, isotropic polymerization produces the 1,6-ionene oligomer (Fig. 33b) [62].

D. Example of an Anionic Polymerized Micelle

Sodium 10-undecenoate is a polymerizable micelle-forming surfactant which has a double bond in the aliphatic chain, producing a T-type polymeric micelle. The radical polymerization was initiated by γ-radiation [75,76] and monitored by

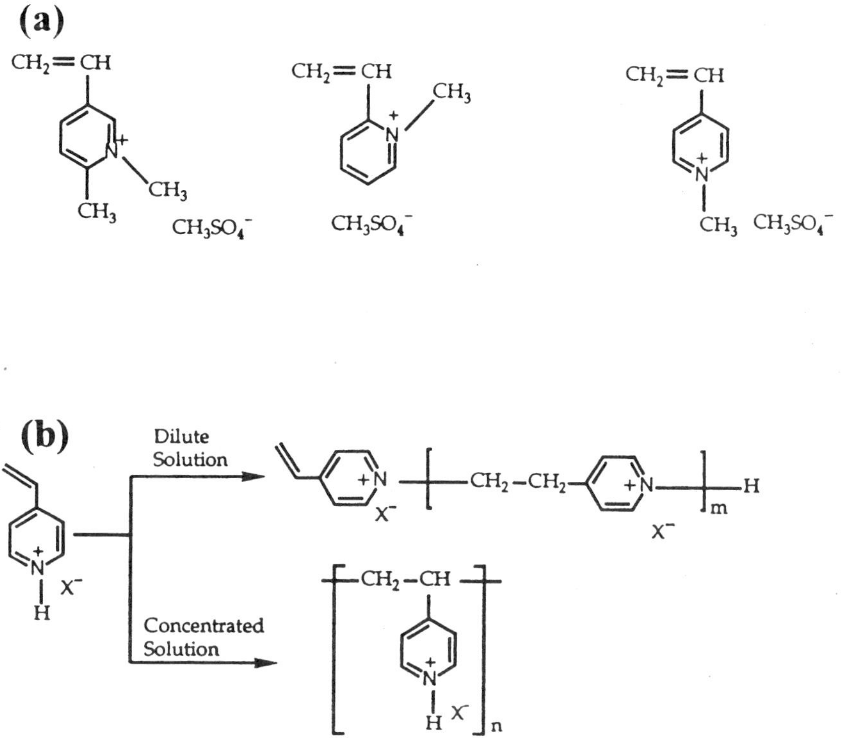

Fig. 33 (a) Structure of vinylpyridinium salts. (b) Different polymerization modes that occurred with protonated salts of 4-vinyl pyridine. Reprinted from Ref. 62 with permission from the author.

proton NMR and infrared (IR) spectroscopy. For such surfactants, NMR usually indicates the disappearance of double-bond peaks around 5.5 ppm, whereas the IR shows the loss of the C–H stretching vibration band at approximately 3100 cm^{-1}. The degree of polymerization of the polymerized micelle produced was determined by vapor-pressure osmometry and was found to be equal to the aggregation number of the monomeric micelle. This suggests that the polymerization is intramicellar and the polymerized micelle is an oligomer of 10 monomer units. Below the CMC, polymerization rates were found to be zero [62].

In addition to intramolecular micelles (e.g., micelles made of only one polymerized micelle with zero CMC), the polymerized sodium 10-undecenoate also forms intermolecular micelles (polymerized micelles that aggregate intermolecularly could have a CMC equal to 10^{-2} M). The data on the intrinsic viscosities of aqueous solutions of the monomeric and polymerized surfactant reveal that the polymerized micelle has a hydrated size equal to the size of the monomeric micelle ($[\eta]$ = 0.048 100 mL/g in both cases) [77,78], whereas the intrinsic viscosities of the intermolecular micelle was equal to 0.078 [75], suggesting a large hydrated size.

Using fluorescence probes [77], it was found that either more water enters inside the polymerized micelle or the pyrene (the fluorescence probe in this case) cannot penetrate into the polymerized micelle as deeply as it penetrates inside the monomeric micelle. This interpretation also agrees with the expected, more compact packing of the aliphatic chains in the case of polymerized micelle [62]. Finally, it was found that when sodium 10-undecenoate is polymerized in the lyotropic liquid-crystalline state, the structure changes from hexagonal closely packed cylinders to a lamellar structure (Fig. 13). It has been suggested that this structure change occurs because of the restricted free rotation and translation of the end group of the aliphatic chain after polymerization [79].

From this brief synopsis, we conclude that polymerized micelles have enhanced stabilities, controllable sizes, enhanced rigidities, and permeabilities. Ideally, polymerized micelles combine the beneficial properties of stable uniform polymers to the fluidities of micelles [80].

E. Applications of Polymerized Micelles

We choose to use polymerized micelles to minimize and/or eliminate dynamic equilibria in normal micelles that do not always allow for adequate discrimination for many racemic mixtures, as well as to reduce the use of high concentrations of surfactant which must often be used in order for some surfactants to be effective in MEKC. When ionic surfactants are used at high CMC, this may result in joule heating which causes problems in CE separations [5]. As mentioned earlier, polymerized micelles have distinct advantages over normal micelles because they have enhanced stabilities, rigidities, and controllable size, because of the covalent bonds formed between surfactant aggregates eliminating the dynamic equilibrium which occurs between the surfactant monomer and the micelle. To our knowledge, only a few articles report the use of chiral [5,51] and achiral polymerized [81] micelles and polyions [82] in separations by the use of CE.

Sodium-Undecylenyl-L-Valinate

In this part of our review, we report on our work involving CE separations by the use of polymerized micelles. Our new polymer micelle, poly(sodium *N*-undecylenyl-L-valinate), poly(L-SUV) has an additional advantage of containing a chiral functional group. This novel chiral polymer micelle was used in MEKC to separate (±)-binaphthol and D,L-laudanosine.

Poly(L-SUV) was synthesized and characterized by spectroscopic technique. In addition, the utility of poly(L-SUV) versus the nonpolymerized L-SUV were compared under similar conditions in MEKC for chiral separation.

The monomer acid, N-undecylenyl-L-valine was prepared from a procedure reported by Lapidot et al. [83]. The N-hydroxysuccinimide ester of undecylenic acid was reacted with L-valine. Undecylenal valine, L-UV was then converted to the sodium salt (L-SUV). Proton NMR showed the indicative multiplet and doublet at approximately 5.8 and 5.0 ppm, respectively, for the terminal vinyl group (Fig. 34a). The CMC of L-SUV was determined by surface tension measurements to be 2.1×10^{-2} M. The polymerization was achieved by ^{60}C γ-irradiation of a 0.05 M surfactant solution. After irradiation, the polymer solution was purified by lyophilization and treated with hot ethanol to extract any of the unreacted monomer [5]. Dialysis was also used as a purification method. A regenerated cellulose membrane with a 2000 Da molecular-weight cutoff was found to be better for purification than the ethanol rinse method. Again, proton NMR was used to follow the completion of polymerization. A noticeable change in the spectrum was the absence of the peaks at 5.8 and 5.0 ppm and the broadening of the remaining peaks (Fig. 34b). Optical rotation measurements ($[\alpha]^{25}_D$) were taken before and after polymerization. The measurements were $[\alpha]^{25}_D$ equal to $-2.9°$ and $-8.19°$ (c = 1.00 methanol and c = 1.00 water), respectively, showing that the chirality of the micelle had not been destroyed [5].

Figure 35 demonstrates the chiral recognition ability of the polymerized micelle. In Fig. 35a, the baseline separation of (±)-1,1'-bi-2-naphthol was achieved using a 25 mM sodium borate buffer (pH 9.0) containing 0.5% (w/v) poly(L-SUV). The (S)-(–)-1,1'-bi-2-naphthol enantiomer eluted faster than the corresponding (R) form, suggesting that the (R) form has a higher affinity with the (S) (L) form of the chiral polymer. Figures 35b and 35c show the electropherograms of 1,1'-bi-2-naphthol when the nonpolymerized surfactant was used at concentrations of 0.5% (w/v) and 1.0% (w/v), respectively. When 0.5% (w/v) of the monomer was used, no separation was achieved, as the concentration is just below the CMC. However, when 1% (w/v) was used, chiral separation (Fig. 35c) was achieved; however, it is noted that the polymerized micelle allowed for better discrimination, hence better chiral separation. This observation reinforces the advantages of polymerized micelles that were outlined earlier. Due to improved interaction of the solute with the polymerized micelle we see how the compactness of the polymerized micelle versus the normal micelle (Fig. 36) enhances separation. In the polymerized micelle, the solute does not penetrate as deeply into the micelle (Fig. 36b) as into the normal micelle (Fig. 36a); thus, interactions between the solute and the polymerized micelle must occur near the surface of the micelle. Consequently, an increase in mass-transfer rate of the solute with the polymerized micelle was noted. Also, an increase in theoretical plates (N) was observed. The N was 102,240 in Fig. 35a and 28,073 in Fig. 35c. A well-known limitation of MLC and MEKC is the peak-broadening associated with slow mass transfer of the solute between the micelle

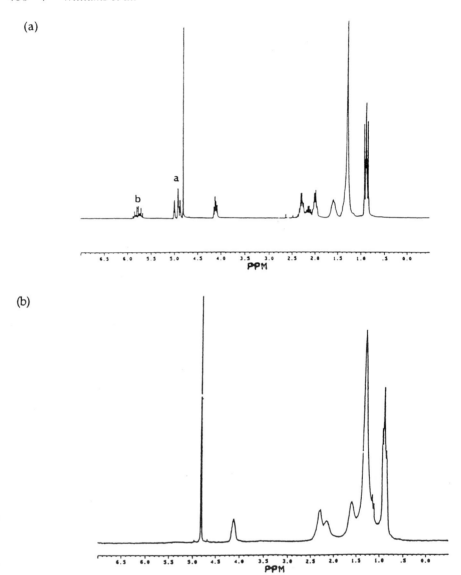

Fig. 34 Proton NMR of (a) L-SUV and (b) poly(L-SUV).

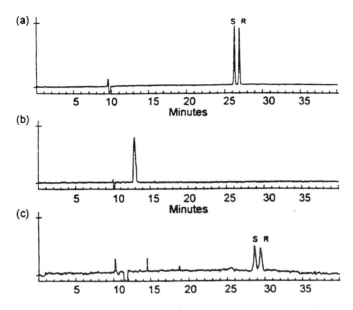

Fig. 35 Comparison between polymerized micelle and nonpolymerized micelle for separation of (+)-1,1′-bi-2-naphthol. (a) 0.5% poly(L-SUV); (b) 0.5% L-SUV; (c) 1% L-SUV. Buffer: 25 mM borate buffer (pH = 9.0); applied voltage 12 kV; current (a) 39 µA, (b) 40 µA, (c) 51 µA; UV detection at 290 nm. Reprinted with permission from I.M. Warner, *Anal. Chem.*, 66:3773 (1994), © 1994 American Chemical Society.

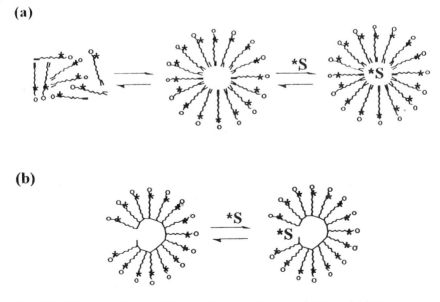

Fig. 36 Schematic diagram of the principle of surfactant, micelle, and solute interactions. (a) Normal (nonpolymerized) micelle. (b) Polymerized micelle. S denotes solute and asterisk the chiral center. Reprinted with permission from I.M. Warner, *Anal. Chem.*, 66:3773 (1994), © 1994 American Chemical Society.

and the bulk solvent [5]. Our observations are in accordance with the spectroscopic data reported earlier [84], in which the authors noted that a polymerized micelle exhibits a compact structure compared to a normal micelle. Consequently, the solute does not incorporate as deeply into the cavity of polymerized micelles as it does in the case of the normal micelles. This phenomenon results in an increased rate of mass transfer of solute in and out of the polymerized pseudophase.

After the initial successful separation, a study was conducted on the effect of poly(L-SUV) concentration. Figure 37 shows the effect of changing the concentration of the poly(L-SUV) in the chiral separation of (±)-1,1'-bi-2-naphthol. In this figure, the concentration of poly(L-SUV) varies from 0.02% to 0.5%. It was noted that the resolution of (±)-1,1'-bi-2-naphthol increased as the concentration of poly(L-SUV) was initially increased. However, at an optimum concentration (0.2–0.5% which is below the CMC), further increases in the concentration of the poly(L-SUV) did not improve the resolution and actually led to a slight decrease in resolution (Fig. 38) [5]. This observation was explained by a chiral separation model proposed by Wren and Rowe [85]. It is assumed that the complexes of individual enantiomers have the same electrophoretic mobility. However, if the two enantiomers have different binding constants with poly(L-SUV), then chiral resolution is achievable because the electrophoretic mobility of the free binaphthol would be different from the binaphthol interacting with the polymer.

The separation power of poly(L-SUV) was also used to resolve D,L-laudanosine. In this separation it was found that pH played an essential role in achieving enantiomer separation and resolution. Figure 39 shows a slight resolution at pH 9.0. However, when the pH was changed to pH 10.0 near baseline resolution (R_s = 1.2) was achieved. Chu and Thomas [86] reported that at lower pHs, negatively charged polymerized micelles have a compact conformation. In contrast, at higher pHs, the highly negatively charged polymerized micelle may have a looser conformation due to electrostatic repulsion. Our data show that at a higher pH, when the polymer conformation is looser, better interactions with the laudanosine enantiomers are achieved [5].

Combination of Polymerized Chiral Micelle with γ-Cyclodextrin

To demonstrate greater utility of our polymerized micelle we added an additional chiral selector to our polymerized micellar system. The motivation for this was to be able to separate a larger variety of enantiomers as well as mixtures of chiral drugs. We choose to use γ-cyclodextrin which was discussed earlier in CZE. The CDs have also been used in MEKC. In CD/MEKC, CDs and micelles are combined into the same buffer. This approach was initially used to separate highly hydrophobic compounds [87] but is now widely used in chiral separations [88–90]. However, when normal micelles are used, there is an inherent disadvan-

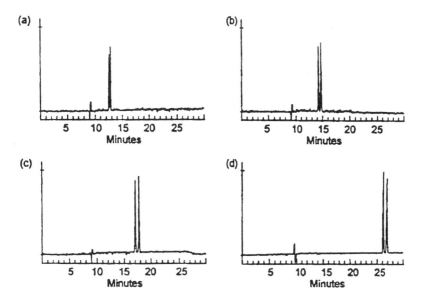

Fig. 37 Effects of poly(L-SUV) concentration on chiral separation of (±)-1,1′-bi-2-naphthol. (a) 0.02% poly(L-SUV); (b) 0.05% poly(L-SUV); (c) 0.1% poly(L-SUV); and (d) 0.5% poly(L-SUV). Buffer, 25 mM borate (pH 9.0); applied voltage 12 kV; detection at 290 nm. Reprinted with permission from I.M. Warner, *Anal. Chem., 66:*3773 (1994), © 1994 American Chemical Society.

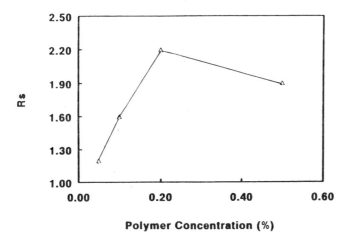

Fig. 38 Influence of poly(L-SUV) concentration on the resolution of (±)-1,1′-bi-2-naphthol. Reprinted with permission from I.M. Warner, *Anal. Chem., 66:*3773 (1994), © 1994 American Chemical Society.

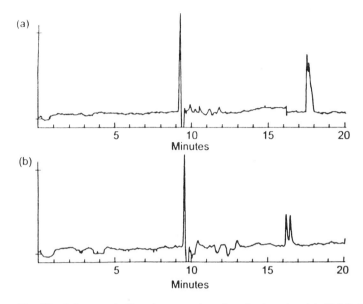

Fig. 39 Influence of pH on the separation of D,L-laudanosine. (a) pH 9.0; (b) pH 10.0 (adjusted by the addition of NaOH). Run buffer: 0.5% poly(L-SUV) in 25 mM borate buffer; UV detection at 254 nm. Other conditions same as Fig. 35. Reprinted with permission from I.M. Warner, *Anal. Chem.*, 66:3773 (1994), © 1994 American Chemical Society.

tage with the CD/MEKC approach for chiral separations. The surfactant monomers in the buffer are partly associated in a complex with the CDs [91,92]. Therefore, in many cases, complexation of surfactant monomers with CDs will interfere with the interactions between the individual enantiomers and the CDs. This interference will directly affect the enantioselectivity of the system and may reduce the observed chiral resolution [51]. Furthermore, monomeric surfactants with high CMC values give high conductivity to the running buffer. This can lead to joule heating, which will cause the separation to deteriorate. Because the polymerized micelle is covalently linked to the surfactant monomers, they can be used in CD/MEKC with elimination of the normal inclusion phenomena associated with surfactant monomers and CDs. Wang and Warner have reported the first chiral separation which employs a combination of a polymerized micelle and γ-CD [51]. Again, the polymer used in this study was poly(L-SUV) and its antipod poly(D-SUV).

Because the CD-modified polymerized chiral micellar system is similar to a normal CD/MEKC system, we can use normal CD/MEKC theory [32] with minor modifications. In our system, the polymerized micelle behaves as a pseudostationary phase and it is assumed that the neutral γ-CD is part of the aqueous phase. In addition, it is assumed that the enantiomers interact independently with the polymerized micelle and the γ-CD. Thus, the capacity factor, k', is defined by Eq. (4), where η_{mp} and η_{aq} are the moles of solute molecules associated with the polymerized micelle and aqueous phase (including CD), respectively. In addition, because the moles of solute in the aqueous phase include not only the moles of free solute, η_f, but also the solute molecules incorporated in CD, η_{CD}, we have

$$\eta_{aq} = \eta_f + \eta_{CD} \tag{15}$$

Based on a partitioning mechanism in this CD/chiral MEKC system, there are two important partitions for an enantiomer, A:

$$[A]_f \rightleftharpoons [A]_{mp}, \qquad K_{mp,A} = \frac{[A]_{mp}}{[A]_f} \tag{16}$$

$$[A]_f \rightleftharpoons [A]_{CD}, \qquad K_{CD,A} = \frac{[A]_{CD}}{[A]_f} \tag{17}$$

where $[A]_f$, $[A]_{mp}$, and $[A]_{CD}$ are the concentration of the enantiomer, A, in the aqueous phase, micelle polymer, and cyclodextrin, respectively. Thus, we obtain

$$\frac{\eta_{mp,A}}{\eta_{f,A}} = K_{mp,A} \frac{V_{mp}}{V_f} \tag{18}$$

$$\frac{\eta_{CD,A}}{\eta_{f,A}} = K_{CD,A} \frac{V_{CD}}{V_f} \tag{19}$$

where K_{mp} and K_{CD} are partition coefficients between the polymerized micelle and the aqueous phase, and between the CD and aqueous phase, respectively, and V_f, V_{mp}, and V_{CD} are the volumes of the aqueous phase, the micellar phase, and the cyclodextrin phase, respectively. Combining the above equations, we obtain the following expression for the capacity factor:

$$k' = \frac{K_{mp} V_{mp}}{V_f + V_{CD} K_{CD}} \tag{20}$$

and the selectivity for an enantiomeric pair can be defined as

$$\alpha = \frac{k'_{app,B}}{k'_{app,A}} \tag{21}$$

where $k'_{app,A}$ and $k'_{app,B}$ are apparent capacity factors for the enantiomeric pair A and B, respectively.

As selectivity (α) is directly related to resolution (R_s), Eqs. (20) and (21) can be combined to give the selectivity for the polymer CD/MEKC system. Therefore, α can be derived:

$$\alpha = \frac{1+\Phi_{CD}K_{CD,A}}{1+\Phi_{CD}K_{CD,B}} \frac{K_{mp,B}}{K_{mp,A}} \tag{22}$$

where Φ_{CD} is the phase ratio of the volume of CD (V_{CD}) to that of the aqueous phase (V_j). For a given plate number N and apparent capacity factor k', the greater the value of α, the higher the resolution. Selectivity must be ≥ 1; therefore, it can be shown that there are only three possible combinations of these parameters: (1) if $K_{CD,A} > K_{CD,B}$ and $K_{mp,B} > K_{mp,A}$, chiral resolution will be superior to that obtained using either CD of the polymerized chiral micelle alone; (2) if $K_{CD,A} > K_{CD,B}$ and $K_{mp,A} > K_{mp,B}$, chiral separation will be poorer than by using γ-CD alone; and (3) if $K_{CD,A} < K_{CD,B}$ and $K_{mp,B} > K_{mp,A}$, then resolution will also be poorer than by use of the polymerized chiral micelle alone. Equation (22) is validated in Fig. 40, where a combination of γ-CD and poly(D-SUV) were used. When using only γ-CD as a chiral selector, R-1,1'-bi-2-naphthol has a high affinity for γ-CD and will migrate faster than the (S) form (Fig. 40a). When using only poly(D-SUV) as a chiral selector, the S-1,1'-bi-2-naphthol interacts stronger with the polymer than the (R) form, and R-1,1'-bi-2-naphthol will migrate through the system faster than the (S) form (Fig. 40b). If γ-CD and poly(D-SUV) are combined, we find that $K_{CD,R} > K_{CD,S}$ and $K_{mp,S} > K_{mp,R}$. Therefore, chiral resolution is greater than by use of either chiral selector alone. This synergistic effect of γ-CD and poly(D-SUV) on the separation of R,S-1,1'-bi-2-naphthol was demonstrated in Fig. 40c. Figure 40d shows that chiral resolution was diminished when γ-CD and poly(L-SUV) were combined, which corresponds to the condition where $K_{CD,R} > K_{CD,S}$ and $K_{mp,R} > K_{mp,S}$.

After combining the chiral polymer and CD/MEKC system for the resolution of R,S-1,1'-bi-2-naphthol, a mixture of four chiral compounds were attempted and successfully separated. A combination of γ-CD and poly(D-SUV) enhanced chiral selectivity as well as extended the migration window, as the neutral CD travels with the electroosmotic flow. Figure 41 shows the separation of the four enantiomeric compounds, (±)-1,1'-binaphthol, (±)-verapamil, (±-BNPO4, and D,L-laudanosine. Using either γ-CD or poly(D-SUV) alone at the concentrations examined, a satisfactory resolution was not obtained (Figs. 41a and 41b). However, when both were used at the same concentrations, three enantiomeric pairs were resolved (Fig. 41c). All four compound were resolved with optimization of the system after closely examining the effect of γ-CD concentration, buffer concentration, and organic solvents (Fig. 41d).

A concentration range of 5–20 mM γ-CD was used to investigate the effect of γ-CD on the separation of the mixture. Resolution generally increased with increasing γ-CD concentration. Figure 42 shows the effect of γ-CD concentration on the four compounds up to a point where maximum separation of the enantiomers was achieved. We found that increases in γ-CD concentration above the optimum concentration resulted in a decrease in resolution. The optimum concentration is

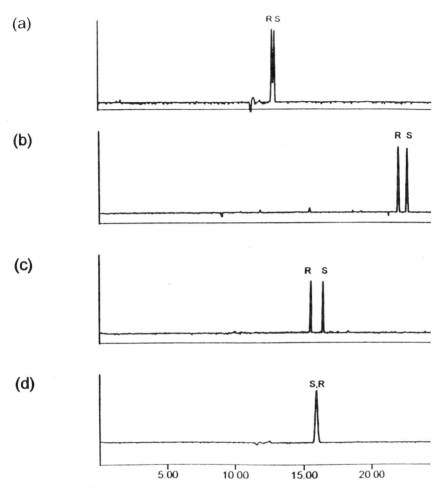

Fig. 40 Chiral separation of (±)-1,1' binaphthol by use of poly(D-SUV) and γ-CD. Experimental conditions: 25 m*M* borate buffer (pH 9), applied voltage 12 kV; UV detection 280 nm. (a) 10 m*M* γ-CD, (b) 0.5% poly(D-SUV), (c) 10 m*M* γ-CD and 0.5% poly(D-SUV), (d) 10 m*M* γ-CD and 0.5% poly(L-SUV). Reprinted from *J. Chromatogr., 711:*297, I.M. Warner, 1995, with permission from Elsevier Science.

dependent on the enantiomeric pair being separated. Qualitative estimates of optimum CD concentrations were 10 m*M* for (±)-BNPO4, 15 m*M* for (±)-1,1'-bi-2-naphthol, and 20 m*M* for (±)-verapamil. For D,L-laudanosine, it was found that an increase in γ-cyclodextrin concentration did not significantly enhance the resolution. This suggests that the interactions between D,L-laudanosine enantiomers and γ-CD are weak, therefore not aiding in the improvement of the resolution [51].

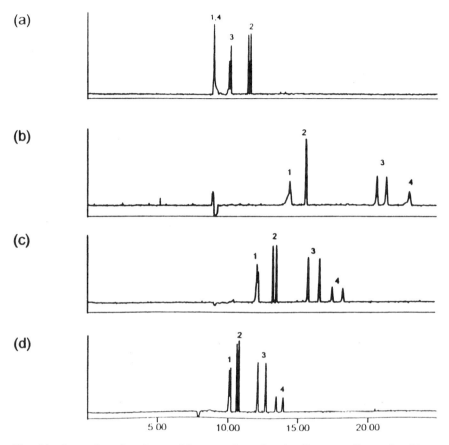

Fig. 41 Separation of a mixture of four enantiomeric pairs. Corresponding peaks: (1) D,L-laudanosine, (2) (±)-BNPO$_4$, (3) (±)-1,1'-binaphthol, (4) (±)-verapamil. Experimental conditions: (a) 10 mM γ-CD, (b) 0.5% poly(D-SUV), (c) 10 mM γ-CD and 0.5% poly(D-SUV); buffer for (a), (b), and (c) is 25 mM borate (pH 9.0), (d) 10 mM g-CD and 0.5% poly(D-SUV), 5 mM borate (pH 9.0); applied voltage 12 kV; UV detection 280 nm. Reprinted from *J. Chromatogr.*, 711:297, I.M. Warner, 1995, with permission from Elsevier Science.

Another factor that was studied and altered in order to optimize the resolution of the mixture was the buffer concentration. As expected, increasing the concentration of borate buffer will reduce the EOF and increase the viscosity of the electrolyte. This, in turn, causes the migration-time window to be expanded in the CD/MEKC system. However, this does not guarantee that an increase in resolution will occur. When we increased the borate buffer concentration from 5 mM to 45 mM, the migration-time window for each compound was lengthened. Again, the resolution for most of the enantiomeric pairs was increased except for

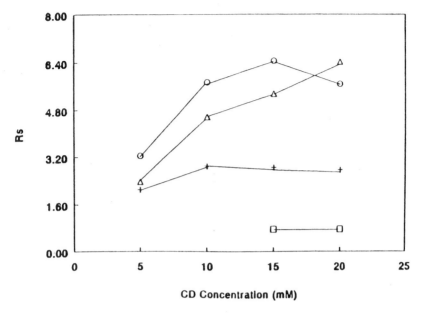

Fig. 42 Effect of γ-CD concentration on the resolution of the enantiomeric mixture. Experimental conditions: 0.5% poly(D-SUV), 25 mM borate (pH 9.0), 5 mM–20 mM γ-CD; applied voltage 12 kV; UV detection 280 nm. Circle: (±)-1, 1′-binaphthol; triangle: (±)-verapamil; +: (±)-BNPO$_4$, square: D,L-laudanosine. Reprinted from *J. Chromatogr.*, *711*:297, I.M. Warner, 1995, with permission from Elsevier Science.

D,L-laudanosine, where the resolution was decreased. At lower buffer concentrations, the charged chiral polymer is more flexible than at higher buffer concentrations. Therefore, the micelle can extend more at lower concentrations which may have induced a large difference in affinities between the individual D- and L-enantiomeric form, thus resulting in better resolution [51].

The effects of organic solvents on chiral resolution is dependent on the type of chiral selectors as well as the properties of the enantiomers. As discussed earlier, the migration time of the solute increases with an increase in organic solvent concentration. In normal MEKC or CD/MEKC, if the concentration of organic solvent is too high, the micelle will decompose into surfactant monomers inhibiting solute–micelle interaction. However, with polymeric micelles, a very high concentration (40% methanol) of organic solvent can be used. In our system, the enantiomeric resolution of the compounds studied was still very good, as shown in Fig. 43. However, D,L-laudanosine showed the same characteristics as with the other optimization techniques. In the presence of a small amount of methanol, separation of D,L-lauanosine was impaired. Moreover, it is interesting to note that in this study the other three enantiomeric compounds respond differ-

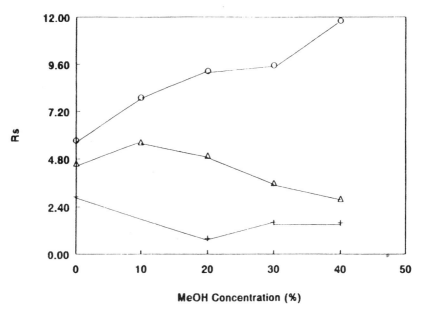

Fig. 43 Effect of methanol concentration on the resolution of the enantiomeric mixture. Experimental conditions: 0.5% poly(D-SUV), 25 mM borate (pH 9.0), 10 mM γ-CD, 0–40% MeOH; applied voltage 12 kV; UV detection 280 nm. Circle: (±)-1,1′-binaphthol; triangle: (±)-verapamil; (±)-BNPO$_4$. Reprinted from *J. Chromatogr.*, *711*:297, I.M. Warner, 1995, with permission from Elsevier Science.

ently to poly(D-SUV) in comparison to their response to γ-CD. For example, the enantiomeric net resolution of (±)-1,1′-bi-2-naphthol is improved by increasing methanol concentration. When methanol was added to the buffer solution containing only poly(D-SUV), the resolution increased as the methanol concentration increased. However, when methanol concentration was increased in the γ-CD buffer solution a slight decrease in resolution was noted. Enantiomeric resolution of (±)-verapamil increased with increasing methanol concentration up to 10%. Above 10% methanol, the resolution starts to decrease. In contrast, enantiomeric resolution of (±)-BNPO4 is decreased with increasing methanol concentration up to 10%, and further addition of methanol increases resolution. These observations show that different effects on the interactions between individual enantiomeric pairs with the chiral selectors poly(D-SUV) and γ-CD are solute dependent when organic solvents (methanol) are added to the system. Similar effects were also noticed when the organic solvent was changed from methanol to acetonitrile. As shown in Fig. 44, the resolution of all of the enantiomers is decreased with an increase in percentage of acetonitrile in the buffer.

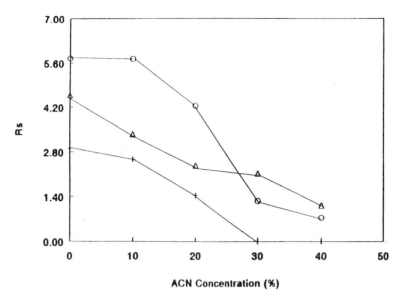

Fig. 44 Effect of acetonitrile (ACN) concentration on the resolution of the enantiomeric mixture. 0.5% poly(D-SUV), 25 mM borate (pH 9.0), 10 mM γ-CD, 0%–40% ACN; applied voltage, 12 kV; detection, 280 nm. Circle: (±)-1,1'-binaphthol, triangle: (±)-verapamil, (±)-BNPO$_4$. Reprinted from *J. Chromatogr., 711*:297, I.M. Warner, 1995, with permission from Elsevier Science.

This is expected, as it is known that acetonitrile can displace solutes from CD cavities [51].

The modified theory discussed above explains the observed synergistic effect on enantioselectivity by the use of two chiral selectors, poly(D-SUV) and γ-CD. Because the micelle is covalently bonded, there is no surfactant monomer interference on the enantioselectivity of the γ-CD. We have found that some effects on the resolution of the enantiomers are very obvious, whereas others are less so. It should be noted that when γ-CD concentration is increased, each pair of racemates reach an optimum resolution at a distinctive point. An increase in buffer concentration causes the resolution to increase except for D,L-laudanosine. Finally, the addition of organic solvents such as methanol is solute dependent. On the other hand, acetonitrile produced a decrease in resolution.

Recently, Dobashi et al. [94] used sodium undecenoyl-L-valinate which they refer to as SUVal in the separation of dinitrobenzoylated (DNB) amino acid isopropyl esters. Note that this is the same polymer micelle used in our research; however, Dobashi et al. used ultraviolet lamps for the polymerization process, whereas we used ^{60}Co γ-irradiation. Figure 45 shows enantiomers separation was achieved; however, it was reported that peak tailing occurred with drifting solute

418 / Williams et al.

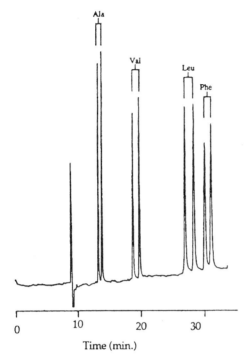

Fig. 45 Separation of a mixture of four (3,5-dinitrobenzoyl)amino acid isopropyl esters with poly(sodium 10-undecenoyl-L-valinate). Experimental conditions: fused silica capillary, 50 cm in length (50 μm i.d.); micellar solution, 0.025 M poly(L-SUV) in 0.025 M borate–0.05 M phosphate buffer (pH 7.0) containing 2 M urea and 0.01 M SDS; total applied voltage, ca. 12.4–12.8 kV; UV detection at 254 nm. Reprinted with permission from A. Dobashi, *Anal. Chem.*, 67:3011 (1995), © 1995 American Chemical Society.

retention. In order to optimize the separation, 2 M urea was added. The addition of urea gave reproducible solute retention but did not aid in the disappearance of peak tailing. Therefore, SDS had to be added to the polymeric micelle to reduce the peak tailing [94].

Dobashi et al. have recently attempted to use our micelle polymer to increase separation factors and/or resolution between the D,L-amino acids using a phosphate–borate buffer containing 2.0 M urea and at a pH above 9.0. It was suggested by Dobashi and co-workers that our separation of enantiomers was simply due to the pH used in our study. They found that our system was not effective for their use. One must remember that the conditions (voltage, concentrations, pH, additives, etc.) most often reported are those that give the best optimization of results for that particular solute. Despite this obvious fact, Dobashi et al. displayed skepticism [94] regarding our assertion that polymeric micelles enhance chiral separations through the elimination of the dynamic equilibrium that exists in such systems. However, from a simple discussion of the basic rules of thermodynamics, one must conclude that the dynamic equilibrium of micelles is detrimental to chromatographic separations, particularly chiral separations. Consider our earlier assertation that conventional (unpolymerized) micelles are polydispersed entities in solution. It is a well-established fact that micelles are in dynamic equilibria with species that are monomeric all the way up to an aggregation number well above the average aggregation number N of the micelle. It is also well established that such polydisperity will cause peak-broadening in chromatography and is therefore detrimental to separations that require the highest chromatographic efficiencies (e.g., chiral separations). We cannot understand how anyone can argue with this point. It is reasonable to conclude that dynamic equilibria and, therefore, polydispersity is detrimental to chromatographic separations. Therefore, we stand by our conclusion that elimination of dynamic equilibria is the primary factor which contributes to improved chiral separations by use of chiral micelle polymers.

V. CONCLUSION

We have attempted to give a conspectus of CZE as related to MEKC. The principles, techniques, and applications of chiral separations in MEKC with monomeric and polymeric chiral surfactants, collected from the literature and from the author's own laboratory, have been summarized in this chapter. The frontier to most recent research implementing CE and MEKC for separation of enantiomers has been briefly and generally compiled in order to provide the background for our novel use of chiral micelle polymers for the separation of chiral compounds. It is safe to say that there is not likely to be one cationic, anionic, or neutral micelle, or polymerized micelle that will separate all enantiomers. Therefore, it is best to find a system that works best for the resolution of particular racemates or a group of racemates.

Polymeric chiral surfactants offer an attractive and alternative venue for chiral separations in MEKC. This is because there continues to be a need for an improvement in efficiency and resolution of racemates. The polymer-based micelle poses advantages in chemical stability, structural rigidity, absence of CMC, lower joule heating, and tolerance to large concentrations of organic solvents. When viewed from these distinct advantages, chiral separations with polymeric micelles represent the beginning of a new and highly efficiency mode of chiral separations.

REFERENCES

1. M. Novotny, H. Soini, and M. Stefansson, *Anal. Chem.*, 66: 646A (1994).
2. J. McMurry, *Organic Chemistry*, 2nd ed., Brooks/Cole Publishing, Pacific Grove, CA, 1988, pp. 258–275.

3. F. Jamali, R. Mehvar, and F. M. Pasutto, *J. Pharm. Sci.*, *78*: 695 (1995).
4. B. Feibush, and N. Grinberg, *The History of Enantiomeric Resolution*, M. Zief and L. J. Crane, Eds., Chromatographic Science Series Vol. 40, Marcel Dekker, Inc. New York, 1988, pp. 1–14.
5. J. Wang and I. M. Warner, *Anal. Chem.*, *66*: 3773 (1994).
6. W. H. Pirkle and T. C. Pochapsky, *Chem. Rev.*, *89*: 347 (1989).
7. R. J. Baczuk, G. K. Landram, R. J. Dubois, and H. C. Dehn, *J. Chromatogr.*, *60*: 351 (1971).
8. S. F. Y. Li, *Capillary Electrophoresis Principles, Practice, and Applications*, Elsevier Science Publishers, Amsterdam, 1992, Vol. 52.
9. R. Kuhn and S. Hoffstetter-Kuhn, *Chromatographia*, *34*: 505 (1992).
10. J. Snopek, J. Jelinak, and E. Smolkova-Kenndler, *J. Chromatogr.*, *438*: 11 (1988).
11. S. Fanali, *J. Chromatogr.*, *545*: 437 (1991).
12. R. Kuhn, F. Stöcklin, and F. Erni, *Chromatographia*, *33*: 32 (1992).
13. A. Guttman, A. Paulus, S. Cohen, N. Grinberg, and B. I. Karger, *J. Chromatogr.*, *448*: 41 (1988).
14. M. Heuermann and G. Blaschke, *J. Chromatogr.*, *648*: 267 (1993).
15. B. A. Ingelse, F. M. Everaerts, J. Sevcik, and Z Stransky, *J. High Resolut. Chromatogr.*, *18*: 348 (1995).
16. Z. Aturki and S. Fanali, *J. Chromatogr*, *680*: 37 (1994).
17. H. Nishi, K. Nakamura, H. Nakai, and T. Sato, *J. Chromatogr.*, *678*: 333 (1994).
18. R. Kuhn, T. Berewer, F. Stöcklin, and F. Erin, *4th International Symposium on HPCE*, Amsterdam, 1992.
19. C. J. Pederson, *J. Am. Chem. Soc.*, *89*: 2495 (1967).
20. E. Höhne, G. J. Krauss, and G. Gübitz, *J. High Resolut. Chromatogr.*, *15*: 699 (1992).
21. R. Kuhn, C. Steinmetz, T. Bereuter, P. Haas, and F. Erni, *J. Chromatogr.*, *666*: 367 (1994).
22. D. W. Armstrong, K. Rundlett, and G. L. Reid, *Anal. Chem.*, *66*: 1690 (1994).
23. T. Tsuda, *J. High Resolut. Chromatogr.*, *10*: 622 (1987).
24. S. Terabe, K. Otsuka, and T. Ando, *Anal. Chem.*, *61*: 251 (1989).
25. K. Otsuka and S. Terabe, *J. Microcol. Sep.*, *1*: 150 (1989).
26. Y. Moroi, *Micelles, Theoretical and Applied Aspects*, Plenum Press, New York, 1992, Chap. 4.
27. H. J. Issaq, *Instrum. Sci. Tech.*, *22*: 119 (1994).
28. J. W. McBain, *Colloid Chemistry, Theoretical and Applied*, J. Alexander, Ed., Reinhold, New York, 1944.
29. P. Debye and E. W. Anacker, *J. Phys. Colloid. Chem.*, *55*: 644 (1951).
30. I. Reich, *J. Phys. Chem.*, *60*: 257 (1956).

31. F. M. Menger, *Angew. Chem. Int. Ed. Engl.*, *30*: 1086 (1991).
32. S. Terabe, K. Otsuka, and H. Nishi, *J. Chromatogr*, *666*: 295 (1994).
33. S. Terabe, *J. Pharm. Biomed. Anal.*, *10*: 705 (1992).
34. S. Terabe, T. Katsura, Y. Okada, Y. Ishihama, and K. Otsuka, *J. Microcol. Sep.*, *5*: 23 (1993).
35. K. Otsuka and S. Terabe, *J. Microcol. Sep.*, *1*: 150 (1989).
36. A. Dobashi, T. Ono, and S. Hara, *J. Chromatogr.*, *480*: 413 (1989).
37. K. Otsuka, J. Kawahara, and K. Tatekawa, *J. Chromatogr.*, *559*: 209 (1991).
38. K. Otsuka and S. Terabe, *Electrophoresis*, *11*: 982 (1990).
39. K. Otsuka and S. Terabe, *J. Chromatogr.*, *515*: 221 (1990).
40. A. Dobashi T. Ono, and S. Hara, *Anal. Chem.*, *61*: 1984 (1989).
41. A. R. Campaneli, S. Canderloro De Sanctic, E. Chressi, M. D'Alagni, E. Giglo, and L. Scaramuzza, *J. Phys. Chem.*, *93*: 1536 (1989).
42. M. J. Sepaniak, R. O. Cole, and W. L. Hinze, *J. High. Resolut. Chromatogr.*, *13*: 579 (1990).
43. S. Terabe, M. Shibata, and Y. Myashita, *J. Chromatogr.*, *480*: 403 (1989).
44. H. Nishi, T. Fukuyama, M. Matsuo, and S. Terabe, *J. Chromatogr.*, *515*: 233 (1990).
45. H. Nishi, T. Fukuyama, M. Matsuo, and S. Terabe, *J. Microcol. Sep.*, *1*: 234 (1989).
46. H. Nishi, T. Fukuyama, M. Matsuo, and S. Terabe, *Chim. Acta*, *236*: 281 (1990).
47. K. Otsuka and S. Terabe, *J. Chromatogr.*, *515*: 221 (1990).
48. K. Otsuka, M. Kahihara, Y. Kawaguchi, R. Koike, T. Hisamitsu, and S. Terabe, *J. Chromatogr.*, *652*: 253 (1993).
49. Y. Ishihama and S. Terabe, *J. Liq. Chromatogr.*, *16*: 933 (1993).
50. D. C. Tickle, G. N. Okafo, P. Camilleri, R. F. D. Jones, and A. J. Kirby, *Anal. Chem.*, *66*: 4121 (1994).
51. J. Wang and I. M. Warner, *J. Chromatogr.*, *711*: 297 (1995).
52. E. A. G. Aniansson, S. N. Wall, M. Almgren, H. Hoffman, I. Kielmann, W. Ulbricht, R. Zana, J. Lang, and C. Tondre, *J. Phys. Chem.*, *80*: 5 (1976).
53. E. A. G. Aniansson and S. N. Wall, *J. Phys. Chem.*, *78*: 1024 (1974).
54. S. N. Wall and E. A. G. Aniansson, *J. Phys. Chem.*, *84*: 727 (1980).
55. K. Shinodal, *Colloidal Surfactants*, Academic Press, New York, 1963, Chap. 1.
56. K. Nagai and H-G. Elias, *Makromol. Chem.*, *188*: 1095 (1987).
57. E. R. Jones and C. R. Bury, *Philos. Mag.*, *4*: 841 (1927).
58. U. Kammer and H-G. Elias, *Kolloids. Z. Z. Polym.*, *250*: 344 (1972).
59. S. Hamid and D. Sherrington, *J. Chem. Soc. Chem. Commun.*, 936 (1986).
60. G. Birrenbach and P. Speiser, *J. Pharm. Sci.*, *65*: 1763 (1976).
61. H. Inaba and S. Kudo, *J. Polym. Chem.*, *23*: 1221 (1985).

62. C. M. Paleos and A. Malliaris, *Macromol. Chem. Phys.*, *C28*: 403 (1988).
63. L. R. Fisher and D. G. Oakenfull, *Chem. Soc. Rev.*, *6*: 25 (1977).
64. H. Hoffmann *Prog. Colloid. Polym. Sci.*, *65*: 140 (1978).
65. H-G. Elias, *Makromoleküle*, Hüthig and Wepf, Basel; 2nd ed., Plenum Press, New York, 1984, pp. 573, 585, 6856, and 699.
66. V. A. Kabanov, *Pure Appl. Chem.*, *15*: 391 (1967).
67. V. A. Kabanov, T. I. Patrikeeva, O. V. Kargina, and V. A. Kargina, *J. Polym. Sci. C*, *23*: 357 (1968).
68. J. C. Salamone, M. V. Mahmud, A. C. Watterson, and P. P. Olson, *J. Polym. Sci. Polym. Chem. Ed.*, *20*: 1153 (1982).
69. V. A. Kabanov, K. V. Aliev, O. V. Karigina, T. I. Patrikeeva, and V. A. Kargin, *J. Polym. Sci. C*, *16*: 1079 (1967).
70. J. C. Salamone, E. J. Ellis, C. R. Wilson, and D. F. Bardoliwala, *Macromolecules*, *6*: 475 (1973).
71. I. Mielke and H. Ringsdorf, *Makromol. Chem.*, *142*: 319 (1971).
72. I. Mielke and H. Ringsdorf, *Makromol. Chem.*, *153*: 307 (1972).
73. V. Martin, W. Sutter, and H. Ringsdorf, *Makromol. Chem.*, *176*: 2029 (1975).
74. V. Martin and H. Ringsdorf, *Makromol. Chem.*, *177*: 89 (1976).
75. C. Tanford, *J. Phys. Chem.*, *76*: 3020 (1972).
76. E. D. Sprague, D. C. Duecher, and C. E. Larrabee, *J. Colloid. Interf. Sci.*, *92*: 416 (1983).
77. C. M. Paleos, C. I. Stassinopoulou, and A. J. Malliaris, *J. Phys. Chem.*, *87*: 251 (1983).
78. C. E. Larrabee, Ph.D dissertation, University of Cincinnati, 1980.
79. R. Thundathil, J. O. Stoffer, and S. E. Friberg, *J. Polym. Sci. Polym. Chem. Ed.*, *18*: 2629 (1980).
80. J. H. Fendle and P. Tundo, *Acc. Chem. Res.*, *17*: 3 (1984).
81. C. P. Palmer, M. Y. Khaled, and H. M. McNair, *J. High Resolut. Chromatogr.*, *15*: 756 (1992).
82. S. Terabe and T. Isemura, *Anal. Chem.*, *62*: 650 (1990).
83. Y. Lapidot, S. Rappoport, and Y. Wolman, *J. Lipid Res.*, *8*: 142 (1967).
84. C. M. Paleos, C. I. Stasslnopoulou, and A. Mallaris, *J. Phys. Chem.*, *87*: 251 (1983).
85. S. A. C. Wren and R. C. Rowe, *J. Chromatogr.*, *603*: 235 (1992).
86. D. Y. Chu and T. K. Thomas, *Macromolecules*, *24*: 2212 (1991).
87. S. Terabe, Y. Miyashita, O. Shibata, E. R. Barnhart, L. R. Alexander, D. J. Patterson, B. I. Karger, K. Hosoya, and N. Tanaka, *J. Chromatogr.*, *516*: 23 (1990).
88. H. Nishi, T. Fukuyama, and S. Terabe, *J. Chromatogr.*, *553*: 503 (1991).
89. T. Ueda, F. Kitamura, R. Mitchell, T. Metcalf, T. Kuwana, and A. Nakamoto, *Anal. Chem.*, *63*: 2979 (1991).

90. H. Nishi, Y. Kokusenya, T. Miyamoto, and T. Sato, *J. Chromatogr.*, *659*: 449 (1994).
91. T. Okubo, H. Kitana, and N. Ise, *J. Phys. Chem.*, *80*: 2661 (1976).
92. V. K. Smith, T. T. Ndou, A. M. LaRena, and I. M. Warner, *J. Incl. Phenom. Mol. Recog.*, *10*: 471 (1991).
93. S. Terabe, K. Otsuka, and H. Nishi, *J. Chromatogr.*, *666*: 295 (1994).
94. A. Dobashi, M. Hamada, Y. Dabashi, and J. Yamaguchi, *Anal. Chem.*, *67*: 3011 (1995).

8
Analysis of Derivatized Peptides Using High-Performance Liquid Chromatography and Capillary Electrophoresis

Kathryn M. De Antonis and Phyllis R. Brown *University of Rhode Island, Kingston, Rhode Island*

I.	INTRODUCTION	426
II.	DERIVATIZATION CHEMISTRIES	426
	A. O-Phthalaldehyde	427
	B. Phenylisothiocyanate	428
	C. 9-Fluoroenylmethyl Chloroformate	429
	D. 5-N-N-Dimethyl Aminonaphthalene-1-Sulfonyl Chloride and 4-Dimethylaminoazobenzene-4-Sulfonyl Chloride	429
	E. Ninhydrin	430
	F. Fluorescamine	430
	G. 6-Aminoquinolyl-N-Hydroxysuccinimidyl Carbamate	431
	H. 3-(4-Carboxybenzoyl)-2-Quinolinecarboxaldehyde	432
	I. Naphthalene-2,3-Dicarboxaldehyde	433
III.	ANALYSIS OF DERIVATIZED PEPTIDES	433
	A. Physiological Samples	435
	B. Peptide Mapping	439
	C. Pharmaceutical Samples	440
	D. Chiral Samples	442
	E. Food Samples	444
	F. Model Synthetic Peptides	446
IV.	CONCLUSION	448
	REFERENCES	449

I. INTRODUCTION

Peptides play a multitude of roles in physiology, pharmaceutics, and foods. They can occur naturally or be synthetically produced. In humans, peptides function as neurotransmitters, growth promoters, coenzymes, and hormones. Microorganisms produce some peptides which have useful antibiotic activity, and plant peptides are used in the food industry as sweeteners, bulking agents, and flavor enhancers. Synthetic peptides are important in the biotechnology industry, and many new pharmaceuticals are peptides. In addition, mixtures of peptide fragments are produced from the enzymatic and chemical cleavage of proteins. Thus, analytical technology is needed to analyze quantitatively peptides which exist in a wide variety of matrices and in a large range of concentrations.

High-performance liquid chromatography (HPLC) and capillary electrophoresis (CE) are the techniques most commonly used for the separation of peptides. They are complementary, each with unique advantages and disadvantages [1]. Because of the importance of determining extremely low concentrations of peptides in a variety of matrices, derivatization now plays a major role in the total HPLC or CE analysis of peptides. Peptides are labeled, mainly on the N-terminal amine group, with absorbing, fluorescing, electroactive, or chemiluminescent tags.

II. DERIVATIZATION CHEMISTRIES

There are a number of essential requirements for successful peptide derivatization. Ideally, the derivatizing agent should produce a single, stable derivative for each target analyte. The derivatization reaction should be rapid and simple, and quantitative yields should be obtained easily. The derivatized peptide should generate a strong signal especially for ultratrace analyses, and the reaction by-products should not complicate the separation and analysis.

There have been many amine derivatizing agents described in the literature. These agents are used to label a variety of amines in addition to peptides (e.g., amino acids, catecholamines, amino sugars, and amine-containing drug compounds).

Since the middle 1950s, biochemists and other scientists have relied heavily on some of the older derivatizing agents. Reagents such as O-phthalaldehyde (OPA), phenylisothiocyanate (PITC), 9-fluorenylmethyl chlorofomate (FMOC), 5-N,N-dimethyl aminonaphthalene-1-sulfonyl chloride (dansyl), 4-dimethylaminoazobenzene-4-sulfonyl chloride (dabs), ninhydrin, and fluorescamine are reliable and have been reported repeatedly in the literature. Many of these older reagents have been marketed as commercial products by major vendors of chromatographic instrumentation. Although they are less familiar to the scientific community, some of the newer reagents have unique advantages. These reagents are 6-aminoquinolyl-N-hydroxysuccinimidyl carbamate (AQC), 3-(4-carboxybenzoyl)-2-quinolinecarboxaldehyde (CBQCA), and naphthalene-2,3-dicarbox-

Table 1 Summary of the Peptide Derivatizing Agents

	Detection mode[a]	Sens.	Deriv. stability	Reaction kinetics	1°/2° amines	Single peak/pep[b]
OPA	F, A, E	fmol	Poor	Rapid	1°	Yes
PITC	A	pmol	Good	Moderate	1°/2°	Yes
FMOC	F, A	fmol	Excellent	Rapid	1°/2°	No
DANSYL	F, A	pmol	Good	Slow	1°/2°	No
DABS	A	pmol	Excellent	Slow	1°/2°	Yes
NIN	A	pmol	N.A.[c]	Rapid	1°/2°	N.A.
FC	F, A	fmol	Poor	Rapid	1°/2°[d]	Yes
AQC	F, A	fmol	Excellent	Rapid	1°/2°	Yes
CBQCA	F, A	amol	Good	Slow	1°	No
NDA	F, A, E, C	amol	Excellent	Slow	1°	Yes

[a]F = Fluorescence; A = absorbance; E = electrochemical; C = chemiluminescence.
[b]Yes = derivatizing agent reaction forms single derivatives for each target analyte; no = multiple peaks are formed for some analytes.
[c]N.A. = not applicable.
[d]2° amines do not form fluorescent derivatives.
Source: Adapted from Ref. 2.

aldehyde (NDA). See Table 1 for a brief summary of some of the characteristics of each reagent [2].

The focus of this chapter is to discuss the major amine derivatizing agents that have been applied to peptide analysis. Other reagents have been used to label the functional groups of specific amino acid side chains. Benzoin has been used to derivatize the guanidinyl group of arginine [3], and the side chain of tyrosine has been labeled by formylation followed by reaction with 4-methoxy-1,2-phenylenediamine [3]. Reagents such as these are not useful for universal peptide detection, however, so further discussion is not included here.

A. O-Phthalaldehyde

O-phthalaldehyde (OPA) was first described for amine derivatization in 1971. It is one of the oldest and at present the most popular derivatizing agent [4].

The derivatization reaction between OPA and primary amines takes place under alkaline conditions (pH 9–11) in the presence of a thiol such as 2-mercaptoethanol, 3-mercapto-1-propanol, or ethanethiol [5]. The reaction is complete within 1 min. Because the chemistry is so rapid, OPA may be employed for either precolumn or postcolumn derivatization. The resulting derivatized peptides are detected using ultraviolet (UV) absorbance (330 nm), fluorescence (excitation at 330 nm and emission at 430 nm), or electrochemical detection [5–7]. OPA does not react with secondary amines and, therefore, will not react with N-terminal proline, a secondary amino acid found in peptides and proteins. These peptides will not be detected unless they contain an internal cysteine, lysine, or hydroxylysine whose side chains are reactive.

The precolumn derivatization procedure with OPA is rapid and produces quantitative yields of strongly fluorescent derivatives. There are four major disadvantages to this reagent. First, peptides with cysteine, lysine, and hydroxylysine at the N-terminus produce weak fluorescence due to the quenching caused by the intramolecular interaction of the two fluorescent labels. However, when lysine occupies a position within the peptide, the ε amine group forms a derivative with 50 times the fluorescence intensity as a peptide without an internal lysine [8]. Second, some peptides will go undetected because the N-terminal peptide group is a secondary amine. Third, the derivatized products are relatively unstable. The accuracy of each analysis is directly affected when the elapsed time prior to injection is not strictly controlled. The use of automated equipment for precolumn derivatization and analysis reduces the variability in elapsed time and can improve accuracy. Auto•Tag OPA™ (Waters Corp., Milford, MA) is a commercially available system which uses OPA derivatization for the analysis of primary amines only, and AminoQuant™ (Hewlett-Packard Co., Avondale, PA) is another product which utilizes OPA for the derivatization of primary amines and FMOC for the derivatization of secondary amines. Fourth, and most importantly for peptide analysis, derivatives of peptides exhibit diminished fluorescence when compared with amino acids. The resulting detection limits for peptides are higher for peptides derivatized with OPA than with other fluorescent derivatizing agents [9].

B. Phenylisothiocyanate

Phenylisothiocyanate (PITC) is a frequently used amine derivatizing agent which was originally described over 30 years ago as a reagent for peptide sequencing in the Edman degradation, and currently it is often used for that purpose. PITC was first reported for amine derivatization in 1982 by Koop et al. [10], and in 1984, Waters Corp. (Milford, MA) marketed a commercial product (Pico•Tag™) based on the work of Heinrikson and Meredith at the University of Chicago [11].

Phenylisothiocyanate reacts with both primary and secondary amines under alkaline conditions to form phenylithiocarbamyl (PTC) derivatives. The reaction is complete in less than 10 min, and the resulting PTC derivatives can be detected by UV absorbance (λ_{max} = 269 nm, but the signal is generally measured at 254 nm for convenience). In Edman sequencing, the N-terminal PTC amino acid is cyclized to the phenylthiohydantoin species (PTH) and cleaved from the remainder of the peptide when exposed to acid in organic solvent.

In addition to the N-terminal amine groups, PITC reacts quantitatively with the side chains of cysteine and lysine, and all derivatives are relatively stable. There are two major drawbacks. First, because PITC is only a UV-absorbing reagent, sensitivity is on the order of picomoles, which is not as good as the sensitivity that can be achieved with the fluorescent reagents. Second, the highly reactive reagent must be removed prior to analysis. The additional steps taken to remove the reagent make the sample preparation more tedious and introduce analytical error. However, once the reagent is removed, only small synthetic impurities are present, which do not interfere with the analysis.

C. 9-Fluoroenylmethyl Chloroformate

Like PITC, FMOC was not used initially for analysis of amines. It was used originally, and is still used, as a protecting group in peptide synthesis. In 1979, FMOC was described for the derivatization and analysis of amines by Moye and Boning [12].

Under basic conditions, FMOC–Cl reacts with both primary and secondary amines to form highly fluorescent, stable derivatives. When done properly, the amine derivatization is complete within 30 s. Excess reagent reacts with water to form FMOC–OH, which is also fluorescent and will cause interference in the chromatogram because it is eluted as a large broad band. Removal is accomplished by extraction with an organic solvent such as pentane or diethyl ether. The resulting FMOC derivatives can be detected using absorbance (254 nm) or fluorescence (the excitation and emission wavelengths are 260 and 310 nm, respectively).

The FMOC-derivatized amino acids show a high degree of stability for up to 13 days [13]. Only diderivatized histidine showed significant losses. A half-life for this derivatized amino acid was 5–6 days. In addition to the potential interferences from the hydrolysis product, the major disadvantage for this reagent is the presence of multiple peaks for this diderivatized amino acid. As was described Section II.A, AminoQuant™ (Hewlett-Packard Co., Avondale, PA) is a commercially available automated product which utilizes OPA for the derivatization of primary amines and FMOC for the derivatization of secondary amines.

D. 5-*N*-*N*-Dimethyl Aminonaphthalene-1-Sulfonyl Chloride and 4-Dimethylaminoazobenzene-4-Sulfonyl Chloride

Both dansyl chloride and dabs chloride have been used for amine derivatization since the 1970s. Because they are analogs, they have several similarities, but dabs has more advantages and has been marketed by Beckman Instruments, Inc. (Bucks, U.K.) as a commercial product called System Gold/Dabsylation Kit.

Dansyl chloride reacts with both primary and secondary amines under basic conditions (pH 9–10) to form derivatives that can be detected by either UV (λ_{max} = 250 nm) or fluorescence (excitation 360 nm and emission 470 nm). The reaction is fairly slow, 30–35 min to completion. In addition to the N-terminal amino acids, dansyl chloride also reacts with the side chains of histidine, lysine, ornithine, cystine, and tyrosine. The dansyl derivatives are stable for up to 7 days if kept frozen at –4°C.

Derivatization with dansyl chloride has a number of disadvantages. First, the derivatized compounds are light sensitive. Second, incomplete derivatization can be a problem. In addition, multiple peaks can be formed from histidine derivatization. Finally, by-products such as dansyl-OH and dansyl-NH_2 are formed during derivatization. These by-products will interfere with the analysis.

Like dansyl chloride, dabs chloride reacts slowly with primary and secondary amines. These derivatives are unusual in that they are detected in the visible region (λ_{max} = 436 nm). The sensitivity of the analysis is in the picomole range. Unlike dansyl derivatives, dabs derivatives are highly stable in the light for up to 1 month. The major drawback of this reagent is that the presence of salts, urea, sodium dodecyl sulfate, phosphate, or ammonium bicarbonate can alter the pH and interfere with the derivatization. In the case of high salt content, precautions must be made to ensure a high degree of buffering capacity.

E. Ninhydrin

Ninhydrin is one of the most widely used amine derivatizing agents. The original automated amino acid analyzer, which was described by Spackman et al. in 1958, employed ninhydrin derivatization [14]. Because all amines produce identical derivatives with ninhydrin, this reagent is exclusively a postcolumn derivatizing agent.

Ninhydrin reacts with both primary and secondary amines. In a multistep reaction mechanism, ninhydrin reacts with primary amines to form Ruhemann's Purple. As is indicated by its name, the product is detected in the visible region at 570 nm. Secondary amines react with ninhydrin to form a yellow product which can be detected at 440 nm. Detection at both wavelengths simultaneously provides structural information about compounds being analyzed. In the postcolumn system, both the chromatographic eluent and the ninhydrin solution are pumped into a heated mixing coil (130–135°C) where the reaction occurs. The reaction is complete before the resulting mixture reaches the detector.

Although it is commonly used, derivatization with ninhydrin has some distinct disadvantages. This reagent is sensitive to light, atmospheric oxygen, and changes in pH and temperature. In addition, the detection limit is in the middle picomole region, which is the highest among detection limits of derivatized amines discussed in this chapter. As a result, ninhydrin is useless for the more sensitive analyses required at present.

F. Fluorescamine

Fluorescamine was first described as an amine derivatizing agent in 1972 [15]; however, it is not commonly used because it is expensive and the derivatives are relatively unstable. Nonetheless, the reagent produces highly fluorescent compounds, and detection limits are on the order of 30–50 fmol for derivatized peptides [16].

Fluorescamine reacts rapidly at alkaline pH (pH~9) with primary amines to form highly fluorescent products. The excitation and emission wavelengths are 390 and 475 nm, respectively. The reaction is complete and the reagent com-

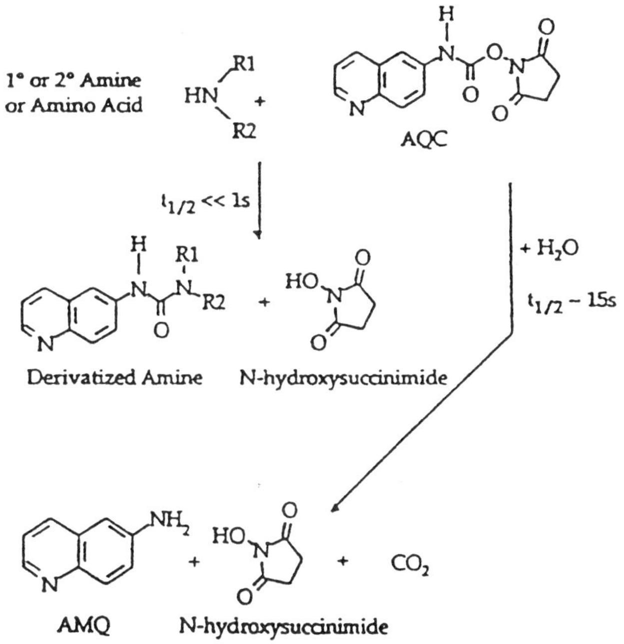

Fig. 1 AQC derivatization chemistry. (From Ref. 19.)

pletely hydrolyzed within 1 min, making this reagent amenable to both precolumn and postcolumn derivatization. Fluorescamine also reacts with secondary amines, but the products are not fluorescent. Both the reaction by-products and the reagent itself are nonfluorescent, so there is no chromatographic or electrophoretic interference from these compounds.

G. 6-Aminoquinolyl-*N*-Hydroxysuccinimidyl Carbamate

First introduced in 1993 by Cohen and Michaud for use in amino acid analysis, AQC is a new amine derivatizing agent [17]. It has since been reported for the analysis of peptides by both HPLC and CE [18,19].

6-Aminoquinolyl-*N*-hydroxysuccinimidyl carbamate reacts quickly and quantitatively with primary and secondary amines under basic conditions to form highly stable, fluorescent derivatives. The derivatization chemistry is shown in Fig. 1. The derivatized products can be detected using either absorbance (254 nm) or fluorescence (excitation wavelength 250 nm, emission wavelength 395 nm). In addition to the amino terminus, AQC also reacts quantitatively with the side chains of lysine residues. Excess reagent reacts with water to form aminoquinoline (AMQ) which has a different fluorescence maximum than the derivatized amine compounds, so it does not need to be removed prior to analysis. All reagent is completely consumed within 1 min.

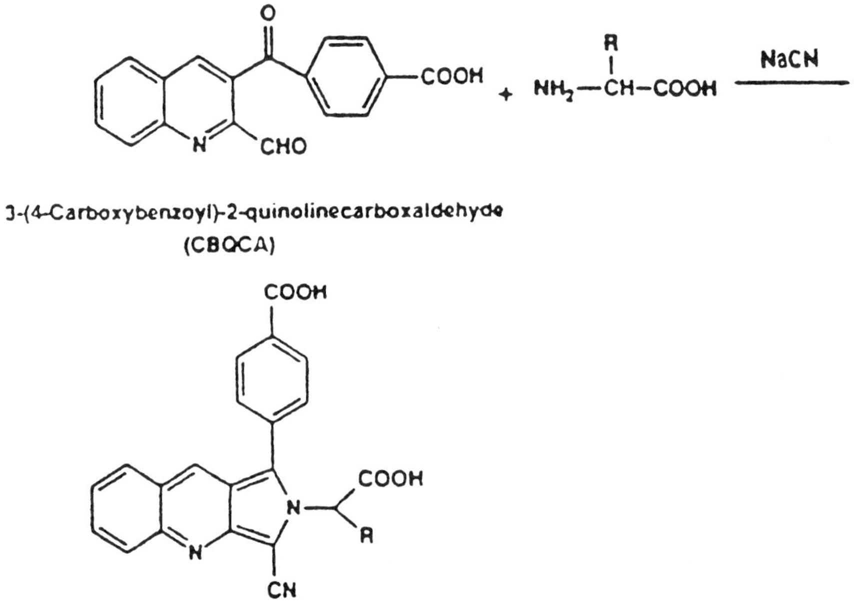

Fig. 2 CBQCA derivatization chemistry. (From Ref. 21.)

A recent comparison between AQC and PITC for amino acid analysis was reported which showed a clear superiority of the AQC methodology in the areas of the derivative stability and low-level reproducibility [20].

H. 3-(4-Carboxybenzoyl)-2-Quinolinecarboxaldehyde

3-(4-carboxybenzoyl)-2-quinolinecarboxaldehyde (CBQCA) was first described for the derivatization of amino acids and peptides by Liu et al. in 1991 [21]. This reagent was structurally designed to be highly fluorescent for analysis of trace levels of biological compounds. Sensitivities on the attomole level have been reported for the CE separation of peptides coupled with laser-induced fluorescence detection [21].

This reagent reacts with primary amines only to form isoindole derivatives. The derivatization chemistry is shown in Fig. 2. The derivatization chemistry is fairly slow, requiring 1 h for completion. Highly sensitive laser-induced fluorescence detection is achieved with an excitation wavelength of 442 nm, which corresponds to the blue line of a helium–cadmium laser, and an emission wavelength of 550 nm.

The derivatives are fairly stable. In solution, derivatives will last for over 24 h, and for over 2 weeks if they are dried and frozen [21]. A major drawback to

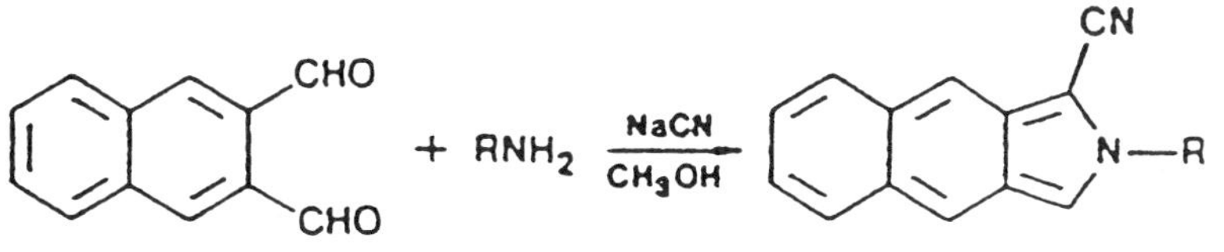

Fig. 3 NDA derivatization chemistry. (From Ref. 22.)

this reagent arises from the fact that peptides with lysine residues form multiple products.

I. Naphthalene-2,3-Dicarboxaldehyde

An analog of OPA, NDA is a relatively new amine derivatizing reagent which was first reported in 1986 by Carlson et al. for the derivatization of amino acids and peptides [22]. It has been particularly useful for the analysis of trace levels of physiological peptides [23,24].

Naphthalene-2,3-dicarboxaldehyde reacts with primary amines in the presence of cyanide ion to yield the highly stable 1-cyano-2-substituted-benz[f]isoindole (CBI) derivatives. Amino acids are derivatized at pH 9.5, but peptides are derivatized at pH 7.0. The derivatization chemistry is shown in Fig. 3. The reaction is complete within 15 min. The highly sensitive CBI products can be detected by measuring UV absorbance (λ_{max} at 250, 420, and 440 nm), fluorescence (excitation at 420 nm and emission at 490 nm), electroactivity, or chemiluminesce. The fluorescence excitation maximum of 420 nm has the advantage that laser-induced fluorescence can be used to enhance sensitivity.

This derivatizing agent has three major disadvantages. First, the multiply derivatized amino acids, lysine and ornithine, exhibit fluorescence quenching due to intramolecular interactions of the fluorescent tags [24]. Second, the peptide derivatization is sensitive to pH. At high pH, an undesirable side reaction takes place in which a cyclic imadazol-3-one, a nonfluorescent compound, is produced, causing a loss in analytical signal [23]. Third, a large excess of reagent can lead to the production of other undesirable by-products [24].

An automated system has been described in which both the derivatization and HPLC analysis are integrated [25]. Automation improves the analytical throughput and is particularly useful for evaluation of physiological samples in which many analyses must be done.

III. ANALYSIS OF DERIVATIZED PEPTIDES

The two principal techniques for peptide analysis today are HPLC and CE. These are complementary techniques because resolution is achieved via orthogonal driving forces. In HPLC, derivatized peptides are most often separated using the reversed-phase mode in which compounds are resolved by hydrophobic interaction between the peptide/derivatization group pair and the stationary phase. In CE, derivatized peptides are most often separated using free-zone electrophoresis or micellar electrokinetic capillary chromatography (MECC). The basic driving force for resolution in these modes is electrophoretic mobility, which is a function of molecular charge and, to a lesser degree, size.

At present, HPLC has a number of advantages over CE. It can be used with gradient elution, but CE cannot. HPLC can be used in a range of scales such as micro, analytical, and preparative scales. CE can be used in only the micro scale. Fraction collection is simpler in HPLC than CE. In addition, a larger number of detectors are commercially available for HPLC, such as UV, fluorescence, electrochemical, mass spectrometry, and others. In CE, the UV detector and, to a lesser degree, fluorescence are the principle modes of detection that are commercially available. However, homemade laser-induced fluorescence detectors [21,26] and mass spectrometers with homemade interfaces have been described in the literature [27–29]. Furthermore, HPLC can be used with multiple detectors in series. HPLC has the advantage that instrumentation is modular, and hardware can be moved from system to system. There is a tremendous need for modular instrumentation in CE as well.

Capillary electrophoresis also has some important advantages over HPLC. Relative to HPLC, CE has higher separation efficiencies, particularly for large compounds, due to the fact that the only source of band-broadening is molecular diffusion. CE also has better mass sensitivity and shorter analysis times than HPLC. Instrumentation is simpler and less expensive to purchase as well as to operate because the consumption of expensive solvents and reagents is minimized, and expensive stationary phases are unnecessary. CE analyses require less sample volume, and sample preparation is significantly reduced, which is particularly useful for complex biological samples [30]. In addition, sensitivity can be improved for CE with the use of sample stacking or stationary-phase packing at the capillary inlet to preconcentrate the sample. A potential advantage for the future is the use of capillary bundles which has been reported for the simultaneous analysis of multiple samples [31].

Because derivatized peptide samples exhibit a wide variety of physicochemical properties, they can be analyzed using one or both techniques. For most applications, one technique is usually sufficient. However, the use of both complementary techniques is beneficial for the analysis of complex samples. For example, both HPLC and CE were used to analyze a tryptic digest of the protein hGH [32]. Good resolution was achieved with each technique. As there was no correlation between the retention and migration times, each technique provided unique information about the peptide, and neither analysis was more valuable than the other. In addition, CE can be used as a quick screening tool for sample purity subsequent to HPLC fraction collection because (1) it operates under a different separation mechanism, (2) it requires as little as 5 nL of solution, (3) it is fast, and (4) it is quantitative. For example, a tryptic digest of β-lactoglobulin A was analyzed using HPLC. Several fractions were collected which appeared to be pure by HPLC but appeared as several peaks by CE. Sequence analysis confirmed the CE result [32].

For analyses of complex samples, multidimensional separations have advantages [33]. Two-dimensional systems such as two coupled orthogonal separation techniques or coupled columns in HPLC have been described. For example, two columns have been used in the analysis of derivatized peptides in biological matrices [26,34]. Column switching was employed to facilitate sample cleanup. The first column was used to separate the peaks of interest from the interferences, and the second column was used to resolve the peaks of interest from each other. In other applications, the automated coupling of HPLC and CE has been described. Yamamoto et al. separated protein mixtures using gel-permeation chromatography (GPC) to separate the proteins first by size, then CE to separate them by charge [35]. Bushey and Jorgenson coupled HPLC and CE for the analysis of derivatized tryptic digests [36]. In all cases, the two-dimensional system has greater resolving power than the two systems used independently.

A. Physiological Samples

Physiological samples, physiological fluids, and tissue extracts are analyzed in basic scientific investigations in order to understand peptide structure and function or to clinically evaluate disease diagnosis. Because the peptides are present at trace levels (on the order of ng/mL) in complex matrices, the analytical method must be highly selective and sensitive.

The derivatizing agent NDA has been frequently reported for the analysis of peptides [24,29,34,37–41]. Enkephalins are an important class of neurotransmitters which are present at ultratrace levels, making analysis difficult. Three NDA derivatized enkephalins, ^{5}leucine-enkephalin (Tyr-Gly-Gly-Phe-Leu), ^{5}methionine-enkephalin (Tyr-Gly-Gly-Phe-Met), and [D-^{2}alanine]-^{5}methionine-enkephalin (Tyr-H-Ala-Gly-Phe-Met), have been analyzed by reversed-phase HPLC and detected at the 200-pmol/mL level in plasma using conventional fluorescence detection [37], and at roughly 25 pmol/g wet tissue in rat striatum [34]. In order to increase the sensitivity of enkephalin detection, both laser-induced fluorescence and chemiluminescence detection have been used [38,39], and electrochemical detection has been employed to increase selectivity [42].

Two approaches to the reversed-phase HPLC analysis of NDA derivatized peptides have been used to eliminate the interference from reagent impurities and synthetic by-products which can complicate highly sensitive analyses. First, some of the interferences can be eliminated by introducing the amino acid taurine as a scavenger to consume the excess reagent [24]. Second, a multidimensional HPLC in which two columns are employed can be used first to separate the peaks of interest from the interferences and, second, to resolve the peaks of interest from each other [26,34]. Neurotensin was detected on the order of pmol/g using a multidimensional HPLC with both fluorescence and chemiluminescence detection. Adequate sensitivity was achieved with both, but a 10-fold increase in sensitivity was achieved with chemiluminescence as can be seen in Fig. 4 [24].

Consalvo et al. studied the two-step mechanism of posttranslational carboxyl terminal amidation of glycyl peptides by the enzymes peptidylamidoglycolate lyase (PAL) and peptidylglycine hydroxylase (PGH) [43]. Samples containing glycyl peptides were derivatized with dansyl chloride and analyzed using C_{18} reversed-phase HPLC with gradient elution and fluorescence detection. As fluorescence detection provides selectivity, little interference from the crude tissue homogenate matrix was observed. The limit of detection was about 1 pmol, and resolution of derivatized peptides was achieved in less than 11 min.

Lewis et al. studied the important physiological activity of the carboxyl terminal sequence of β endorphin, referred to as the melanotropin potentiating factor, and related peptides [44]. They investigated the activity of these peptides in cerebral spinal fluid (CSF) and rat brain tissue. The CSF samples required no prior sample pretreatment, but the tissue extracts required cleanup with size exclusion chromatography on a Sephadex-25 column. The samples were derivatized with FMOC and analyzed using C_6 reversed-phase HPLC with gradient elution followed by fluorescence detection, fraction collection, and peak identification using mass spectrometry. The detection limit for this method was 25 fmol.

Advis et al. [45] measured in vivo levels of leutenizing hormone–releasing hormone (LHRH), β endorphin, and neuropeptide Y in ewe median eminence. The peptides were derivatized with fluorescamine and separated using CE with fluorescence detection. As only small amounts of sample could be collected from the hypothalamus, CE was the best choice for this application. No sample preparation was required. Analysis was done using a 75 μm i.d. × 125 cm long capillary and a 0.05 M borate buffer (pH 8.3) which contained 0.025 M lithium chloride. The separation was run at 25 kV for 1 h. A linear response for LHRH was measured for injections of 6.7–108 nmol.

Paroni et al. used HPLC with fluorescence detection to analyze total, oxidized, and protein-bound glutathione (γ-L-glutamyl-cysteinyl-glycine) in blood, plasma, and tissue [46]. Glutathione acts as a redox buffer through the sulfhydryl group on cysteine for protection against oxidative tissue damage. Prior to analysis, N-ethylmaleimide was used to block sulfhydryl groups in order to evaluate

Fig. 4 Detection of phenylalanine–neurotensin in human plasma using multidimensional HPLC system. (a) Fluorescence detection. Column: 10 cm × 4.6 mm i.d., 5 mm CN (ES Industries). Mobile phase: methanol/50 mM sodium acetate, pH 4.0 (1 : 1 v/v). Flow rate: 1 mL/min. Detection (Shimadzu RF-530): excitation wavelength = 420 nm, emission wavelength = 490 nm. (b) Chemiluminescence detection. Column: 15 cm × 4.6 mm i.d., 5 mm TSK-ODS C18. Mobile phase: acetonitrile/10 mM imidazole, pH 7.0 (38 : 62 v/v). Postcolumn chemiluminescent reagents: 0.5 mM bis(2,4,6-trichlorophenyl)oxalate/50 mM hydrogen peroxide, added 1 mL/min. Chemiluminescence detection (Atto Corp. Biomonitor AC 2220) followed addition of chemiluminescent reagents. Peak A = phenylalanine–neurotensin. (From Ref. 24.)

(a)

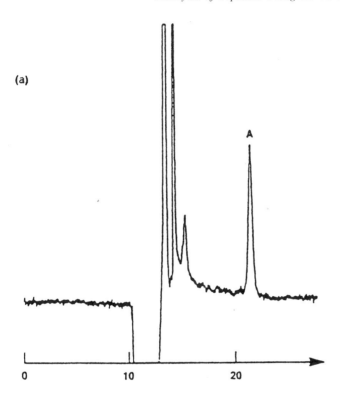

(b)

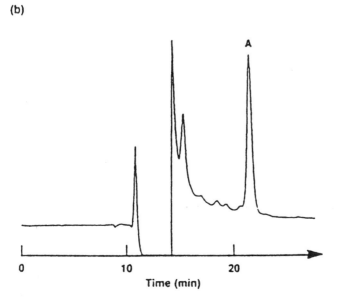

Time (min)

the oxidation state of the cysteine. The peptide was then derivatized with OPA and injected. The limit of detection was less than 0.05 pmol.

In a recent review, Rissler summarizes the reported methodologies for the determination of low levels of substance-P-related peptides which perform a variety of functions in the vertebrate brain and central nervous system [47]. Separation of these compounds is performed using reversed-phase HPLC with either isocratic or gradient elution. Low-UV detection is employed occasionally, but it is generally not sensitive enough for analysis of small amounts of sample. Derivatization with NDA, OPA, and CBQCA has been reported with fluorescence detection. However, at the present time, HPLC in conjunction with radioimmunoassay is found to be the most sensitive and the least influenced by matrix interferences.

Yajima et al. used a dansylated peptide substrate to evaluate the enzymatic activity of carboxypeptidase H (CPH) [48]. CPH activity was measured in rat sciatic nerves and dorsal root ganglia. A mixture of the substrates (dansyl-gly-lys and dansyl-gly-arg), and product (dansyl-gly) were injected on a reversed-phase HPLC system with fluorescence detection. Resolution was achieved within 5 min, and as little as 30 fmol of dansyl-gly was detected. An advantage of this method over the traditional radiometric assay is that no radioactive materials are involved.

In a similar report, Ishiguro et al. used a dabsylated peptide to determine farnesyl–protein transferase (FPTase) activity [49]. The substrates were a dabsylated nonapeptide which corresponds to the C-terminus of human N-RAS p21 and farnesyl diphosphate. FPTase catalyzed the binding of the farnesyl group to the C-terminal cysteine. A reversed-phase (C_{18}) HPLC assay was used with detection at 436 nm to determine overall activity of the enzyme and to determine inhibitors to the activity. A limit of detection (LOD) of 2 pmol/100 µL was reported. This method is useful for analysis of crude materials such as tissue homogenates from clinical samples and has the advantage that there are no radioactive probes involved.

Gilman and Ewing analyzed the contents of single cells using CE with laser-induced fluorescence detection [50]. Individual rat pheochromocytoma cells were captured at the injector end of a capillary and lysed. The amines inside were derivatized with NDA, and the entire contents were injected. Two derivatized peptides were included as internal standards. The amounts of five amino acids and dopamine were determined at levels as low as 180 amol for aspartic acid. CE was the method of choice for this application because only a small amount of sample was available.

Capillary electrophoresis with laser-induced fluorescence was used to do near-real-time analysis of physiological neurotransmitters in anesthetized, or awake free-moving rats [51]. Samples from rat brain were collected, microdialysed, derivatized with NDA, and injected. Quantities of aspartate and glutamate were monitored with time. The detection limit was 0.1 µM. CE is better than HPLC for this application because of the small sample volume required.

B. Peptide Mapping

Highly selective analyses must be developed to resolve the complex mixtures of peptides in peptide mapping. As there is a need to analyze trace amounts of protein, these analyses must be highly sensitive as well. In peptide mapping, a protein is enzymatically or chemically cleaved into small peptide fragments, and the resulting mixture is separated by reversed-phase HPLC or CE to generate a peptide map or fingerprint which can be used to identify the protein. This technique provides a wealth of information about protein structure, including expression errors, mutations, location of glycosylation, disulfide linkages, and structural identification of newly discovered recombinant proteins. In addition, it can be used routinely for the separation of peptides for sequence analysis [52].

Colilla et al. analyzed the structures of three proteins using peptide mapping [52]. Human α-immunoglobulin (L-chain), bovine β-lactoglobulin, and α-lactoglobulin were digested with trypsin and derivatized with PITC prior to C_{18} reversed-phase HPLC analysis. As is customary for peptide mapping, an acetonitrile/trifluoroacetic acid gradient was employed for resolution followed by UV detection at 254 nm. Injection of 5 pmol of digested protein produced peptide maps with adequate sensitivity and baseline resolution although this amount was close to the limit of detection. Fractions were collected from some samples, and the peptides were sequenced using Edman sequencing.

Mendez et al. studied two almost identical cytotoxic proteins, mitogellin and restricosin [53]. These proteins differed by only one amino acid. Following a trypsin digest, the fragmented proteins were derivatized using both dansyl chloride and OPA. The OPA derivatives were prepared and injected using the Auto•Tag OPA™ system. Use of this system eliminated the effect of derivative decay because the samples were derivatized immediately prior to injection. It also provided improved reproducibility of sample preparation. The dansyl derivatives were prepared manually. The derivatized mixtures were then separated using reversed-phase HPLC with a trifluoroacetic acid (TFA)/acetonitrile gradient and detected by fluorescence. The chromatograms generated using both derivatizing agents had a high degree of peptide resolution. However, only the peptide map of the OPA derivatized tryptic digest resolved the peptides that contained the different amino acid. The peptides that contained the different amino acid were coeluted in the peptide map of the dansylated tryptic digest; 10–30 pmol of peptide were routinely detected with both derivatizing agents.

O-phthalaldehyde, fluorescamine, and CBQCA were used for the derivatization of tryptic digests of cytochrome C [54]. The resulting derivatized peptide mixtures were resolved using MECC with conventional fluorescence detection for OPA and fluorescamine, and laser-induced fluorescence detection for CBQCA derivatives. A homemade instrument with either a conventional fluorescence detector or a helium–cadmium laser for laser-induced fluorescence was employed. The capillary was 50 μm i.d. and 45–90 cm long, and the voltage was 20 kV. The eluent was a borate buffer (pH 9.50) which contained sodium dodecyl sulfate micelles to facilitate the resolution of peptides with similar charges. In addition, α- and β-cyclodextrins were added to the eluent to sharpen the peaks and improve the fluorescence quantum yield of the CBQCA derivatives. The addition of 20 mM β-cyclodextrin increased the fluorescence signal by an order of magnitude. A high degree of selectivity was achieved for all of the derivatizing agents, and separations were complete within 30 min.

In a related article, Novotny and his group [21] reported the use of CBQCA for the derivatization of tryptic digests of β-casein. Peptide maps were generated using the same homemade instrument with laser-induced fluorescence described in the previous paragraph. The capillary was 50 μm i.d. by 90 cm long. The eluent was 50 mM borate buffer (pH 9.5) with 20 mM α-cyclodextrin as an additive to improve selectivity and sensitivity. The separation was complete within 20 min. Peptide maps of as little as 17 fmol gave good sensitivity and selectivity.

Bushey and Jorgenson described an automated, coupled HPLC and CE instrument for the multidimensional analysis of tryptic digests of ovalbumin [36]. The digested protein was derivatized using fluorescamine, and the mixture of fragments was analyzed. The multidimensional system consisted of an HPLC system whose effluent was pumped into a CE injector port. Frequent injections of the effluent stream were made on the CE which was configured with fluorescence detection. A three-dimensional "chromatoelectropherogram" graphically demonstrated the results (Fig. 5). Retention time was on the x axis, migration time was on the y axis, and fluorescence signal was on the z axis. Greater resolution was achieved with the combination of techniques than could have been achieved with either of the two alone.

C. Pharmaceutical Samples

Peptide drugs play an important role in the pharmaceutical and biotechnology industries. For thorough drug evaluation, these compounds must be analyzed in both formulations and physiological matrices after dosing. Because the bulk drug must be highly pure, free from both racemic impurities and synthetic impurities such as deletion sequences, highly accurate analyses must be developed to determine synthetic purity, and difficult chiral analyses (which will be discussed in the next section) must be used to determine optical purity. In order to evaluate biological activity, these compounds must also be measured in physiological matrices with high sensitivity and selectivity.

Guzman et al. described a CE analysis for peptides and proteins in biotechnological formulations [55]. The formulations included the drug products recombinant leukocyte A interferon or humanized anti-TAC monoclonal antibody, and the excipient amino acids arginine and glycine. Human serum albumin was included as an internal standard. The samples were derivatized using fluorescamine prior to injection on a CE instrument with UV detection (absorbance was monitored at 200, 214, and 280 nm). Electropherograms were generated us-

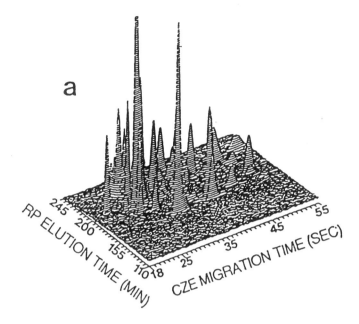

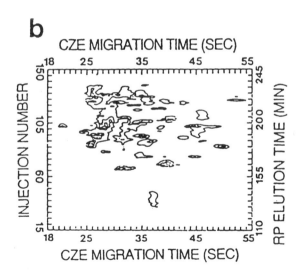

Fig. 5 (a) Surfer-generated chromatoelectropherogram of fluorescamine-labeled tryptic digest of ovalbumin: C1 flow rate, 10 µL/min; gradient 0–10 min, isocratic in buffer A; 10–175 min., 0–30% acetonitrile; 175–300 min., 30–90% acetonitrile; data acquisition is 5 points/s; all points from 18 to 55 s are plotted. (b) Countour plot of the same data set. Tick marks on injection number axis represent 5 injections and 5 min each. (From Ref. 36.)

ing phosphate and borate buffers in the pH range of 7.0–8.3 which contained lithium chloride, and the analysis was complete within 30 min. See Fig. 6 for electropherograms of formulations containing recombinant leukocyte A interferon and humanized anti-TAC monoclonal antibody. UV absorbance at 280 nm of the derivatized peptides (fluorescence detection is more sensitive for this derivatizing agent) provided much better sensitivity than low-UV detection of the peptides without derivatization. In addition, successful derivatization was achieved for both the singly derivatized amino acids and the multiply derivatized large peptides, and derivatization improved the peak resolution as well.

In a related article Guzman et al. described the effects of various buffer constituents and analytical conditions on the CE separation and quantification of a humanized monoclonal antibody in bulk form and in a typical therapeutic formulation [56]. The samples were derivatized with fluorescamine and analyzed using CE with fluorescence detection as described above. The addition of alkylamines and/or zwitterions to the eluent improved the performance of separation.

Boppana et al. described a method for the analysis of two synthetic hexapeptides (manufactured by SmithKline Beecham Pharmaceuticals, Swedeland, PA) which are known to release the growth hormone in some species [57]. The peptide of interest was designated SK&F110697 and the other peptide, SK&F110910, was included as an internal standard. These peptides were extracted from plasma using a weak cation-exchange solid-phase extraction cartridge. An aliquot of the resulting extract was loaded into the autosampler which automatically derivatized the peptides with fluorescamine and injected the derivatives on a reversed-phase (C_{18}) HPLC system with gradient elution and fluorescence detection. The detection limit of the method was 30–50 fmol.

Boppana and Miller-Stein also developed a method for the analysis of a novel synthetic hematoregulatory peptide (SmithKline Beecham Pharmaceuticals), SB107647 {(S)5-oxo-1-prolyl-L-α-flutamyl-L-α-aspartyl-N8-(5-amino-1-carboxypentyl)-8-oxo-N7-[N-[5-oxo-L-prolyl-L-α-glutamyl]-L-α-aspartyl]-L-threo-2,7,8-triaminooctanoyl-L-lysine} [58]. This drug was measured in the plasma of dogs and rats in support of toxicokinetic studies. It was extracted from plasma using solid-phase extraction and analyzed using reversed-phase ion-pair chromatography followed by postcolumn OPA derivatization and fluorescence detection. Postcolumn OPA derivatization was chosen in order to minimize degradative losses of the derivatives. The limit of detection of SB107647 in 0.25 mL of plasma was 10 ng/mL.

D. Chiral Samples

All peptides except those containing only glycine are chiral. Chiral analyses are important in the pharmaceutical and biotechnology industries because racemates of chiral pharmaceuticals often possess different efficacy, toxicity, and pharmacokinetic properties. Therefore, techniques must be developed which are accurate, reproducible, and sensitive for the determination of enantiomeric purity of

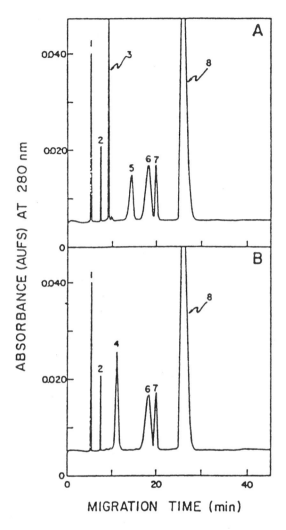

Fig. 6 Capillary electrophoresis profile of fluorescamine derivatized formulation mixture components. (A) Electropherogram of a formulation mixture containing human leukocyte A interferon; peaks: 1 = acetone, 2 = L-arginine, 3 = ammonia, 5 = interferon, 6 = human serum albumin, 7 = glycine, 8 = fluorescamine reagent. (B) Electropherogram of a formulation mixture containing humanized anti-TAC monoclonal antibody; peaks: 4 = anti-TAC monoclonal antibody, all other peaks as in (A). The separation buffer consisted of 0.05 M lithium chloride. The concentration of analytes, per 100 μL reaction mixture, used in this experiment were as follows: L-arginine, 5 μg (28.7 nmol); anti-TAC monoclonal antibody, 215 μg (1.4 nmol); interferon, 145 μg (7.4 nmol); glycine, 2.1 μg (28.7 nmol); and human serum albumin, 125 μg (1.8 nmol). (From Ref. 55.)

synthetic peptides. Generally, chiral resolution of derivatized peptides is achieved in one of two ways, either through the use of a chiral stationary phase (in HPLC only) or with a mobile phase or eluent containing a chiral recognizing agent. Although some success has been achieved for the separation of small model chiral peptides, analysis of real samples is still generally done by hydrolyzing the peptides into constituent amino acids (special effort must be taken to prevent racemization during hydrolysis) and analyzing the resulting mixtures [59–61].

Zukowski et al. described the enantioseparation of FMOC derivatized amino acids and peptides [62]. FMOC derivatized di- and tripeptides were analyzed using a chiral column with α-, β-, and γ-cyclodextrin-bonded stationary phases and fluorescence detection [62]. Multiple analyses were performed with mobile phases of water, acetonitrile, methanol, triethylamine, and acetic acid in various proportions. It was determined that chiral resolution was best with mobile phases that had no aqueous content. Under aqueous conditions, the fluorene group competed with the chiral centers for inclusion in the cyclodextrin cavity. Work was done with di- and tripeptides which had either one or two chiral centers with the additional amino acid(s) being glycine which is achiral. For each center, two peaks were observed.

In related articles, Armstrong et al. described the enantioseparation of AQC derivatized amino acids and di- and tripeptides with one or two chiral centers [63,64]. These derivatized compounds were separated using α, β, γ, R,S-hydroxypropyl acetylated β-, s-naphthylethyl carbamated, and R-naphthylethyl carbamated cyclodextrin-bonded stationary phases. In the case of the amino acids, the D enantiomer eluted before the L. The detection limit was 10 fmol, making this method useful for trace and ultratrace analyses. Detection of 0.049% D-leucine was measured in an L-leucine standard. As with the FMOC derivatives, water diminished the enantioselectivity as is demonstrated in Fig. 7. However, the use of functionalized β-cyclodextrin stationary phases greatly enhanced the enantioselectivity of both the amino acids and peptides because of additional interactions between the analytes and the cyclodextrin substituents.

E. Food Samples

Peptides are used in the food industry as artificial sweeteners, flavor enhancers, and bulking agents. Stringent requirements are placed on the use of peptides in food formulations because many peptides can be harmful at high levels. Rugged and accurate methods are needed for the analysis of these formulations.

Aspartame, L-α-aspartyl-L-phenylananine methyl ester (APM), is a common sweetener used in diet sodas as an alternative to sugar. It has 180 times the sweetness of sucrose. However, high levels of aspartame may cause illness, so precise monitoring is important. APM degrades in solution to L-α-aspartyl-L-phenylananine (AP) and 2,5-disubstituted-diketopiperazine (DKP). AP further degrades to aspartic acid and phenylalanine. In stability studies, it has been shown that only 30–40% of APM remains after 6 months of storage at ambient

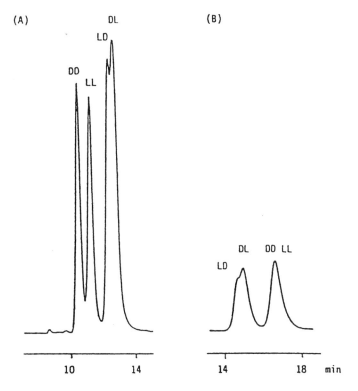

Fig. 7 Enantiomeric resolution of AQC-Leu-Leu obtained on (R,S)-hydroxypropyl derivatized β cyclodextrin stationary phase using (A) nonaqueous mobile phase: 475 acetonitrile + 25 methanol + 3 acetic acid + 6 triethylamine (v/v/v/v); (B) aqueous mobile phase consisting of 5% acetonitrile in 95% triethylammonium acetate buffer (1% pH 7.1). Fluorescence detection (λ_{ex} = 250 nm, λ_{em} = 395 nm) was used. (From Ref. 64.)

temperature [65]. Methods of analyses of underivatized APM have detection limits on the order of 100 pmol. Because APM generally is present in beverages at the micromolar level, these methods are not sensitive enough. In order to reach the desired sensitivity, samples were derivatized using NDA and analyzed using reversed-phase HPLC with fluorescence detection [23,24]. Both the APM and the degradation products were easily detected in the beverage sample.

Aspartame and glutamate (monosodium glutamate), another commonly used food additive, were quantified in diet soda and dried-soup samples, respectively [66]. Like APM, high levels of monosodium glutamate can cause illness. These compounds were derivatized with fluorescamine and analyzed by reversed-phase C_{18} HPLC with isocratic elution and fluorescence detection. For the soda samples, no pretreatment was required, but the soup samples required reconstitution, filtration, dialysis, and preconcentration prior to injection. Results for the chromatographic analysis were more accurate than those obtained previously using spectrofluorimetry alone. The limit of detection were 0.04 and 0.02 µg/mL for the APM and monosodium glutamate, respectively.

Aspartame and its degradation products, AP, aspartic acid, and phenylalanine were analyzed in soft drinks by Hayakawa et al. [67]. These amine compounds were derivatized with NDA and analyzed using C_{18} reversed-phase HPLC with gradient elution and fluorescence detection. Sample preparation included degassing and solid-phase extraction on C_{18} to remove the APM prior to derivatization. APM was removed and analyzed separately to minimize the effect of degradation to AP and DKP. Good compositional analyses were generated for all of the components, and detection limits were in the subpicomole range.

F. Model Synthetic Peptides

Model synthetic peptides are used to evaluate a method prior to applying that method to a particular sample type.

The use of AQC derivatization was reported for both the CE and HPLC analyses of 13 analogs of the prothrombin leader peptide (Ala-Asn-Lys-Gly-Phe-Leu-Glu-Glu-Val) [18,19]. These peptides were identical except for the amino acid at the carboxyl terminus. HPLC analysis was done using C_{18} reversed-phase HPLC with gradient elution and fluorescence detection. The peptides were eluted in order of the relative hydrophobicity of the carboxyl terminal amino acid. Resolution was achieved for most of the derivatized peptides (see Fig. 8), and the detection limit was 19 fmol. Because of its high degree of sensitivity and selectivity, this method is being further evaluated for physiological peptide analysis and peptide mapping.

The CE analysis of the same series of peptides was done using a borate buffer (pH 9.5) and UV detection. Retention was governed by the absolute charge on the peptides. Under the electrophoretic conditions used, derivatized peptides have charges of –1 to –3, and underivatized peptides have charges from –2 to –4. These charges were indicative of their relative mobilities. Despite the fact that UV detection was employed, better sensitivity was achieved for the derivatized peptides than for the underivatized peptides under optimal conditions at 185 nm. The mass limit of detection for the derivatized peptides was 4 fmol.

The derivatization of CBQCA was used for the analysis of 10 model peptides by CE [21]. The CE instrument was homemade with a helium cadmium laser for laser-induced fluorescence detection. The capillary was 50 µm i.d. by 90 cm long, and the voltage was 20 kV. The eluent was a borate buffer (pH 9.50) with 20 mM cyclodextrin added for improved sensitivity and selectivity. The separation of the peptides was complete in 15 min, and the detection limits for two model peptides, Val-Ala-Ala-Phe and Gly-Gly-Tyr-Arg, were 4.6 and 13.8 amol, respectively, making CBQCA one of the most sensitive derivatizing agents.

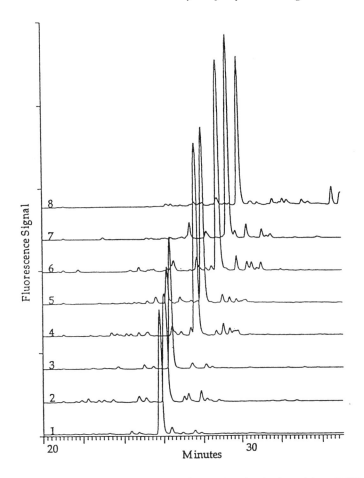

Fig. 8 Overlaid chromatograms of several derivatized peptides. (1) PT-Asp, (2) PT-Ser, (3) PT-Thr, (4) PT-Val, (5) PT-Met, (6) PT-Ile, (7) PT-Phe, and (8) PT-Trp. The separation was done using a Novapak C_{18} column (15 cm × 4.6 mm i.d.) with a linear gradient from 0% to 60% acetonitrile in 1 h. The initial eluent was an acetate buffer (140 mM sodium acetate, 17 mM triethylamine, 3.0 mM ethylenediaminetetraacetic acid, pH 5.05 with phosphoric acid). The flow rate was 1.0 mL/min, and the column temperature was 37°C. (From Ref. 18.)

Grunau and Swaider developed a separation for 99 model amine-containing compounds [68]. These compounds were chosen because they are physiological compounds which occur naturally in plant tissues. These compounds were mostly amino acids, amino acid analogs, and small peptides, including homocarnosine, carnosine, and anserine. Separation of the native compounds was done using ion-exchange chromatography with lithium-containing eluents in the gradient. Separation was followed by postcolumn derivatization with ninhydrin and visible detection at 570 nm. Separation was complete within 2 h. The advantage for ion-exchange chromatography for this application is that it is less sensitive than reversed-phase chromatography to matrix interferences other than proteins which can be easily removed. In addition, ninhydrin provides the advantage of reactivity with both primary and secondary amines.

Nussbaum et al. studied the electrochemical characteristics of NDA derivatives of 18 model amino acids and 15 model peptides (enkephalin fragments and other small peptides which were 2–4 amino acids long) [69]. After derivatization, these compounds were analyzed using both cyclic voltammetry and reversed-phase HPLC with electrochemical detection. Several advantages of electrochemical detection over fluorescence were demonstrated. First, the oxidation potentials of different peptides vary greatly and are largely affected by the presence of arginine. Second, multiply derivatized peptides, such as those which contain lysine, are more readily oxidized than singly derivatized ones. Although the CBI derivatives of lysine show reduced fluorescence due to fluorescence quenching, electroactivity is enhanced by the presence of multiple derivatizations [70,71]. Finally, selectivity can be controlled by varying pH and electrode potential, as some compounds will be more readily oxidized at one pH over another.

Krisjansson et al. developed a method for the analysis of NDA derivatized Pro^2-Lys peptides which are common in physiologic samples [72]. Substance P (Arg-Pro-Lys-Pro-Gln-Phe-Phe-Gly-Leu-Met-NH$_2$) and analogs were chosen as the model compounds. These derivatives were analyzed using both fluorimetry and reversed-phase HPLC with fluorescence detection. The existence of diminished fluorescence was caused by multiple derivatization of peptides with an internal lysine residue. Chromatographic analysis of these derivatives can be improved by selectively derivatizing only the ε amino group of lysine residues. It was demonstrated that the derivatization rate for ε amino group is 100-fold faster than for the α amino group at pH 9–10, and selective derivatization of only the ε amino group was achieved. Derivatization yield was linear for five solutions between 50 and 500 nM (r^2 = 0.998).

IV. CONCLUSION

The demands on peptide sensitivity are becoming more rigorous. Analysis of peptides without derivatization is not sensitive enough to measure trace levels of naturally occurring or synthetic peptides in physiological samples, or to generate a peptide map of a trace amount of protein.

There are new derivatizing agents which can fill the high sensitivity and selectivity needs of peptide analyses. Many of the older derivatizing agents can achieve sensitivity in the middle femtomole to low picomole range, and those that are fluorescent also provide selectivity. The newer reagents have shown even better sensitivity. For example, CBQCA and NDA provide sensitivity in the attomole range using laser-induced fluorescence, and AQC provides sensitivity in the low femtomole range with conventional fluorescence detection. In addition, OPA and NDA can be used with electrochemical detection, and NDA can be used with chemiluminescence detection; both are good alternatives to obtain high selectivity and sensitivity.

Both HPLC and CE are useful techniques for the analysis of derivatized peptides. They are complementary techniques, each with unique advantages and disadvantages. They both provide highly accurate, reproducible, and rugged analyses. However, at the present time, HPLC with fluorescence detection is the technique most often used. As more CE detectors become available commercially, CE will be used more widely for the analysis of derivatized peptides. To date, several references to laser-induced fluorescence detection have been made; however, homemade CE instruments had to be used. Although the CE analysis of underivatized peptides with low-UV detection is still widely used, it is just a matter of time before the CE analysis of derivatized peptides is widely accepted.

REFERENCES

1. H. J. Issaq, G. M. Janini, I. Z. Atamna, and G. M. Muschik. *J. Chromatog.*, *14*(5): 817 (1991).
2. S. M. Lunte and O. S. Wong. *Current Separations*, *10*: 19 (1990).
3. K. A. Cobb and M. V. Novotny, *Anal. Biochem.*, *200*: 149 (1992).
4. M. Roth. *Anal. Chem.*, *43*: 880 (1971).
5. J. A. White and R. J. Hart in *Food Analysis by HPLC*, L. M. L. Nollet, Ed., Marcel Dekker, Inc., New York, 1992, pp. 53–74.
6. W. A. Jacobs. *J. Chromatogr.*, *392*: 435 (1987).
7. L. A. Allison, G. S. Mayer, and R. E. Shoup. *Anal. Chem.*, *56*: 1089 (1984).
8. M. P. Polo, D. G. de Llano, and M. Ramos in *Food Analysis by HPLC*, L. M. L. Nollet, Ed., Marcel Dekker, Inc., New York, 1992, pp. 117–140.
9. B. G. Matuszewski and R. S. Givens, *Anal. Chem.*, *59*: 1102 (1987).
10. D. R. Koope, E. T. Morgan, G. E. Tarr, and M. J. Coon, *J. Biol. Chem.*, *257*: 8472 (1982).
11. R. L. Heinrikson and S. C. Meredith, *Anal. Biochem.*, *136*: 65 (1984).
12. H. A. Moye and A. J. Boning, *Anal. Lett.*, *12*: 25 (1979).
13. S. Einarsson, B. Josefsson, and S. Lagervist, *J. Chromatogr.*, *282*: 609 (1983).
14. D. H. Spackman, W. H. Stein, and S. Moore, *Anal. Chem.*, *30*: 1190 (1958).
15. M. Wiegele, J. F. Bluont, and J. P. Tengi, *J. Am. Chem. Soc.*, *94*: 4052 (1972).

16. V. K. Boppana, C. Miller-Stein, J. F. Politowski, and G. R. Rhodes, *J. Chromatogr.*, *548*: 319 (1991).
17. S. A. Cohen and D. P. Michaud, *Anal. Biochem.*, *211*: 279 (1993).
18. K. M. De Antonis, P. R. Brown, and S. A. Cohen, *Anal. Biochem.*, *223*: 191 (1994).
19. K. M. De Antonis, P. R. Brown, Y.-F. Cheng, and S. A. Cohen, *J. Chromatogr.*, *661*: 279 (1994).
20. D. J. Strydom and S. A. Cohen, *Anal. Biochem.*, *222*: 19 (1994).
21. J. Liu, Y.-Z. Hsieh, D. Wiesler, and M. Novotny, *Anal. Chem.*, *63*: 408 (1991).
22. R. G. Carlson, K. Srinivasachar, R. S. Givens, and B. K. Matuszewski, *J. Organ. Chem.*, *51*: 3978 (1986).
23. S. M. Lunte and O. S. Wong, *Current Separations*, *10*: 19 (1990).
24. S. M. Lunte and O. S. Wong, *LC–GC*, *7*(11): 908 (1989).
25. H. Konig, H. Wolf, K. Venema, and J. Korf, *J. Chromatogr.*, *533*: 171 (1990).
26. T. M. Jeffries, C. M. Riley, S. C. Crowley, and J. Stobaugh, *Pharmaceut. Res.*, *5*: S-15 (1988).
27. W. G. Kuhr and C. A. Monnig, *Anal. Chem.*, *64*: 389R (1992).
28. Z. Deyl and R. Struzinsky, *J. Chromatogr.*, *569*: 63 (1991).
29. F. Y. L. Hsieh, J. Cai, and J. Henion, *J. Chromatogr.*, *679*: 206 (1994).
30. K. D. Altria and D. D. Rogan, *J. Pharmaceut. Biomed. Anal.*, *8*: 1005 (1990).
31. H. H. Lauer and J. B. Ooms, *Anal. Chim. Acta*, *250*: 45 (1993).
32. P. D. Grossman, J. C. Colburn, H. H. Lauer, R. G. Nielsen, R. M. Riggin, G. S. Sittampalam, and E. C. Rickard, *Anal. Chem.*, *61*: 1186 (1989).
33. J. G. Dorsey, J. P. Foley, W. T. Cooper, R. A. Barford, and H. G. Barth, *Anal. Chem.*, *64*: 353R (1992).
34. M. Mifune, D. K. Krehbiel, J. Stobaugh, and C. M. Riley, *J. Chromatogr.*, *496*: 55 (1989).
35. H. Yamamoto, T. Manabe, and T. Okuyama, *J. Chromatogr.*, *515*: 659 (1990).
36. M. M. Bushey and J. W. Jorgenson, *Anal. Chem.*, *62*: 978 (1990).
37. L. A. Sternson, J. F. Stobaugh, J. Ried, and P. de Montigny, *J. Pharmaceut. Biomed. Anal.*, *6*: 657 (1988).
38. L. Nicholson, T. Jeffries, J. Stobaugh, K. Dave, A. Thakur, H. Patel, D. Jencen, and C. Riley, *13th International Symposium on Column Liquid Chromatography* 1989, p. L-016.
39. D. A. Jencen, J. F. Stobaugh, C. Riley, and R. S. Givens, *14th International Symposium on Column Liquid Chromatography*, 1990, p. 328.
40. J. F. Stobaugh, P. de Montigny, S. C. Crowley, and A. Thakur, *Pharmaceut. Res.*, *5*: S-15 (1988).

41. J. F. Stobaugh and S. C. Crowley, *Pharmaceut. Res.*, 5: S-15 (1988).
42. L. Zech and O. Wong, *Pharmaceut. Res.*, 5: S-15 (1988).
43. A. P. Consalvo, S. D. Young, and D. J. Merkler, *J. Chromatogr.*, 617: 25 (1992).
44. J. R. E. Lewis, J. S. Morley, and R. F. Venn, *J. Chromatogr.*, 615: 57 (1993).
45. J. P. Advis and N. A. Guzman, *J. Liq. Chromatogr.*, 16: 2129 (1993).
46. R. Paroni, E. De Vecchi, G. Cighetti, C. Arcelloni, I. Fermo, A. Grossi, and P. Bonini, *Clin. Chem.*, 41(3): 448 (1995).
47. K. Rissler, *J. Chromatogr. B*, 665: 233 (1995).
48. R. Yajima, T. Chikuma, and T. Kato, *J. Chromatogr. B*, 667: 333 (1995).
49. H. Ishiguro, S. Kawata, E. Yamasaki, Y. Matsuda, S. Fujii, and Y. Matsuzawa, *J. Chromatogr. B*, 663: 35 (1995).
50. S. D. Gilman and A. Ewing, *Anal. Chem.*, 67: 58 (1995).
51. S. Y. Zhou, H. Zou, J. F. Stobaugh, C. E. Lunte, and S. M. Lunte, *Anal. Chem.*, 67: 594 (1995).
52. F. J. Colilla, S. P. Yadav, K. Brew, and E. Mendez, *J. Chromatogr.*, 548: 303 (1991).
53. E. Mendez, R. Matas, and F. Soriano, *J. Chromatogr.*, 323: 373 (1985).
54. J. Liu, K. A. Cobb, and M. Novotny, *J. Chromatogr.*, 519: 189 (1990).
55. N. A. Guzman, J. Moschera, C. A. Bailey, K. Iqbal, and W. Malick, *J. Chromatogr.*, 598: 123 (1992).
56. N. A. Guzman, J. Moschera, K. Iqbal, and A. W. Malick, *J. Chromatogr.*, 608: 197 (1992).
57. V. K. Boppana, C. Miller-Stein, J. F. Politowski, and G. R. Rhodes, *J. Chromatogr.*, 548: 319 (1991).
58. V. K. Boppana and C. Miller-Stein, *J. Chromatogr. A*, 676: 161 (1994).
59. H. Brückner, T. Westhauser, and H. Godel, *J. Chromatogr. A*, 711: 201 (1995).
60. J. Ermer, J. Gerhardt, and M. Siewert, *Arch. Pharm. (Weinheim)*, 328: 635 (1995).
61. D. R. Goodlett, P. A. Abuaf, P. A. Savage, K. A. Kowalski, T. K. Mukherjee, J. W. Tolan, N. Corkum, G. Goldstein, and J. B. Crowther, *J. Chromatogr. A*, 707: 233 (1995).
62. J. Zukowski, M. Pawlowska, M. Nagatkina, and D. A. Armstrong, *J. Chromatogr.*, 629: 169 (1993).
63. M. Pawlowska, S. Chen, and D. W. Armstrong, *J. Chromatogr.*, 641: 257 (1993).
64. S. Chen, M. Pawlowska, and D. W. Armstrong, *J. Liq. Chromatogr.*, 17(3): 483 (1994).
65. S. Motellier and I. W. Wainer, *J. Chromatogr.*, 516: 365 (1990).
66. F. G. Sanchez and A. A. Gallardo, *Anal. Chim. Acta*, 270: 45 (1992).

67. K. Hayakawa, T. Schlipp, K. Imai, T. Higuchi, and O. S. Wong, *J. Agric. Food Chem.*, *38*: 1256 (1990).
68. J. A. Grunau and J. M. Swiader, *J. Chromatogr.*, *594*: 165 (1992).
69. M. A. Nussbaum, J. E. Przedwiecki, D. U. Staerk, S. M. Lunte, and C. M. Riley, *Anal. Chem.*, *64*: 1259 (1992).
70. M. A. Nussbaum, J. E. Przedwiecki, D. U. Starek, S. M. Lunte, and C. M. Riley, *Anal. Chem.*, *64*: 1259 (1992).
71. S. M. Lunte, T. Mohabbat, O. S. Wong, and T. Kuwana, *Anal. Biochem.*, *178*: 202 (1989).
72. F. Krisjansson, A. Thakur, and J. F. Stobaugh, *Anal. Chim. Acta*, *262*: 209 (1992).

Index

Absorbance ratioing, 13-15
Additives used in CIEF, 300-301
Affinity capillary electrophoresis (ACE), 261-265
Amine derivatizing agents for peptide analysis, 427-433
   6-aminoquinolyl-*N*-hydroxysuccinimidyl carbamate, 426, 431-432
   3-(4-carboxybenzoyl)-2-quinoline-carboxaldehyde, 426, 431-432
   4-dimethylaminoazobenzene-4-sulfonyl chloride, 426, 429
   5-*N*-*N*-dimethyl aminonaphthalene-1-sulfonyl chloride, 426, 429
   fluorescamine, 426, 430
   9-fluoroenylmethyl chloroformate, 426, 429-430
   naphthalene-2,3-dicarboxaldehyde, 426-427, 433
   ninhydrin, 426, 430
   phenylisothiocyanate, 426, 427
   O-phthalaldehyde, 426-428
6-Aminoquinolyl-*N*-hydroxysuccinimidyl carbamate (AQC), 426, 431-432
Analyte collection method in SFE-GC, 168
Anionic polymerized micelle, 402-404

Aromatic molecules, separation on PGC of, 124-129, 152
Artificial sweeteners, 444
Atrazine, 217
Automated systems for SPE, 214-215

Bence Jones proteinuria, CE in analysis of, 335-336
Benign monoclonal gammopathy, 333
Benzodiazepines, 216
Benzoyleconine, 216
Beta-induced fluorescence (BIF) detection, 38-39
Bile salts (chiral surfactants used in MEKC), 386-392
   structure of, 389
Biller–Biemann method of peak purity assessment, 21
Biomedical applications for SPE-GC, 215-216
Biotechnological peptides and protein formulations, analysis of, 440-442
Buffer additives to minimize protein-wall interactions, 241-243
Bulking agents, 444

453

Capillary electrochromatography (CEC) 367
Capillary electrophoresis (CE) (*see also* Peptide analysis using HPLC and CE; Polymers for chiral separation in CE):
　fundamentals of, 367-378
Capillary electrophoresis (CE) of proteins, 237-361
　capillary isoelectric focusing, 282-310
　　applications, 303-310
　　capillary IEF optimization, 295-303
　　CIEF process, 284-295
　capillary zone electrophoresis, 248-282
　　affinity capillary electrophoresis, 261-265
　　analysis of milk and dairy products, 270-276
　　analysis of protein folding, 269-270
　　cereal proteins, 280-282
　　chiral separations, 265-269
　　enzyme assays, 258-261
　　glycoproteins, 279-280
　　metalloproteins, 276-279
　　sample preconcentration techniques, 249-258
　clinical applications, 323-352
　　cerebrospinal fluid, 323, 339-340
　　enzymes or isoenzymes, 323, 345-350
　　hemoglobin, 323, 341-345
　　lipoproteins, 351-352
　　peptide hormones, 323, 350
　　serum proteins, 323-334
　　soluble mediators from immune cells, 323, 351
　　urine proteins, 323, 335-339
　sieving separations, 310-323
　　analysis of native proteins, 311
　　analysis of SDS-protein complexes, 312-323
　strategies for reducing protein-wall interactions, 240-248
　　capillary coatings, 243-248
　　operation at pH extremes, 240-241
　　use of buffer additives, 241-243
Capillary gel electrophoresis (CGE), 367
Capillary isoelectric focusing (CIEF or IEF) for proteins, 282-310, 367

Capillary zone electrophoresis (CZE) of proteins, 245-282, 367, 368-371
　affinity capillary electrophoresis, 261-265
　analysis of milk and dairy products, 270-276
　analysis of protein folding, 269-270
　cereal proteins, 280-282
　chiral separations, 265-269, 371-378
　enzyme assays, 258-261
　glycoproteins, 279-280
　metalloproteins, 276-279
　sample preconcentration techniques, 249-258
Carbohydrates, separation on PGC of, 138-140, 156
Carbon-based packing materials for LC, 73-162
　correlation of retention on graphite with physical properties, 99-116
　　factor analysis approach, 113-116
　　single parameter approach, 104-113
　development and production, 75-78
　　comparison of properties of porous graphites, 78
　　properties of ideal HPLC packing materials, 75-76
　enantiomer separations, 132-138, 154-155
　　based on inclusion complexation, 136, 155
　　based on ion-pair formation, 136-138, 155
　　based on metal complexation, 138, 155
　　based on nonbonding interactions, 134-136, 154
　　separations of diastereoisomers, 133-136, 154
　existing theories of retention by graphite in reversed phase chromatography, 93-99
　performance of PGC/Hypercarb, 79-81
　PGC and its distinguishing characteristics, 122-123
　residue analysis, 142-147, 157-158
　　polychorinated biphenyls 142-143, 157

[Carbon-based packing materials for LC]
  solid-phase extraction and water
    analysis, 143-147, 158
  tissue residues, 143, 157-158
  retention by graphite in LC
    major features (seen in 1988), 81-85
    retention studies post-1988, 85-92
  separation of geometric isomers and
    related compounds, 123-132,
    152-153
    natural products, 129-132, 152-153
    pesticides, 132, 153
    pharmaceuticals, 129-132, 152-153
    simple aromatic molecules, 124-129,
      152
  separations of ionized and other highly
    polar compounds, 147-151, 159
    ion-exchange separations, 151, 159
    ionized solutes, 147, 159
    water-soluble un-ionized solutes,
      147-151, 159
  separations of sugar, carbohydrates,
    and glucuronides, 138-142, 156
  structure of PGC/Hypercarb, 78-79
3-(4-Carboxybenzoyl)-2-quinoline-
  carboxaldehyde (CBQCA), 426,
  432-433
Casein analysis by CE, 272
Cationic polymerized micelle, 402
Cereal proteins, 322
  CZE in separation of, 280-282
Cerebrospinal fluid (CSF), CE in analysis of,
  323, 339-340
Chemical mobilization of focused zones in
  CIEF process, 287-290
Chemical warfare agents in water, SPE/GC
  procedure for, 218
Chiral compounds, 364
  analysis of, 442-444
  proteins used in HPLC for separation
    of, 265-269
  separation using CZE, 371-378
    crown ethers, 374-375
    host-guest complexations, 371-
      374
    ligand exchange complexations,
      376-377
    macrocyclic antibiotics, 377-378

Chiral separations with polymeric micelles
  (see Polymers for chiral separation
  in CE)
Chiral surfactants:
  for MEKC separation of enantiomers,
    383-397
  MEKC theory with, 379-383
  micelles and, 378-379, 380
Chromatographic ratioing, 15-16
Cocaine, 216
Collagens, 322
Commercially available SPME fibers, 220
Coupling SFE/GC, techniques for, 166-168
Crown ethers for chiral separation, 374-375
Curve fitting technique for peak purity
  assessment, 8-9
Cyclodextrins (CDs):
  combination of polymerized chiral
    micelle with γ-cyclodextrin,
    408-419
  for enantiomeric separation, 371-374

Dairy products, CZE in analysis of, 270-276
  casein analysis, 272
  peptide analysis, 273-274
  whey analysis, 272, 273
  whole milk analysis, 274-276
Derivative techniques for peak purity
  assessment, 5-6
Derivatized peptides, analysis of, 432-448
  chiral samples, 442-444
  food samples, 444-446
  model synthetic peptides, 446-448
  peptide mapping, 439-440
  pharmaceutical samples, 440-442
  physiological samples, 435-438
Detection strategies for peak purity
  assessment:
  operational considerations, 22-23
  statistical considerations, 23-25
Diastereoisomers, separations on PGC of,
  133-134, 154
Diastereomers, 365
Digitonin (chiral surfactant used in MEKC),
  392-394
4-Dimethylaminoazobenzene-4-sulfonyl
  chloride, 426, 429-430
5-*N*-*N*-Dimethyl aminonaphthalene-1-
  sulfonyl chloride, 426, 429-430

Directly coupled supercritical fluid
    extraction (SFE)/GC, 163-204
  applications, 198
  construction of SFE-GC instrumentation,
    174-178
  external trapping of the analytes,
    168-170
  internal accumulation of analytes,
    171-174
  optimization of extraction conditions
    for SFE-GC, 191-198
  optimization of SFE-GC
    chromatography, 178-187
    column stationary-phase thickness,
      186-187
    column trapping temperature, 182-186
    extraction flow rate, 178-182
  quantitative considerations for SFE-GC,
    187-191
  techniques for coupling SFE-GC, 166-
    168
Dyes, CIEF in analysis of, 309
Dynamic-state fluorescence detection, 46-66
  frequency-domain lifetime detection,
    59-66
  time domain resolution and lifetime
    detection, 50-58

Electroosmotic mobilization of focused
    zones in CIEF process, 287
Emission wavelength:
  filter channel selection of, 31-32
  monochromator selection of, 32
Enantiomers, 364-365
  CE in separation of, 365-366
  cyclodextrins in separation of, 371-374
  monomeric chiral surfactants for
    MEKC separation of, 383-397
    bile salts, 386-392
    digitonin, 392-394
    glucopyranoside, 394-397
    saponins, 394, 395
    sodium-N-dodecanoyl-L-valinate,
      384-385
  PGC in separation of, 132-138, 154-155
    based on inclusion complexation,
      136, 155
    based on ion-pair formation, 136-138,
      155

[Enantiomers]
    based on metal complexation, 138,
      155
    based on nonbonding interactions,
      134-136, 154
    separations of diastereoisomers,
      133-136, 154
Environmental applications:
  SPE/GC for, 216, 217
  SPME/GC for, 229-231
Enzymes:
  CE in analysis of, 323, 345-350
  CZE in analysis of enzyme assays,
    258-262
Excitation-emission matrix (EEM)
    detection, 34-36, 37
External accumulation devices used in SFE-
    GC, 169

Factor analysis (FA), 18-19
  explanation of graphite retention
    properties by, 113-116
  FA/GC techniques for peak purity
    assessment, 18-19
Ferguson plots, analysis of SDS-protein
    complexes with, 321-322
Filter channel selection of emission
    wavelength, 31-32
Five-zone serum protein separation, CZE in,
    324-325
Flavor enhancers, 444
Fluorescamine, 426, 430-431
Fluorescence-detected circular dichroism
    (FDCD), 41-42
Fluorescence detectors in HPLC, 29-71
  dynamic-state fluorescence detection,
    46-66
    frequency-domain lifetime detection,
      59-66
    time-domain resolution and lifetime
      detection, 50-58
  innovations in detector technology and
    methodology, 37-46
    beta-induced fluorescence detection,
      38-39
    fluorescence-detected circular
      dichroism detection, 41-42
    fluorescence detection combined
      with other detectors, 46

[Fluorescence detectors in HPLC]
  indirect detection, 44-45
  laser-induced fluorescence
    detection, 45-46, 47-48
  on-column fluorescence detection,
    42-44
  selective modulation, 37-38
  sequentially excited fluorescence
    detection, 39-40
  supersonic jet laser fluorometric
    detection, 41
  two-photon-excited fluorescence
    detection, 40
  overview of fluorescence detection,
    30-31
  steady-state fluorescence intensity and
    spectral detection, 31-37
    excitation-emission matrix
      detection, 34-36, 37
    filter channel selection of emission
      wavelength, 31-32
    monochromator selection of emission
      wavelength, 32
    rapid scanning and array detection,
      32-33
    synchronous fluorescence spectral
      detection, 37
9-Fluoroenylmethyl chloroformate FMOC),
  427, 428-429
Fluoxetine, 216
Food industry, peptides in, analysis of, 444-
  446
Forensic applications for SPME/GC, 231
Frequency-domain techniques:
  for lifetime detection, 59-66
  for peak purity assessment, 9-12
Fungicides, 217

Gel-filled capillaries, analysis of SDS-
  protein complexes with, 314
Geometric isomers, separation on PGC of,
  124-132, 152-153
  natural products, 129-132, 152-153
  pesticides, 132, 153
  pharmaceuticals, 129-132, 152-153
Glomerular proteinuria, CE in analysis of,
  336-339
Glucopyranoside (chiral surfactant used in
  MEKC), 394-397

Glucuronides, separation on PGC of, 141-
  142, 156
Glycated hemoglobulin, CE in analysis of,
  345
Glycoform, CIEF in analysis of, 307-308
Glycoproteins, CZE in separation of, 279-
  280
Government-approved methods for SPE/GC,
  216, 217

Hemoglobin:
  CE in analysis of, 323, 341-345
  CIEF in analysis of variants, 303-306
High-performance liquid chromatography
  (HPLC) (see Fluorescence
  detectors in HPLC; Peptide analysis
  using HPLC and CE)
High resolution serum protein separation
  and identification, 325-333
High-throughput capillary electrophoresis,
  334
Horváth solvophobic theory of graphite
  retention, 94-95, 98
Host–guest complexes for chiral separation,
  371-374
Hydraulic mobilization of focused zones in
  CIEF process, 287, 290-293
Hydrophilic polymers used in CIEF, 301-
  302
Hypoproteinemic serum protein patterns,
  328-329

Immunoglobulins, CIEF in analysis of, 306-
  307
Indirect fluorometric detection, 44-45
Industrial applications for SPME/GC, 229
Internal accumulation devices used in SFE-
  GC, 169, 171-174
Ion-exchange separations on PGC, 151, 159
Ionized solutes, separation on PGC of, 147,
  159
Ion-pair formation, separation of
  enantiomers on PGC based on,
  136-138, 155
Isoenzymes, CE in analysis of, 323, 345-
  350
Isotachophoresis (ITP), 252-255
  preconcentration using ITP with
    hydraulic counterflow, 256-257

Laser-induced fluorescence (LIF) detection, 45-46, 47-48
Ligand-exchange electrophoresis (LEE) for chiral separation, 376-377
Lipoproteins, CE in analysis of, 351-352

Macrocyclic antibiotics for chiral separation, 377-378
Malignant monoclonal gammopathy, 333
Martire and Boehm unified theory of graphite retention, 96, 99
Metal complexation, separation of enantiomers on PGC based on, 138, 155
Metalloproteins, CZE in separation of, 276-279
Micellar electrokinetic capillary chromatography (MEKC or MECC), 367, 434
  concepts for, 378-397
    MEKC theory with chiral surfactants, 379-383
    monomeric chiral surfactants for MEKC separation of enantiomers, 383-397
    surfactants and micelles, 378-379, 380
Micelles:
  polymeric, 398-419
    anionic polymerized micelle, 402-404
    applications, 404-408
    cationic polymerized micelle, 402
    combination of polymerized chiral micelle with γ-cyclodextrin, 408-409
    concept of surfactant/micelle equilibria, 397-399
    factors affecting polymerization rate and polymer structure, 400-401
  surfactants and, 378-379, 380
Milk and dairy products, CZE in analysis of, 270-276
  casein analysis, 272
  peptide analysis, 273-274
  whey analysis, 272, 273
  whole milk analysis, 274-276
Milk proteins, 322-323
Model synthetic peptides, 446-448
Monochromator selection of emission wavelength, 32

Monoclonal γ-globulin identification, CZE analysis in, 333-334
Monomeric chiral surfactants for MEKC separation of enantiomers:
  bile salts, 386-392
  digitonin, 392-394
  glucopyranoside, 394-397
  saponins, 394, 395
  sodium-N-dodecanoyl-L-valinate, 384-385
Multichannel techniques for peak purity assessment, 3, 4, 13-22
  absorbance/spectral ratioing, 13-15
  chromatographic ratioing, 15-16
  factor analysis/principal component analysis, 18-19
  other assessment techniques, 21
  spectral correlations, 16-17
  spectral libraries matching, 19-21
Multiharmonic Fourier transform phase-modulation fluorometer (MHF), 61-66

Naphthalene-2,3-dicarboxaldehyde (NDA), 426, 433
Natural products:
  CIEF in analysis of, 311
  separation on PGC of, 129-132, 152-153
Ninhydrin, 426, 430
Nippon/Toyoh carbon, 77
  physical and chromatographic properties of, 78
Nonbonding interactions, separation of enantiomers on PGC based on, 134-136, 154
Norfluoxetine, 216

On-column fluorescence detection, 42-44
O-phthalaldehyde (OPA), 426-428
Optimization of SFE-GC chromatography, 178-187
  column stationary-phase thickness, 186-187
  column trapping temperature, 182-186
  extraction conditions, 191-198
  extraction flow rate, 178-182
Organochlorine pesticides, 217
Organophosphorous pesticides, 217
Oxazepam, 216

Packing materials for LC (*see* Carbon-based packing materials for LC)
Peak purity assessment techniques, 1-28
  detector's view, 3, 4
  multichannel techniques, 3, 4, 13-22
    absorbance/spectral ratioing, 13-15
    chromatographic ratioing, 15-16
    factor analysis/principal component analysis, 18-19
    other assessment techniques, 21
    spectral correlations, 16-17
    spectral libraries matching, 19-21
  peak purity detection strategies
    operational considerations, 22-23
    statistical considerations, 23-25
  single-channel techniques, 3, 4, 5-12
    curve fitting, 8-9
    derivatives, 5-6
    frequency-domain techniques, 9-12
    profile shape analysis, 7-8
Peptide analysis using HPLC and CE, 273-274, 426-452
  analysis of derivatized peptides, 433-448
    chiral samples, 442-444
    food samples, 444-446
    model synthetic peptides, 446-448
    peptide mapping, 439-440
    pharmaceutical samples, 440-443
    physiological samples, 435-438
  derivatization chemistries, 426-433
    6-aminoquinolyl-N-hydroxy succinimidyl carbamate, 426, 431-432
    3-(4-carboxybenzoyl)-2-quinoline-carboxaldehyde, 426, 432-433
    4-dimethylaminoazobenzene-4-sulfonyl chloride, 426, 429-430
    5-N-N-dimethyl aminonaphthalene-1-sulfonyl chloride, 426, 429
    fluorescamine, 427, 430
    9-fluoroenylmethyl chloroformate, 426, 429
    naphthalene-2,3-dicarboxaldehyde, 426, 433
    ninhydrin, 426, 430
    phenylisothiocyanate, 426, 428-429
    O-phthalaldehyde, 426-428

Peptide hormones, CE in analysis of, 323, 350
Peptide mapping, 438-440
Peptides:
  CE in analysis of, 273-274
  CIEF in analysis of, 309
Pesticides:
  separation on PGC of, 132, 153
  SPC/GC procedure for, 217
pH, operation at pH extremes to minimize protein-wall interactions, 240-241
Pharmaceuticals:
  analysis of, 440-443
  separation on PGC of, 129-130, 152-153
Phenylcyclidine, 216
Phenylisothiocyanate (PITC), 427, 428
Phenylthiohydantoin (PTH) amino acid separation by MEKC, 392-394, 396
Physiological samples, analysis of, 435-438
Polar retention effect (PREG), 99-116
  theories for, 93-99
Polyacrylamide (PA) coatings, 244-246
Polychlorinated biphenyls (PCBs), separation on PGC of, 142-143, 157
Polymers for chiral separations in CE, 363-423
  fundamentals of capillary electrophoresis, 367-378
    chiral separation using CZE, 371-378
  MEKC concepts, 378-397
    MEKC theory with chiral surfactants, 379-383
    monomeric chiral surfactants for MEKC separation of enantiomers, 383-397
    surfactants and micelles, 378-379, 380
  polymeric micelles, 398-419
    anionic polymerized micelle, 402-404
    applications, 404-408
    cationic polymerized micelle, 402
    combination of polymerized chiral micelle with γ-cyclodextrin, 408-419
    concept of surfactant/micelle equilibria, 397-399
    factors affecting polymerization rate and polymer structure, 400-401

Porous graphitic carbon (PGC) (*see also* Carbon-based packing materials for LC), 76-78, 122-124
   applications, 124
   distinguishing characteristics, 122-123
   performance of PGC/Hypercarb, 79-81
   physical and chromatographic properties, 78
   structure of PGC/Hypercarb, 78-79
Preconcentration techniques in CZE, 249-258
   to improve detection, 250-258
     conductivity gradients, 255
     electrical force distribution and/or mobility manipulation, 251-252
     isotachophoresis, 252-255
     physical barriers, 250-251
     preconcentration through adsorption, 257
     preconcentration to regulate sample conductivity, 258
     preconcentration using ITP with hydraulic counterflow, 256-257
     sample focusing, 255-256
Principal component analysis (PCA), 18-19
Profile shape analysis of chromatographic peak, 7-8
Proteins, capillary electrophoresis of, 237-361
   capillary isoelectric focusing, 282-310
     applications, 303-310
     capillary IEF optimization, 295-303
     CIEF process, 284-295
   capillary zone electrophoresis, 248-282
     affinity capillary electrophoresis, 261-265
     analysis of milk and dairy products, 270-276
     analysis of protein folding, 269-270
     cereal proteins, 280-282
     chiral separations, 265-269
     enzyme assays, 258-261
     glycoproteins, 279-280
     metalloproteins, 276-279
     sample preconcentration techniques, 249-258
   clinical applications, 323-352
     cerebrospinal fluid, 323, 339-340

[Proteins, capillary electrophoresis of]
     hemoglobin, 323, 341-345
     lipoproteins, 351-352
     peptide hormones, 323, 350
     serum proteins, 323-334
     soluble mediators from immune cells, 323, 351
     urine proteins, 323, 335-339
   sieving separations, 310-323
     analysis of native proteins, 322
     analysis of SDS-protein complexes, 312-323
   strategies for reducing protein-wall interactions, 240-248
     capillary coatings, 243-248
     operation at pH extremes, 240-241
     use of buffer additives, 241-243

Quantitative considerations for SFE-GC, 187-191

Racemates, 364
Rapid scanning of emission spectra, 32-33
Recombinant human bone morphogenetic protein (rhBMP), 280
Residue analysis, separation on PGC of, 142-147, 157-158
   polychorinated biphenyls, 142-143, 157
   solid-phase extraction and water analysis, 143-147, 158
   tissue residues, 143, 157-158
Retention by graphite in HPLC:
   major features of (as seen in 1988), 81-85
   retention studies post-1988, 85-92
   theories of retention in reversed-phase chromatography, 93-99

Sample preparation for GC with solid-phase extraction (SPE) and solid-phase microextraction (SPME), 205-236
   conventional SPE, 206-218
     batch processing, 214-215
     main features, 232, 233
     method development, 207-213
     SPE/GC applications, 215-218
   SPME, 218-231
     applications, 229-231
     background, 218-220

[Sample preparation for GC with solid-phase
    extraction (SPE) and solid-phase
    microextraction (SPME)]
  main features, 232, 233
  optimizing SPME sampling, 226-229
  principles of SPME, 220-225
Saponins (chiral surfactants used in MEKC),
    394, 395
Secondary monoclonal gammopathy, 333
Selective modulation, 37-38
Sequentially excited fluorescence detection,
    39-40
Serum proteins, CZE in analysis of, 323-334
  five-zone protein separation, 324-325
  high-resolution serum protein
      separation and identification,
      325-333
  high-throughput capillary
      electrophoresis, 334
  monoclonal γ-globulin identification,
      333-334
Sieving separations, CIEF in, 310-323
  analysis of native proteins, 311
  analysis of SDS-protein complexes,
      312-323
Single-channel techniques for peak purity
    assessment, 3, 4, 5-12
  curve fitting, 8-9
  derivatives, 5-6
  frequency-domain techniques, 9-12
  profile shape analysis, 7-8
Single parameter approach to explanation of
    graphite retention properties, 99,
    104-113
Single-step CIEF, 292-295
Snyder theory of graphite retention, 96-97,
    99
Sodium-(N)-dodecanoyl-L-valinate (chiral
    surfactant used in MEKC), 384-385
Sodium dodecylsulfate (SDS)-protein
    complexes, analysis of, 312-323
  applications, 322-323
  Ferguson analysis, 321-322
  gel-filled capillaries, 314
  polymer solutions, 314-321
Sodium-undecylenyl-L-valinate, 404-408
Solid-phase extraction (SPE), 206-218
  batch processing, 214-215
  four basic steps in SPE process, 208

[Solid-phase extraction (SPE)]
  introduction, 206-207, 208
  main features, 232, 233
  method development, 207-213
    optimizing wash and elution steps,
        211-213
    sorbent capacity, 209-211
    sorbent selectivity, 208
    testing SPE procedure, 211
  SPE/GC applications, 215-218
    environmental, 216, 217
    toxicology and biomedical, 215-216
Solid-phase microextraction (SPME), 206,
    218-231
  applications, 229-231
    environmental, 229-231
    industrial, 229
    toxicology and forensic, 231
  background, 218-220
  commercially available SPME fibers, 220
  main features, 232, 233
  optimizing sampling, 226-229
    achieving clean blank run, 228
    preparation of GC injector, 226-
        228
    preparation of standards and
        samples, 226
    selecting sampling parameters, 228
  principles of, 220-225
Soluble mediators from immune cells, CE in
    analysis of, 323, 351
Spectral correlations, 16-17
Spectral libraries matching, 19-21
Spectral ratioing, 13-15
Steady-state fluorescence detection
    technique, 31-37
  excitation-emission matrix
      detection, 34-36, 37
  filter channel selection of emission
      wavelength, 31-32
  monochromator selection of emission
      wavelength, 32
  rapid scanning and array detection,
      32-33
  synchronous fluorescence spectral
      detection, 37
Sugars, separation on PGC of, 138-140, 156
Supercritical fluid extraction (SFE) systems
    (*see* Directly coupled SFE/GC)

Supersonic jet laser fluorometric detection, 41
Surfactant-coated capillaries, 247-248
Synchronous fluorescence spectral detection, 37

Thalassemias, CIEF analysis of, 303-306
Thermodynamics of micellization, 398-399
Time-domain resolution and lifetime detection, 50-58
Tissue residue, separation on PGC of, 143, 157-158
Tonen carbon, 77
    physical and chromatographic properties of, 78
Toxicology applications:
    SPE/GC for, 215
    SPME/GC for, 231
Triazine pesticides, 217
Tubular proteinuria, CE in analysis of, 339
Two-photon-excited fluorescence detection, 40

Urea herbicides, 217
Urine proteins, CE in analysis of, 323, 335-339
    Bence Jones proteinuria, 335-336
    glomerular proteinuria, 336-339
    tubular proteinuria, 339
UV absorption detection, fluorescence detection combined with, 46

Wall-coated capillaries to minimize protein-wall interactions, 243-248
    one-layer wall modifications, 244
    two-layer wall modifications, 244-248
Water:
    chemical warfare agents in, 218
    solid-phase extractions, on PGC, from, 143-147, 158
Water-soluble un-ionized solutes, separation on PGC of, 147-151, 159
Whey analysis by CE, 272, 273
Whole milk analysis by CE, 274-276

Zwitterion as buffer additive for CE, 241-243